Mosaic Landscapes and Ecological Processes

Mosaic Landscapes and Ecological Processes

Edited by

Lennart Hansson

Professor of Population Ecology
Swedish University of Agricultural Sciences
Uppsala, Sweden

Lenore Fahrig

Associate Professor of Biology
Carleton University
Ottawa, Canada

and

Gray Merriam

Professor of Biology
Carleton University
Ottawa, Canada

CHAPMAN & HALL
London · Glasgow · Weinheim · New York · Tokyo · Melbourne · Madras

Published by Chapman & Hall, 2–6 Boundary Row, London SE1 8HN, UK

Chapman & Hall, 2–6 Boundary Row, London SE1 8HN, UK

Blackie Academic & Professional, Wester Cleddens Road, Bishopbriggs, Glasgow G64 2NZ, UK

Chapman & Hall GmbH, Pappelallee, 3, 69469 Weinheim, Germany

Chapman & Hall USA, One Penn Plaza, 41st Floor, New York, NY 10119, USA

Chapman & Hall Japan, ITP-Japan, Kyowa Building, 3F, 2–2–1 Hirakawacho, Chiyoda-ku, Tokyo 102, Japan

Chapman & Hall Australia, Thomas Nelson Australia, 102 Dodds Street, South Melbourne, Victoria 3205, Australia

Chapman & Hall India, R. Seshadri, 32 Second Main Road, CIT East, Madras 600 035, India

First edition 1995

© 1995 Chapman & Hall

Phototypeset in 10/12pt Palatino by Intype, London
Printed in Great Britain by St Edmundsbury Press, Bury St Edmunds, Suffolk

ISBN 0 412 45460 2

Contents

Contributors

Henrik Andrén
Grimsö Wildlife Research Station,
Department of Wildlife Ecology,
Swedish University of Agricultural
Sciences,
S–730 91 Riddarhyttan,
Sweden

Graham Arnold
CSIRO Division of Wildlife and
Ecology,
LMB 4 Midland,
Western Australia 6056,
Australia

A. Joy Belsky
Oregon Natural Resources Council,
522 Southwest Fifth Avenue,
Suite 1050,
Portland, Oregon 97204,
USA

Judith L. Bronstein
Department of Ecology and
Evolutionary Biology,
University of Arizona,
Tuscon, Arizona 85721,
USA

Lenore Fahrig
Department of Biology,
Carleton University,
1125 Colonel By Drive,
Ottawa, Ontario,
Canada K1S 5B6

Ilkka Hanski
Department of Zoology,
Division of Ecology,
PO Box 17, P. Rautatiekatu 13,
University of Helsinki, FIN–00014
Finland

Lennart Hansson
Department of Wildlife Ecology,
Swedish University of Agricultural
Sciences,
PO Box 7002,
S–750 07
Uppsala,
Sweden

Susan Harrison
Division of Environmental Studies,
University of California,
Davis, California 95616,
USA

Rolf A. Ims
Division of Zoology, Department of
Biology,
University of Oslo,
PO Box 1050 Blindern,
N–0316 Oslo,
Norway

Carol A. Johnston
Natural Resources Research
Institute,
University of Minnesota
Duluth,
Minnesota 55812,
USA

Michał Kozakiewicz
Department of Biology,
Warsaw University,
ul. Krakowskie Przedmieście
26/28,
00–927 Warsaw,
Poland

David E. McCauley
Department of Biology,
Vanderbilt University,
Nashville, Tennessee 37235,
USA

Gray Merriam
Department of Biology,
Carleton University,
1125 Colonel By Drive,
Ottawa, Ontario,
Canada K1S 5B6

Douglas W. Morris
Lakehead Centre for Northern
Studies,
Lakehead University,
Thunder Bay, Ontario,
Canada P7B 5E1

Reed F. Noss
Department of Fisheries and
Wildlife,
Oregon State University,
Corvallis, Oregon 97331
USA

John A. Wiens
Department of Biology,
Colorado State University,
Fort Collins, Colorado 80523,
USA

Andrew Young
Centre for Plant Biodiversity
Research,
Australian National Herbarium,
Division of Plant Industry,
CSIRO,
GPO Box 1600, Canberra,
ACT 2601,
Australia

Series foreword

This series presents studies that have used the paradigm of landscape ecology. Other approaches, both to landscape and landscape ecology are common, but in the last decade landscape ecology has become distinct from its predecessors and its contemporaries. Landscape ecology addresses the relationships among spatial patterns, temporal patterns and ecological processes. The effect of spatial configurations on ecological processes is fundamental. When human activity is an important variable affecting those relationships, landscape ecology includes it. Spatial and temporal scales are as large as needed for comprehension of system processes and the mosaic included may be very heterogeneous. Intellectual utility and applicability of results are valued equally. The International Association for Landscape Ecology sponsors this series of studies in order to introduce and disseminate some of the new knowledge that is being produced by this exciting new environmental science.

Gray Merriam
Ottawa, Canada

Foreword

This is a book about real nature, or as close to real as we know – a nature of heterogeneous landscapes, wild and humanized, fine-grained and coarse-grained, wet and dry, hilly and flat, temperate and not so temperate. Real nature is never uniform. At whatever spatial scale we examine nature, we encounter patchiness. If we were to look down from high above at a landscape of millions of hectares, using a zoom lens to move in and out from broad overview to detailed inspection of a square meter we would see that patterns visible at different scales overlay one another. The land is an almost endless series of mosaics upon mosaics. Which scale or level of resolution we choose depends largely on the questions we ask. If we are interested in the ecotones between biomes, we had better use a wide-angle lens; if we wish to study the heterogeneity of regeneration sites for plants in a meadow, we must zoom in closely. No one scale is fundamental.

As the chapters in this book attest, ecological patterns determine – just as they are determined by – a rich suite of ecological processes, from the coldly abiotic to the intricate interactions between individual organisms and with their environments. Yes, pattern and process are intertwined, making scientific study of landscape mosaics challenging. What causes what? Do mechanisms (processes) at one scale determine patterns in another, only to feed back and alter the original process? How can we predict anything about the ecology of landscape mosaics when those mosaics never sit still?

For those people even vaguely familiar with the topics explored in this book, it is difficult to believe that until recently the typical response of ecologists to ecological mosaics was to pretend they do not exist. Sure, there were exceptions. W. S. Cooper, writing in 1913 about the forests he studied on Isle Royale in Lake Superior, described 'a complex of windfall areas of differing ages, the youngest made up of dense clumps of small trees, and the oldest containing a few mature trees with little young growth beneath. . . . The result in the forest in general is a mosaic or patchwork which is in a state of continual change.' And we should not forget the 'space–time mosaics' of British heathlands studied by A. S. Watt. But for the most part ecologists went out of their way to ignore heterogeneity. Textbooks in vegetation science and plant ecology instructed students to stratify vegetation cover into recognizable entities, then sample characteristic plots within the interiors of these relatively homogeneous stands. Ecotones and mosaics should be avoided at all costs.

We must forgive those ecologists for seeking the elusive homogeneous stand, the archetype of nature. Science progresses by simplification, and mosaics are not simple. Now that landscape ecology has embraced heterogeneity, how much closer are we to reality? We can be sure that real ecological mosaics are much more complex than our most complicated models suggest. Perhaps, as Frank Egler reminded us, ecosystems are more complex than we can ever think. But at least ecologists have been trying to think.

As a conservation biologist, I am not very interested anymore in knowledge for its own sake; losing the purity of intellectual pursuit is one of the hazards of doing applied work. I am interested in solving problems. Thus, I see the greatest contribution of landscape ecology in the way it has revolutionized thinking and practice in real-life conservation. I write this with an admitted North American bias, knowing full well that landscape ecology has had a tremendous influence on land-use planning and sometimes biological conservation for many decades in much of Europe. But here in North America, the idea of setting aside tidy little museum specimens of nature for conservation purposes has given way only recently to the idea of managing biodiversity and ecological processes on the scale of regions. We have learned that many species are distributed as metapopulations (either naturally or due to habitat fragmentation) and cannot be managed site by site. Although the long-term dynamics of these metapopulations elude us, landscape ecology gives us the 'bit picture' overview we need to solve these problems. Many metapopulation models remain abstract, but increasingly we are modeling and mapping the distributions and movements of organisms as they really occur. We can thank our science for this breakthrough, but also (I hesitantly admit, as a confirmed Luddite) the technologies of powerful computers, satellite imagery, and geographic information systems. We can now view, map and model nature at scales once thought impossibly big.

This book is hardly the last word in the ecology of heterogeneous landscapes. The field progresses rapidly. As John Wiens notes in his introductory chapter, landscape ecology has remained largely descriptive and has lagged behind other areas in ecology in developing coherent theory and applying hypothetico-deductive methods of analysis. Yet, the descriptive and correlative aspects of this science have proven useful. We have learned that managing species site by site, or managing ecosystems species by species, are not very promising approaches. Landscape ecology is perhaps above all a perspective, a way of seeing and interpreting problems in a new light. Sites do not exist in isolation from one another and their matrix. Organisms, energy, nutrients, water and disturbances flow through complex mosaics; hence we cannot manage or conserve them one by one or out of context. Again, perhaps my applied orientation biases me, but I do not think the absence of clearly articulated theory and rigorous hypothesis testing has doomed landscape ecology to irrelevance. I do not suggest we forget about theory or predictability, but only that we take what we do know and apply it where needed.

Finally, my hope is that the science of landscape ecology will not only become more rigorous, but that it will be increasingly linked with an ethic of landscape ecology, or what Aldo Leopold called simply a 'land ethic'. Some ways of treating land are right and other ways are wrong. When we come to appreciate the beauty and integrity of natural landscapes, and to understand the threats that unrestrained human activities pose to these landscapes, then perhaps we can learn humility. We are powerful, but not all that bright. When by learning more about ecological complexity we come to acknowledge our profound ignorance, then we will have achieved something greater than any knowledge. Maybe then we will treat the land with proper respect.

Reed F. Noss
Idaho, USA

Preface

A marked change has occurred recently within the science of ecology. Previously, ecological processes commonly were assumed to proceed within homogeneous environments, and usually within populations of randomly distributed individuals. Recently it has been widely recognized that environments are not homogeneous and organisms are usually clumped into patchy populations, and that this heterogeneity has significant effects on ecological processes. Early precedents for these ideas were present in both the empirical literature (e.g. Andrewartha and Birch, 1954; Rowe, 1961; Wegner and Merriam, 1979) and the theoretical literature (e.g. den Boer, 1968; Levins, 1969; Roff, 1974; Levin, 1976). Concern with effects of heterogeneity on ecological processes continued throughout the 1980s (e.g. Chesson, 1981; Wilcox and Murphy, 1983; Holt, 1984; Fahrig and Merriam, 1985; Kareiva, 1987; Harrison, Murphy and Ehrlich, 1988). However, the idea that heterogeneity is a dominant factor influencing ecological processes did not become widely appreciated until around 1990 when most theorists finally and rather suddenly accepted the reality that nature is heterogeneous and that models required severe modification to even crudely represent reality (for example, see recent volumes such as *Quantitative Methods in Landscape Ecology* (Turner and Gardner, 1990), and *Ecological Heterogeneity* (Kolasa and Pickett, 1991), and special journal issues such as *Philosophical Transactions of the Royal Society of London*, **330** (1990), *Biological Journal of the Linnean Society*, **42** (1991), and *Theoretical Population Biology* **41**(3) (1992). Thus in the last few years, many publications have presented new models and analyses of behavior, population dynamics, genetics and community dynamics, all occurring in heterogeneous environments. This new approach became known as landscape ecology and the young International Association for Landscape Ecology (IALE) became a major global ecological society.

Applied ecologists and field biologists have long realized that ecological systems were heterogeneous (Naumov, 1936; Frank, 1956). However, they have relied largely on theories and models assuming homogeneity because they had no alternative, more realistic, but still simplifying way to make the complex tractable. Focusing on critical ecological processes rather than the structural detail of the system was a first step (e.g. Stenseth and Hansson, 1981). Clearly, in a heterogeneous system composed of spatially separated resource patches, if there is any differential in distribution of the resources among the patches, there will be gradients along which resources can flow between patches. The same will be true for temporally separated resource patches. Likewise, heterogeneously distributed individuals will form patch populations with potential for flows of both individuals and genetic variants among them. Ecological processes that operate at relatively large scales and which are critical to these heterogeneous systems could lead to predictive ability and also to management interventions. Development of these capabilities should have been helped by the new theoretical insights but there has been only weak collaboration of theorists and empiricists in landscape ecology until very recently. Thus the voluminous theoretical work has not been seen by empiricists as testable until recently, and theorists have not found generalities in the work of empiricists to incorporate into improved theory. This volume is intended to move the two bodies of work closer together.

This book is focused primarily on the results of field work in mosaic landscapes. Our main approach was to attract authors able to outline theoretical ideas and predictions pertaining to environmental heterogeneity and the landscape scale and at the same time to examine these ideas in relation to recent field studies that have included more than one habitat and where interactions between habitats have been considered. This approach may provide not only a means of determining the relevance of various theoretic attempts, but may also supply generalizations on ecological mechanisms that differ in their operation between homogeneous and heterogeneous environments. This book is thus primarily focused on the results of field work in mosaic landscapes, including some instructive case studies. The book is therefore a synthesis of recent theoretical and empirical work in landscape ecology, which will be useful for both basic and applied work in general ecology.

We start with an introductory review on the treatment of heterogeneity in ecological theory. We then examine the influence of species on landscape pattern and, conversely, the influence of landscape pattern on individuals and populations. This includes both proximate behavioral responses to different landscape patterns as well as adaptations for securing dispersed resources to maximize lifetime reproductive output. In the next section we present analyses of the effects of landscape pattern on genetic pattern and interactions of genetics and population dynamics in mosaic landscapes. In the fourth section we examine landscape effects on species interactions. Chapters include effects on competitive interactions, local effects of landscape composition on predation rates, and effects of spatio-temporal pattern of plant–pollinator landscapes on plant–pollinator interactions. In the final section we review implications of landscape pattern for population conservation and conservation programs as a whole.

The papers in this volume were reviewed by 36 leading specialists in the various topic areas, from eight nations. The editors could not have accomplished their task without the creative and constructive help of these referees but responsibility for final content lies with the editors.

REFERENCES

Andrewartha, H. G. and Birch, L. C. (1954) *The Distribution and Abundance of Animals*, Chicago University Press, Chicago.

Chesson, P. L. (1981) Models for spatially distributed populations: the effects of within-patch variability. *Theor. Pop. Biol.* **19**, 288–325.

Den Boer, P. J. (1968) Spreading the risk and stabilization of animal numbers. *Acta. Biotheor.* **18**, 165–94.

Fahrig, L. and Merriam, G. (1985) Habitat patch connectivity and population survival. *Ecology,* **66**, 1762–8.

Frank, F. (1956) Grundlagen. Möglichkeiten und Methoden der Sanierung von Feldmausplagegebieten. *Nachrichtenblatt des deutsches Pflanzenschutzdienstes,* **8**, 147–58.

Gilpin, M. E. and Hanski, I. (1991) *Metapopulation Dynamics.* Biol. J. Linn Soc., **72**, 1–323.

Harrison, S., Murphy, D. D. and Ehrlich, P. R. (1988) Distribution of the Bay checkerspot butterfly, *Euphydryas editha bayensis*: evidence for a metapopulation model. *Am. Nat.,* **132**, 360–82.

Holt, R. D. (1984) Spatial heterogeneity, indirect interactions, and the coexistence of prey species. *Am. Nat.,* **124**, 377–406.

Kareiva, P. (1987) Habitat fragmentation and the stability of predator–prey interactions. *Nature,* **326**, 388–90.

Kolasa, J. and Pickett, S. (eds) (1991) *Ecological Heterogeneity,* Springer Verlag, New York.

Levin, S. A. (1976) Population dynamic models in heterogeneous environments. *Ann. Rev. Ecol. Syst.,* **7**, 287–310.

Levins, R. (1969) Some demographic and genetic consequences of environmental heterogeneity for biological control. *Bull. Entomol. Soc. Am.,* **15**, 237–40.

Naumov, N. P. (1936) On some peculiarities of ecological distribution of mouse-like rodents in southern Ukraine. *Zoologizheskij Zhurnal,* **15**, 675–96 (in Russian).

Roff, D. A. (1974) Spatial heterogeneity and the persistence of populations. *Oecologia,* **15**, 245–58.

Rowe, J. A. (1961) The level-of-integration concept and ecology. *Ecology,* **42**, 420–7.

Stenseth, N. C. and Hansson, L. (1981) The importance of population dynamics in heterogeneous landscapes: management of vertebrate pests and some other animals. *Agro-Ecosystems,* **7**, 187–211.

Turner, M. G. and Gardner, R. (1990) *Quantitative Methods in Landscape Ecology,* Springer Verlag, New York.

Wegner, J. F. and Merriam, G. (1979) Movements by birds and small mammals between a wood and adjoining farmland habitats. *J. App. Ecol.,* **16**, 349–57.

Wilcox, B. A. and Murphy, D. D. (1983) Conservation strategy: the effects of fragmentation on extinction. *Am. Nat.,* **125**, 879–87.

Landscape mosaics and ecological theory

1

John A. Wiens

1.1 INTRODUCTION

To most people, the term 'landscape' refers to the scenery of fields and forests, of mountains and streams. Gardeners apply the same term to their plantings of flower beds, shrubs, trees and lawns. Although the scales are different, the essence of 'landscape' in both cases is heterogeneity: more than a single element is present, and the more appealing landscapes are usually varied and spatially complex.

The settings that ecologists study are no different. Whether they are natural or altered by human activities, or are defined at fine or broad spatial scales, environments are spatially heterogeneous. To an ecologist, the beauty of landscapes lies not only in their reality, but in the challenge of understanding how their complex spatial structure affects ecological patterns and processes. This is the focus of the science of landscape ecology.

The underlying premise of this discipline is that the explicit composition and spatial form of a landscape mosaic affect ecological systems in ways that would be different if the mosaic composition or arrangement were different. The chapters in this volume indicate something of the scope of these effects. They address questions such as 'How does landscape structure affect movement patterns or foraging dynamics?', or 'How do species affect landscape patterns?', or 'How does landscape structure affect the demography or genetics of populations, or species interactions such as predation, competition or pollination?', or 'How should an understanding of spatial processes affect our use or management of resources?'

Mosaic Landscapes and Ecological Processes.
Edited by Lennart Hansson, Lenore Fahrig and Gray Merriam.
Published in 1995 by Chapman & Hall, London. ISBN 0 412 45460 2

These are relatively new questions, but they have generally been derived by coupling the observation that landscape mosaics have a spatial structure with topics that have interested ecologists for a long time. Their foundation is more empirical than theoretical. Indeed, much of the recent progress in landscape ecology has been in the description and analysis of landscape patterns (Wiens, 1992a). Even though spatial phenomena have attracted some attention from theoretical ecologists, the development of formal, predictive theory that deals explicitly with mosaic effects on ecological processes has lagged far behind the more descriptive approaches (Merriam, 1988; Turner, 1989; Gosz, 1991; Wiens et al., 1993). In contrast to many areas of ecology, few questions in landscape ecology are derived from theoretical postulates. Hypothesis-testing is infrequent, perhaps because clearly defined, quantitative predictions about how landscapes affect ecological phenomena are so scarce.

I contend that we are not likely to progress very far in understanding how mosaic structure influences ecological phenomena without a body of theory that deals specifically with such effects. To see what such landscape-ecology theory might look like, we must first consider some general aspects of what theory is and how it is used in ecology, and then examine how ecologists have dealt with spatial variation in their theoretical work.

1.2 THE ROLE OF THEORY IN ECOLOGY

When ecologists speak of 'theory', they have in mind a variety of things, from impromptu speculations to formal statements based on underlying principles such as thermodynamics. These notions may be expressed in different forms (e.g. verbal models, graphs, computer simulations, mathematical formulations) and they may be intended to apply to specific situations or to cover a general class of phenomena. This diversity of meanings has contributed to disagreement among ecologists about what we should expect theory to do and has generated a lot of writing about theory in ecology (e.g. Haila and Järvinen, 1982; Roughgarden, 1983; Stenseth, 1984; Kingsland, 1985; Haila, 1986; Fagerström, 1987; Loehle, 1987; Caswell, 1988; Wiens, 1989a; Peters, 1991).

Regardless of the approach, theory represents a means of simplifying a complex reality so that we can achieve some understanding and make reliable predictions. Understanding natural phenomena requires first that we develop an explanation, a statement that ties the observed pattern to an underlying process or processes (Cale, Henebry and Yeakley, 1989). Most ecological theories incorporate assumptions about cause and effect. Niche theory, for example, posits that ecological divergence between ecologically similar species where they occur in sympatry (i.e. character displacement) is caused by interspecific competition (Grant, 1972; Schluter and McPhail, 1993). In such cases, a pattern-process linkage is contained within the theory (Figure 1.1). When a pattern that is observed in nature matches that contained in a theory, one may then look within the theory to discover a likely underlying process. This process may then be offered as an explanation for the observed pattern (Figure 1.1).

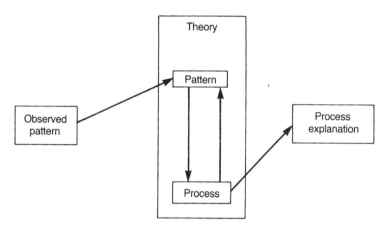

Figure 1.1 The relationship between observation, theory and explanation. Ecologists frequently observe a pattern in nature and then compare it with the pattern predicted in theory. If a match is found, the process contained within the pattern-process linkage of the theory may then be used as a process explanation of the observed pattern. In the absence of independent validation of the pattern-process linkage, such an explanation rests on inference.

Understanding natural phenomena also requires that we determine whether a suggested causal explanation is likely to be correct. The theory must be tested. More specifically, we must (1) test the predictions of the theory and (2) determine whether the essential assumptions of the theory are satisfied. We must also (3) assess the tightness of the pattern-process linkage contained in the theory; if other processes can lead to the same pattern, this linkage is loose, and one must test multiple, alternative hypotheses about causation (Hilborn and Stearns, 1982; Wiens, 1989b).

Testing predictions is the hallmark of hypothetico-deductive science (Popper, 1959; Strong, 1982). Although not all ecological theories are predictive or are framed in operational (i.e. testable) terms (Haila, 1986), there is little question that testing of theoretically derived predictions has contributed greatly to ecological progress. The logical chain leading from theory to hypothesis to prediction to test is the foundation of experimental research (Wiens, 1989b; Underwood, 1990). In a sense, the theoretical model becomes a simplified surrogate for a more complex and muddled reality, and we study the model performance (by testing its predictions) rather than the nature it represents (Haila, 1989).

Theory is therefore a way of deriving deductions about nature, which may then be used as a foundation for tests. In the absence of independent tests, explanations that are logically consistent with a given theory are little more than formalized speculations. The testing phase itself is not theory (although additional theory can be employed in the testing process), but good tests require good theory. Good theory, in turn, is characterized by internally consistent logic, testability, and predictive power. Often we strive for generality in theory, but that is not essential. Predictive theory is obviously much desired as a management tool.

Theory also plays an important role in generating new questions that give direction to research activities. While it is true that such questions may often arise, in the form of speculations, from non-theoretical musings, they frequently emerge from the structure of a theory itself or from tests of predictions.

A good deal of ecological theory is expressed in mathematical terms. By formalizing theoretical propositions in mathematical expressions, ecological problems are clearly specified in simple terms, predictions become clear-cut and quantitative, the logical structure of the theory is strengthened, and the assumptions (or at least some of them) can be stated precisely. If a solution to a mathematical theory is found, the range of values over which that solution holds can be determined, and this establishes the domain of generality of the solution (Fahrig, 1991). Generality is also enhanced by the simplification that accompanies mathematical formulations, since the idiosyncrasies of particular situations are ignored. Mathematical theory is appealing to those who seek general principles in ecology.

It has been said that science is the search for solutions (Judson, 1980). By posing problems clearly and by generating testable predictions, theory encourages solutions. Science is just as much the quest for questions, however, and theory may also contribute to progress by generating questions that lead to new areas of inquiry. This role may be especially important in developing disciplines such as landscape ecology.

1.3 HOW HAVE ECOLOGISTS VIEWED SPACE?

There is a natural progression in how one may view spatial patterns (Figure 1.2). The simplest approach, of course, is to ignore spatial variation and treat space as homogeneous. There are two ways to introduce spatial complexity (and realism) into this view of nature. One approach is simply to view space as variable or heterogeneous, without regard to the spatial pattern. Any area that is not homogeneous is, by definition, heterogeneous. Alternatively, the specific form of nonhomogeneous spatial patterning may be considered. Perhaps the simplest view of spatial pattern is that of patches set in a background matrix, akin to islands in

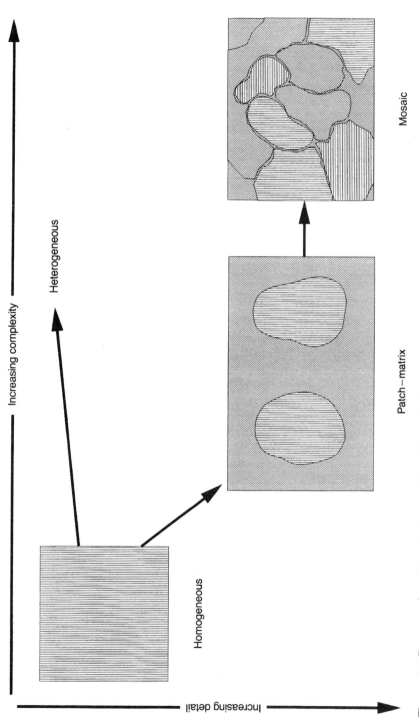

Figure 1.2 Patterns of spatial variation considered in ecological theory. Heterogeneity expresses complexity of spatial patterning or spatial variance, and includes all nonhomogeneous patterns.

a featureless sea. The patches are generally internally homogeneous and similar to one another, and the matrix is ecologically neutral. This view of nature as patches and matrix may be extended further, to consider spatial variation in terms of a mosaic of patches. Here, patches are arrayed in a particular configuration and the explicit structure of spatial patterns becomes important. Although I do not consider them here, gradients are another form of spatial variation in which patch boundaries are indistinct. Gradients have received considerable empirical and analytical attention from plant ecologists (e.g. Whittaker, 1960; Gauch, 1982; Keddy, 1991), and they are closely related to passive diffusion and epidemiological models (Bailey, 1975; Okubo, 1980; O'Neill *et al.*, 1992). Discontinuities in diffusion rates may lead to boundaries and patchiness (Haggett, Cliff and Frey, 1987; Wiens, 1992b).

Here, I comment on theoretical aspects of homogeneity, heterogeneity, and patch–matrix patterns, deferring a discussion of mosaic patterns to the next section.

1.3.1 HOMOGENEITY

To a large degree, this progression of increasingly detailed attention to pattern also represents an historical progression in ecology. The formal ecological theory that developed prior to, say, 1960, generally emphasized spatial homogeneity. Dealing with spatial variation was analytically intractable and beyond the capacity of most simulation tools, and a major step toward simplification could be made by ignoring space altogether. Thus, theories dealing with population dynamics, predator–prey interactions, competition, life-history variation, ecosystem energy flow, nutrient cycling and the like, generally did not include spatial terms. Researchers looking to theory for guidance found little to direct their attention to spatial variation, even though early descriptive work by plant ecologists had focused on spatial dispersion patterns (Grieg-Smith, 1979) and animal ecologists such as Andrewartha and Birch (1954) had made strong statements about the importance of patchy distributions in the 1950s. To test theory (which did not explicitly include space), one was directed to seek internally homogeneous study areas (e.g. MacArthur, 1972). Watersheds, which were functionally thought of as contained, homogeneous areas, became the model for ecosystem research (e.g. Bormann and Likens, 1979). Tests of population theory generally dealt with numbers of individuals, not the space they occupied. As Paul Keddy (1991) has observed, once a world view based on equilibrium and homogeneous environments was adopted, heterogeneity became an obstacle to scientific progress.

Because theories that assumed homogeneity developed early in the history of ecology, they have had a powerful and persistent effect on how we view ecological systems. In their efforts to develop theory that would have general applicability to real situations, ecologists sometimes forgot that the assumption of homogeneity enhanced computational and theoretical tractability, but was not otherwise justified. As a result, this body of work was sometimes regarded as a demonstration of how nature is and how it works, rather than a statement of what might be if nature were indeed homogeneous.

1.3.2 HETEROGENEITY

But nature is not homogeneous. One way of dealing with spatial variation in habitat, resource distribution, predation risk and the like, emphasized heterogeneity *per se*. Spatial heterogeneity was regarded as a measurable expression of the overall spatial complexity or variety of an area. Thus, MacArthur and MacArthur (1961) initially related the diversity of bird species occupying a habitat to the vertical heterogeneity of the vegetation, and subsequent workers (e.g. Wiens, 1974; Roth, 1976) extended this approach to small-scale, within-habitat horizontal heterogeneity. At a broader spatial scale, variations in ß-diversity were related to differences among areas in the mixture of habitats present (between-habitat heterogeneity) (Whittaker, 1965; Cody, 1975). (Harrison (1993) has recently noted the ways in which dispersal differences among species may also influence ß-diversity.) The so-called 'intermediate disturbance hypothesis' (Harper, 1969; Huston, 1994) is yet another version of a relationship between spatial heterogeneity and diversity, as heterogeneity is generally greatest at intermediate levels of disturbance, when diversity is also high.

Research on heterogeneity has generally involved relatively little formal theory (but see Kolasa and Pickett, 1991). Instead, empirical relationships have been derived between measures of heterogeneity and ecological features of interest. Heterogeneity simply refers to variation in space. In a statistical sense, heterogeneity (spatial variance > 0) contrasts with homogeneity (spatial variance = 0).

1.3.3 PATCHINESS

While some ecologists were focusing on heterogeneity, others were building spatial variation into theory in the form of patches. Much of this work focused on population genetics and population dynamics, although patch theory also developed in studies of foraging behavior. Some ecologists studying succession adopted a patch perspective in their focus on gap dynamics (Brokaw, 1985). Island biogeography theory can perhaps be considered a form of community-level patch theory.

Various patch configurations were modeled: two patches imbedded in a matrix, an archipelago of islands, a series of stepping-stones, or islands linked to an adjacent mainland. To retain analytical tractability, the patches were usually assumed to be internally homogeneous and equivalent in size and quality, and all patches were either equally accessible to dispersing individuals or dispersal was restricted to adjacent patches. Each subpopulation was modeled using traditional expressions (e.g. logistic growth, Lotka-Volterra predator–prey equations), but terms were included linking the dynamics of each subpopulation to that of the other(s). No specific dynamics were assigned to the matrix. Thus, for example, changes in gene frequencies in local populations were modeled not only as a consequence of selection, but were affected by gene flow from another patch that depended on the degree of isolation between the patches (Slatkin, 1985). Similarly, the dynamics of a large population that was subdivided into a number of patches (i.e. a metapopulation; Levins, 1970) were modeled as the aggregate of the dynamics of individual subpopulations that were largely independent of one another but were linked by dispersal (Hanski, 1991).

Not surprisingly, this work showed that patchiness makes a difference. The changes in local effective population size and gene flow produced by population subdivision generally decrease genetic variance within local subpopulations but increase variance among populations, and the heterozygosity of the overall metapopulation may decline (Slatkin, 1985). Things become much more complex, however, if patch populations undergo local extinction and recolonization (Gilpin and Hanski, 1991; McCauley, 1993). The spatial subdivision of a variable population tends to increase the probability of population persistence when the local patch populations have asynchronous dynamics and dispersal is limited (Levins, 1970; Hanski, 1985; Lande, 1987; Chapter 12). Similarly, coexistence of competitors (Horn and MacArthur, 1972; Slatkin, 1974; Hanski, 1983) or of predator–prey or parasitoid–host systems (Crowley, 1981; Reeve, 1988; Taylor, 1988, 1990; Sabelis, Diekmann and Jansen, 1991) is enhanced when differential dispersal rates of the interacting species among patches lead to a decoupling of their local population dynamics.

Why did theory dealing with patchiness develop more than that dealing with heterogeneity? Two reasons come to mind. First, patchiness refers to a particular spatial pattern – bounded elements in a background matrix. Although the specific arrangement of patches may take a variety of forms (see Pickett and White, 1985; Harrison, 1991), the basic structure is the same, and it is well-defined. In contrast, any form of spatial variation, from an unbounded gradient to a collection of various patch types (and including a simple patch–matrix system) is heterogeneous. The concept of 'heterogeneity' is diffuse, and it does not lend itself to precise theoretical development, especially using analytical models. Second, because of the unitary nature of patches, patch theory could be developed by analogy with traditional population dynamics models in ecology, with 'patches' substituting for 'individuals' (Hastings, 1991). It thus involved an extension of familiar theory into a spatial domain, rather than the generation of entirely new theory.

1.3.4 RECENT DEVELOPMENTS

Over the past two decades or so, interest in what might be called 'spatial ecology' has grown. Some of this growth has led to increased sophistication in the description of spatial patterns, aided by more powerful spatial statistics (Haining, 1990; Turner *et al.*, 1991; Rossi *et al.*, 1992), techniques for detecting patch boundaries (Gosz, 1991), and technological tools such as Geographic Information Systems (GIS) (Burke *et al.*, 1990; Schimel, Davis and Kittel, 1993). As computational power has increased, modelers have developed simulation models that incorporate spatial variation by allowing individual cells in a spatial grid to undergo dynamics that are spatially linked in various ways (Costanza, Sklar and White, 1990). Cellular automata (Phipps, 1992) are one version of such models, and some individual-based models incorporate a spatial dimension as well (e.g. Hyman, McAninch and DeAngelis, 1991; DeAngelis and Gross, 1992).

These tools all enhance our ability to test spatially explicit theory. Much of the recent growth in spatial theory has involved elaboration and extension of patch-based population models, especially metapopulation theory (Gilpin and Hanski, 1991). The emergence of conservation biology as a coherent discipline and increased concern about how human activities such as habitat fragmentation affect the spatial structure of populations have enhanced interest in metapopulation models, and 'metapopulation' has become something of a buzzword in the conservation literature.

Patch-based population theory has developed in several ways. A variety of patch arrangements and interactions has been incorporated into models (Harrison, 1991). Source–sink models, in which some patches exhibit a positive net recruitment and export dispersers while other patches have negative net recruitment and absorb dispersers, have attracted particular attention (Pulliam, 1988). Hanski (1991) has incorporated local patch dynamics into metapopulation models, while others have explored the consequences of varying patch size, quality or configuration (e.g. Verboom, Metz and Meelis, 1993). Hanski's work (1991) suggests that metapopulation persistence is enhanced by variation in patch size and quality, while other studies indicate that (as intuition would suggest) increasing patch clumping or patch size also increases the probability of population survival (Chapter 12). Other theoretical analyses have indicated that temporal variation in patch structure (i.e.

disturbance frequency) may have greater effects than spatial subdivision alone on population persistence (Fahrig, 1991). However, it has also become apparent that metapopulation theory may only apply in certain situations. If patches are too large and/or too close together, dispersal may mix their dynamics and erode the stabilizing influences of population subdivision. On the other hand, if patches are too small, too few, or too distant, dispersal may be inadequate to recolonize patches in which local extinction has occurred, and the system may proceed (stochastically) to global extinction (Hanski, 1991; Opdam *et al.*, 1993).

Despite these modifications, the basic structure of the theory remains much the same. Populations are subdivided into discrete patches separated by an inert matrix. Although the characteristics of neighboring patches may (through emigration) affect population density in a given patch, immediate patch context has no effect. Within-patch variation (in space, or among individuals) is generally not considered (i.e. local populations are 'well-mixed'). Increasingly, it is becoming clear that movement or dispersal is the key to the dynamical behavior of such patch-population models (Stenseth and Hansson, 1981; Hansson, 1991). In fact, De Roos, McCauley and Wilson (1991) have demonstrated that limited mobility of individuals in predator–prey systems can produce patchiness in predator and prey distributions and enhance system stability, irrespective of the underlying environmental heterogeneity. In a similar vein, Caswell and Cohen (1991) have shown that temporal variation and decoupling in disturbance rates, dispersal, and competitive exclusion can produce spatial variation among patches in the absence of any pre-existing spatial differentiation in the environment. The physical pattern of an area occupied by populations may also be affected by the populations. Such reciprocal patch–population interactions complicate attempts to generate simple, analytical theory.

1.4 HOW HAVE ECOLOGISTS DEALT WITH SPATIAL MOSAICS?

Despite all of the attention given to spatial patterns, heterogeneity and patchiness by ecologists, relatively little work has focused on the structure of spatial mosaics and their effects on ecological systems. This is (or should be) the focus of landscape ecology. The emergence of landscape ecology as a scientific discipline has drawn attention to ecological mosaics, and has generated a lot of questions and a fair amount of empirical research. Much of this research has focused on describing landscape structure or patterns, especially through the use of GIS. Several studies, however, have documented the importance of landscape-level effects on ecological phenomena such as community assembly or population structure (e.g. Drake *et al.*, 1993; Scribner and Chesser, 1993). Because theory is not well developed, formal tests of predictions about the effects of mosaic structure are rare. Indeed, reading papers published in the journal *Landscape Ecology* gives one the impression that landscape ecology is not yet a particularly quantitative discipline, nor is it much concerned with the formalities of hypothesis-testing (Wiens, 1992a). To some degree, this situation reflects the difficulty of replicating samples and conducting experiments at the kilometers-wide spatial scales of landscapes as we normally perceive them (Forman and Godron, 1986). One solution to the problem is to use smaller-scale 'experimental model systems' (Ims and Stenseth, 1989; Wiens and Milne, 1989) to study mosaic effects.

This is not to say that theory (or at least concepts) are entirely lacking. Attention on theory that is explicitly related to landscapes has focused especially on two areas: 'ecological flows', i.e. the movement of individuals, materials or disturbances through mosaics (Hansen and di Castri, 1992; Vos and Opdam, 1993), and scaling (O'Neill *et al.*, 1986; Wiens, 1989c; Allen and Hoekstra, 1992; Levin, 1992).

1.4.1 ECOLOGICAL FLOWS

One dominant theme in studies of ecological flows has emphasized boundaries or ecotones between mosaic elements (Holland, Risser and Naiman, 1991). Some of this work has dealt with edge effects, such as the influence of predation or abiotic factors across edges among neighboring patches (Angelstam, 1992). Other studies have focused on the dynamics of movements across boundaries and on how boundaries control the strength of interactions among elements in a mosaic (Wiens, Crawford and Gosz, 1985). Both areas lead to the prediction that movement across boundaries may be reduced as boundary sharpness or contrast among elements increases.

As ecotones generally restrict movement, corridors linking similar elements in a landscape enhance it. Corridors have received considerable attention in conservation biology, where they are viewed as a way to counteract the damaging effects of habitat fragmentation (Noss, 1991; Saunders and Hobbs, 1991; Soulé, 1991; but see Simberloff and Cox, 1987). Computer simulation studies (Lefkovitch and Fahrig, 1985; Merriam, 1988; Merriam, Henein and Stuart-Smith, 1991; Soulé and Gilpin, 1991) confirm the intuitive predictions that movement of organisms through a corridor depends on the width of the corridor, the strength of the edge effect, the shape and relative linearity of the corridor, the spatial pattern of interconnections among patches and corridors, and (of course) the mobility of the organisms (see also Harrison, 1992).

A somewhat more esoteric way of modeling corridor effects on landscape flows involves the use of percolation theory, which was originally developed to deal with the physics of particle movement through systems

(Stauffer, 1985). Percolation models describe the probability that a particle (or an organism, or a disturbance event) will move across a landscape composed of two types of interdigitating elements, as a function of the relative proportions of elements enhancing or restricting movement. Gardner and his colleagues (Gardner *et al.*, 1987, 1989, 1991; Turner *et al.*, 1989) have applied this approach to a variety of ecological questions. One prediction of this theory is that the relationship between movement probability and coverage of the permeable cover type is nonlinear: as coverage decreases, there is a threshold of connectivity beyond which movement through the landscape suddenly becomes unlikely. Although the theory is abstract and simplified, the prediction of a threshold effect has obvious implications for assessing habitat fragmentation effects and the use of corridors as management tools.

In these theories or concepts, the effects of structural features of landscapes (boundaries, corridors, percolation networks) are mediated by movement. Understanding how movement occurs among landscape elements is therefore critical (May and Southwood, 1990; Wiens *et al.*, 1993; Chapter 4), but very little is known about how individuals respond to boundaries, use corridors, or move through a mosaic (Opdam, 1991). There has none the less been a good deal of modeling of animal movement, some incorporating only dispersal distance (e.g. Fahrig and Paloheimo, 1988), others employing variations on diffusion, random walk or correlated random walk algorithms (e.g. Okubo, 1980; McCulloch and Cain, 1989; Bovet and Benhamou, 1991; Cain, 1991; Johnson, Milne and Wiens, 1992; Crist *et al.*, 1992; Vail, 1993), and some using empirically derived movement pathways (With and Crist, unpublished; Crist and Wiens, unpublished). This work clearly indicates that there is an interplay between movement and landscape structure; for example, the spatial arrangement of patches is more likely to have important effects on a population if dispersal distances are intermediate rather than short or long, relative to the spacing of suitable patches (Fahrig, 1988).

1.4.2 SCALING

Perhaps because it is usually practiced at a broader scale of resolution than the local habitats historically studied by ecologists, landscape ecology has become closely associated with ecological scaling (Urban, O'Neill and Shugart, 1987; Turner, 1989; Wiens, 1992a; Allen *et al.*, 1993). Scaling issues, however, are not confined to landscape studies; they affect all ecological investigations (Wiens, 1989c). Indeed, the problem of interrelating processes occurring at different scales may be 'the central problem of theoretical biology' (Levin, 1992, 1993).

It has been clear for a long time to those describing spatial patterns that the nature of a pattern (e.g. the dispersion of individuals) changes with changes in the scale of analysis (Grieg-Smith, 1979). What is a boundary or a patch or a corridor at one scale may disappear or take on a different structure at another scale (Gosz, 1991). Several empirical studies have demonstrated clearly that patterns in community organization (e.g. Wiens, Rotenberry and Van Horne, 1987) or landscape complexity (e.g. Krummel et al., 1987) differ as the scale of analysis is changed. Exactly how patterns or processes change with changes in scale, however, is still unresolved. If patterns and processes change continuously with scale, then results at one scale can tentatively be extrapolated to other scales, once the scaling function is known. On the other hand, if ecological phenomena change discontinuously with scale, then extrapolation is only possible within scale 'domains' in which certain pattern–process relationships hold (Wiens, 1989c). Translating among scales then becomes a major problem (Turner, Dale and Gardner, 1989; King, 1991).

Although a formal 'theory of scaling' does not yet exist in ecology, hierarchy theory (Allen and Starr, 1982; O'Neill et al., 1988; Allen and Hoekstra, 1992) is closely related to scaling issues. Hierarchy theory is a way of ordering observational scales in a way that draws attention to the mechanisms and constraints that operate at a given level and how these change among levels. Gosz (1993), for example, has proposed a hierarchy of ecotones, from individual plants to populations to patches to landscapes to biomes. The factors affecting plants differ across this hierarchy, and much of the variation that is produced by constraints at fine scales is integrated or averaged out at broader scales, where only broad-scale constraints such as climate or topography may be important. The levels in a hierarchy such as this define the scales at which one must study across-scale influences.

Hierarchies, however, are artificial constructs that we impose on nature: we categorize phenomena into levels that are logically related. These categories may not reflect the actual scaling of natural systems. Hierarchical levels may be seemingly clearly defined when we are dealing with units such as cells, organisms or populations, but they are not intuitively apparent when we are dealing with spatial variation. The value of hierarchy theory may be more heuristic than operational.

Other studies have approached the scaling problem by examining the mechanisms underlying scaling. Individual movement may be particularly important in determining how the effects of spatial patterns on ecological phenomena are scaled (Wiens et al., 1993). For example, Gardner et al. (1991) have used percolation models to show how variation in the mean and shape of the dispersal function determines how a population will be affected by landscape fragmentation at a given scale. Using a cell-based stochastic simulation approach, De Roos, McCauley and Wilson (1991) have modeled how movement by predators affects the dynamics of predator–prey systems. If movement is limited, the dynamics at small scales are determined by biological interactions and can be portrayed by traditional density-limited, spatially homogeneous models. At large scales, different portions of space are so far apart that the limited movement only weakly couples them and their dynamics are out of phase. The magnitude of movement imposes a 'natural' scale on the spatial domain: observations at scales smaller or larger than this characteristic scale are entirely different. Because movement is so closely tied to mosaic structure, analyses of movement patterns may be particularly well-suited to examinations of scaling dynamics, as these studies suggest.

1.5 DEVELOPING THEORY IN LANDSCAPE ECOLOGY

If the traditional, hypothetico-deductive model of science applies to landscape ecology, we will make greatest progress in understanding how spatial mosaics influence (and are influenced by) ecological systems by developing theories that generate testable predictions. What form might such theory take? There are really two parts to this question. Before addressing these, however, I should comment briefly on the problem of defining and measuring landscape mosaics, since this will determine how (or whether) such theory can be tested or applied.

1.5.1 WHAT IS A LANDSCAPE MOSAIC?

The term 'mosaic' implies discreetness of elements, the existence of clear boundaries between neighboring patches. 'Landscapes', of course, may also be composed of elements that grade gradually into one another or in which the boundaries are blurred. If we are to make progress in developing landscape-level theory, however, it seems best to begin with the multi-patch, mosaic extension of patch–matrix approaches rather than to attempt to deal with the full range of possible landscape expressions.

To study mosaics, then, we must be able to define and map boundaries. In environments that are heavily modified by human activities, such boundaries are often sharp, but in natural settings they may be less evident. In either case, methods of image analysis (Overington and Greenway, 1987; Brunt and Conley, 1990) may be useful in detecting patch edges. Once the boundaries have been determined, mosaic structure can be quantified through a variety of measurements (Figure 1.3). Obviously, the scale on which mosaics are mapped and the parameters that are measured will depend on the organisms studied and the questions asked (Wiens, 1989c; Wiens et al., 1993). To be useful, theory should be insensitive to the choice of scale or (to some degree) parameters.

Even if consideration of landscapes is confined to well-defined mosaics, the variety of possible mosaic structures is still virtually unlimited. In contrast to the simple spatial structure considered in patch–matrix theory,

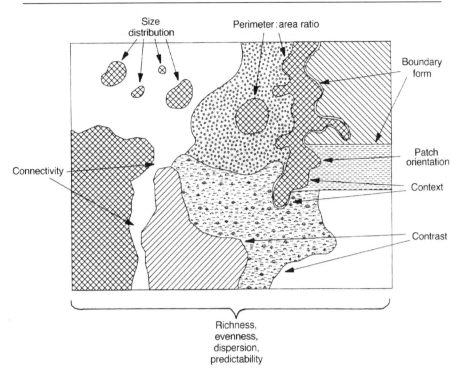

Figure 1.3 A hypothetical landscape mosaic, illustrating various parameters that may be measured. Measures such as richness, evenness, dispersion and predictability are features of the mosaic as a whole, whereas other parameters refer to specific configurations within the mosaic. (From Wiens *et al.* (1993), who provide further details, reproduced with permission.)

there is no single, general mosaic pattern about which theory can be generated. As a result, both empirical and theoretical studies may become bound to the idiosyncracies of the particular mosaic pattern considered, with little hope of generalization. It may be inappropriate to expect theory to provide global generalizations. Instead, it may be more useful to seek 'generic' generalizations. Thus, we might categorize species according to ecological or life-history characteristics (e.g. Keddy, 1991; Hansen, Urban and Marks, 1992; Hansen and Urban, 1992; Wiens *et al.*, 1993), or define groupings of landscape types based on various indices of landscape structure (e.g. O'Neill *et al.*, 1988), or assess the consequences of habitat fragmentation based on a relation between the scale of fragmentation and the scale of an organism's movements or population dynamics (e.g. Hassell, Godfray and Comins, 1993). Such approaches might serve to define 'domains' for the operation of particular theories.

1.5.2 WHAT ISSUES SHOULD LANDSCAPE THEORY ADDRESS?

In my view, theoretical efforts in landscape ecology may be most reward-
ing if they are focused on four areas. A start has been made on thinking
about ecological flows and scaling, and this work should certainly continue. Flows are the linking processes in mosaics; if flows are not affected
by landscape structure, then it is doubtful that the mosaic pattern will
have much effect on other ecological phenomena. Because movement is
so important in determining flows, it merits particular attention. If we
understand how movement patterns are affected by the configuration of
mosaics, we may be able to develop a mechanistic foundation for landscape ecology (Wiens et al., 1993).

A second area of emphasis is scaling. Calls for developing a 'theory
of scaling' in ecology are becoming more frequent (Meentemeyer and
Box 1987; Wiens, 1989c; Levin, 1992, 1993). As this theory develops,
attention should center on defining domains of scale and dealing with
the problem of translating pattern–process relationships across scales.
Because we know so little about scaling thresholds or domains, it may
be necessary to gather empirical observations over a range of scales in
different systems to determine where thresholds lie and where extrapolations are bounded (Gardner et al., 1989). Fractal analysis may be useful
in detecting such thresholds (Krummel et al., 1987; Palmer, 1988; Wiens,
1989c).

A third focus for theory is on the effects of spatial context on ecological
phenomena. The central premise of landscape ecology is that the structure of mosaics, not just patches, affects ecological patterns and processes.
Patch–matrix theory will continue to be useful for addressing problems
in which the island analogy is apt, such as a subdivided population in
which dispersal is unaffected by the habitat between patches. In many
situations, however, patch models may be inadequate to capture the
complexity of spatial interactions. Here, also, empirical studies may be
necessary to determine the sorts of patch-context effects that should
be incorporated into theory.

Finally, it may be useful to begin looking at landscape mosaics in
different ways. We often think of landscapes in terms of the physical
patterns that we (or satellites) perceive. In terrestrial landscapes, these
patterns are usually determined by vegetation and topography. This
perception may be convenient, but it may not reflect the spatial patterns
of factors that influence other organisms. Perhaps elements in a landscape
mosaic would better be viewed in terms of various costs and benefits.
Spatial locations may differ in such features as physiological stress, food
availability, predation risk, mating probability, and the like. Different
patches in a mosaic may then have different aggregate cost–benefit values

to an organism, and the mosaic as a whole can be viewed in terms of spatial cost–benefit contours (Wiens et al., 1993). If organisms moved through this cost–benefit landscape optimally, populations would tend to aggregate at the locally available 'peaks' in the contoured surface. However, costs and benefits are often separated in time, the cost–benefit surface changes through time, and individuals may be more concerned about minimizing costs than about maximizing benefits ('satisficing'; Ward, 1992). For these reasons, the 'optimal' distributions may rarely be attained. In any event, it may be useful to devote some theoretical attention to landscapes as cost–benefit surfaces.

1.5.3 HOW SHOULD THEORY BE DEVELOPED?

We should recognize that, even though nature usually exists as heterogeneous mosaics, not all problems in ecology require a spatially explicit analysis or solution. Models that assume spatial homogeneity may be applicable in some situations (e.g. De Roos, McCauley and Wilson (1991)), patch–matrix theory in others. Fahrig (personal communication) has used simulations to determine when spatial pattern might have important effects on population persistence. She concluded that spatially explicit models might not be necessary when the favored habitat is abundant, movement distances are large relative to inter-patch distances, movement patterns do not differ strongly among elements in the landscape, or habitats are dynamic and ephemeral, i.e. when various processes effectively 'homogenize' the landscape. Whatever the approach, the choice of models should be based on biological, not mathematical, considerations. Clearly, spatially homogeneous theories are no longer the default in ecology.

When mosaic theory is needed, it should be kept simple. One may be tempted to use all of the computational power at one's disposal or to push GIS to the limits of resolution. But bigger is not necessarily better. By the same token, individual-based models offer the potential to develop tremendously detailed bookkeeping analyses of movements and fates (Levin, 1993). But such models may become hopelessly complex and cumbersome if they simply aggregate individual dynamics over a varied spatial array (Kingsolver, Huey and Kareiva, 1993). Developing mechanistic insights about landscape processes is important, but reductionism must be judicious (Wiens et al., 1993).

When we think of theory, we generally think of mathematical, analytical models. Even if landscape effects are simplified to a few general forms, however, it is still likely that their complexity will exceed the limits of analytical tractability. We must therefore turn to simulation models to play the role usually occupied by analytical models. Although they may lack the beauty and elegance of analytical models, properly formulated simulation models may generate quantitative, testable hypotheses that can be used to enhance our understanding of mosaic-level interactions (Fahrig, 1991). It may also be possible to use simulation and analytical models in a coupled, iterative fashion. Simulation models, for example, could be used to conduct sensitivity tests for a large number of parameters, filtering out variables that might not need to be included in a simplified analytical model. A simplified analytical model could, in turn, suggest tests that might be best performed in a simulation model. The challenge of dealing with mosaic complexity may promote a synthesis of approaches to modeling that have often been regarded as mutually exclusive by ecologists.

In the end, we must ask whether it is reasonable to expect theory to play the same roles in landscape ecology as it has in other areas of ecology. Perhaps the array of complex spatial patterns that unfolds when we begin to consider mosaics is too daunting. Analytical approaches may be intractable, and simulation models may be bound too closely to specific parameter sets. The premise of landscape ecology, that spatial context makes a difference in ecological patterns and processes, is still more an article of faith than an empirically validated fact. Our approach to developing landscape theory thus far has focused on particular aspects of landscape structure – corridors, boundaries and the like. In an attempt to develop theory, we have intellectually fragmented landscape ecology. Perhaps we must develop theory inductively, by gathering enough empirical information about mosaic effects to permit us to generate testable propositions. The contributions in this volume provide a good perspective on what we know and what we need to know to take this step.

ACKNOWLEDGEMENTS

This chapter was written during a sabbatical leave at the University of British Columbia. I thank Charley Krebs, Tony Sinclair, and an International Scientific Exchange Award from the Natural Sciences and Engineering Research Council of Canada for making my stay in Vancouver possible. Peter Abrams and Beatrice Van Horne offered insightful comments on an early draft, and Tom Crist, Bruce Milne and Nils Chr. Stenseth provided an earlier impetus to think about theory in landscape ecology. My research on spatial ecology has been supported by the US National Science Foundation, most recently through Grant DEB–9207010.

REFERENCES

Allen, T. F. H. and Hoekstra, T. W. (1992) *Toward a Unified Ecology*, Columbia University Press, New York.

Allen, T. F. H. and Starr, T. B. (1982) *Hierarchy: Perspectives for Ecological Complexity*, University of Chicago Press, Chicago.

Allen, T. F. H., King, A. W., Milne, B. T. *et al.* (1993) The problem of scaling in ecology. *Evol. Trends Plants*, **7**, 3–8.

Andrewartha, H. G. and Birch, L. C. (1954) *The Distribution and Abundance of Animals*, University of Chicago Press, Chicago.

Angelstam, P. (1992) Conservation of communities – the importance of edges, surroundings and landscape mosaic structure, in *Ecological Principles of Nature Conservation* (ed. L. Hansson), Elsevier, London, pp. 9–70.

Bailey, N. T. J. (1975) *The Mathematical Theory of Infectious Diseases and its Applications*, Hafner, New York.

Bormann, F. H. and Likens, G. E. (1979) *Pattern and Process in a Forested Ecosystem*, Springer, New York.

Bovet, P. and Benhamou, S. (1991) Optimal sinuosity in central place foraging movements. *Anim. Behav.*, **42**, 57–62.

Brokaw, N. V. L. (1985) Treefalls, regrowth, and community structure in tropical forests, in *The Ecology of Natural Disturbance and Patch Dynamics* (eds S. T. A. Pickett and P. S. White), Academic, New York, pp. 53–69.

Brunt, J. W. and Conley, W. (1990) Behavior of a multivariate algorithm for ecological edge detection. *Ecol. Modelling*, **49**, 179–203.

Burke, I. C., Schimel, D. S., Yonker, C. M. *et al.* (1990) Regional modeling of grassland biogeochemistry using GIS. *Landsc. Ecol.*, **4**, 45–54.

Cain, M. L. (1991) When do treatment differences in movement behaviors produce observable differences in long-term displacements? *Ecology*, **72**, 2137–42.

Cale, W. G., Henebry, G. M. and Yeakley, J. A. (1989) Inferring process from pattern in natural communities. *BioScience*, **39**, 600–5.

Caswell, H. (1988) Theory and models in ecology: a different perspective. *Ecol. Modelling*, **43**, 33–44.

Caswell, H. and Cohen, J. E. (1991) Disturbance, interspecific interaction and diversity in metapopulations. *Biol. J. Linn. Soc.*, **42**, 193–218.

Cody, M. L. (1975) Towards a theory of continental species diversities: bird distributions over mediterranean habitat gradients, in *Ecology and Evolution of Communities* (eds M. L. Cody and J. M. Diamond), Harvard, Cambridge, pp. 214–57.

Costanza, R., Sklar, F. H. and White, M. L. (1990) Modeling coastal ecosystem dynamics. *BioScience*, **40**, 91–107.

Crist, T. O., Guertin, D. S., Wiens, J. A. and Milne, B. T. (1992) Animal movement in heterogeneous landscapes: an experiment with *Eleodes* beetles in shortgrass prairie. *Funct. Ecol.*, **6**, 536–44.

Crowley, P. H. (1981) Dispersal and the stability of predator–prey interactions. *Am. Nat.*, **118**, 673–701.

DeAngelis, D. L. and Gross, L. J. (eds) (1992) *Individual-based Models and Approaches in Ecology*, Chapman & Hall, London.

De Roos, A. M., McCauley, E. and Wilson, W. G. (1991) Mobility versus density-limited predator–prey dynamics on different spatial scales. *Proc. R. Soc. London, B*, **246**, 117–22.

Drake, J. A., Flum, T. E., Witteman, G. J. *et al.* (1993) The construction and assembly of an ecological landscape. *J. Anim. Ecol.*, **62**, 117–30.

Fagerström, T. (1987) On theory, data and mathematics in ecology. *Oikos*, **50**, 258–61.

Fahrig, L. (1988) A general model of populations in patchy habitats. *Appl. Math. Comput.*, **27**, 53–66.

Fahrig, L. (1991) Simulation methods for developing general landscape-level hypotheses of single-species dynamics, in *Quantitative Methods in Landscape Ecology* (eds M. G. Turner and R. H. Gardner), Springer, New York, pp. 417–42.

Fahrig, L. and Paloheimo, J. (1988) Determinants of local population size in patchy habitats. *Theor. Pop. Biol.*, **34**, 194–213.

Forman, R. T. T. and Godron, M. (1986) *Landscape Ecology*, Wiley, New York.

Gardner, R. H., Milne, B. T., Turner, M. G. and O'Neill, R. V. (1987) Neutral models for the analysis of broad-scale landscape pattern. *Landsc. Ecol.*, **1**, 19–28.

Gardner, R. H., O'Neill, R. V., Turner, M. G. and Dale, V. H. (1989) Quantifying scale-dependent effects of animal movement with simple percolation models. *Landsc. Ecol.*, **3**, 217–27.

Gardner, R. H., Turner, M. G., O'Neill, R. V. and Lavorel, S. (1991) Simulation of the scale-dependent effects of landscape boundaries on species persistence and dispersal, in *Ecotones. The Role of Landscape Boundaries in the Management and Restoration of Changing Environments* (eds M. M. Holland, P. G. Risser and R. J. Naiman), Chapman & Hall, New York, pp. 76–89.

Gauch, H. G., Jr (1982) *Multivariate Analysis in Community Ecology*, Cambridge University Press, Cambridge.

Gilpin, M. and Hanski, I. (eds) (1991) *Metapopulation Dynamics: Empirical and Theoretical Investigations*, Academic Press, London.

Gosz, J. R. (1991) Fundamental ecological characteristics of landscape boundaries, in *Ecotones. The Role of Landscape Boundaries in the Management and Restoration of Changing Environments* (eds M. M. Holland, P. G. Risser and R. J. Naiman), Chapman & Hall, New York, pp. 8–30.

Gosz, J. R. (1993) Ecotone hierarchies. *Ecol. Appl.*, **3**, 369–76.

Grant, P. R. (1972) Convergent and divergent character displacement. *Biol. J. Linn. Soc.*, **4**, 39–68.

Grieg-Smith, P. (1979) Pattern in vegetation. *J. Ecol.*, **67**, 755–79.

Haggett, P., Cliff, A. and Frey, A. (1977) *Locational Analysis in Human Geography*, Edward Arnold, London.

Haila, Y. (1986) On the semiotic dimension of ecological theory: the case of island biogeography. *Biol. Philos.*, **1**, 377–87.

Haila, Y. (1989) Ecology finding evolution finding ecology. *Biol. Philos.*, **4** 235–44.

Haila, Y. and Järvinen, O. (1982) The role of theoretical concepts in understanding the ecological theater: a case study on island biogeography, in *Conceptual Issues in Ecology* (ed. E. Saarinen), D. Reidel, Boston, pp. 261–78.

Haining, R. (1990) *Spatial Data Analysis in the Social and Environmental Sciences*, Cambridge University Press, Cambridge.

Hansen, A. J. and di Castri, F. (eds) (1992) *Landscape Boundaries: Consequences for Biological Diversity and Ecological Flows*, Springer, New York.

Hansen, A. J. and Urban, D. L. (1992) Avian response to landscape pattern: the role of species life histories. *Landsc. Ecol.*, **7**, 163–80.

Hansen, A. J., Urban, D. L. and Marks, B. (1992) Avian community dynamics: the interplay of human landscape trajectories and species life histories, in *Landscape Boundaries: Consequences for Biological Diversity and Ecological Flows* (eds A. J. Hansen and F. di Castri), Springer, New York, pp. 170–95.

Hanski, I. (1983) Coexistence of competitors in a patchy environment. *Ecology*, **64**, 493–500.

Hanski, I. (1985) Single-species spatial dynamics may contribute to long-term rarity and commonness. *Ecology*, **66**, 335–43.

Hanski, I. (1991) Single-species metapopulation dynamics: concepts, models and observations. *Biol. J. Linn. Soc.*, **42**, 17–38.

Hansson, L. (1991) Dispersal and connectivity in metapopulations. *Biol. J. Linn. Soc.*, **42**, 89–103.

Harper, J. L. (1969) The role of predation in vegetational diversity. *Brookhaven Symp. Biol.*, **22**, 48–62.

Harrison, R. L. (1992) Toward a theory of inter-refuge corridor design. *Conserv. Biol.*, **6**, 293–5.

Harrison, S. (1991) Local extinction in a metapopulation context: an empirical evaluation. *Biol. J. Linn. Soc.*, **42**, 73–88.

Harrison, S. (1993) Species diversity, spatial scale, and global change, in *Biotic Interactions and Global Change* (eds. P. M. Kareiva, J. G. Kingsolver and R. B. Huey), Sinauer, Sunderland, MA, pp. 388–401.

Hassell, M. P., Godfray, H. C. J. and Comins, H. N. (1993) Effects of global change on the dynamics of insect host–parasitoid interactions, in *Biotic Interactions and Global Change* (eds P. M. Kareiva, J. G. Kingsolver and R. B. Huey), Sinauer, Sunderland, MA, pp. 402–23.

Hastings, A. (1991) Structured models of metapopulation dynamics. *Biol. J. Linn. Soc.*, **42**, 57–71.

Hilborn, R. and Stearns, S. C. (1982) On inference in ecology and evolutionary biology: the problem of multiple causes. *Acta Biotheor.*, **31**, 145–64.

Holland, M. M., Risser, P. G. and Naiman, R. J. (eds) (1991) *Ecotones. The Role of Landscape Boundaries in the Management and Restoration of Changing Environments*, Chapman & Hall, New York.

Horn, H. S. and MacArthur, R. H. (1972) Competition among fugitive species in a harlequin environment. *Ecology*, **53**, 749–52.

Huston, M. A. (1994) *Biological Diversity: The Coexistence of Species*, Cambridge University Press, Cambridge.

Hyman, J. B., McAninch, J. B. and DeAngelis, D. L. (1991) An individual-based simulation model of herbivory in a heterogeneous landscape, in *Quantitative Methods in Landscape Ecology* (eds M. G. Turner and R. H. Gardner), Springer, New York, pp. 443–75.

Ims, R. A. and Stenseth, N. C. (1989) Divided the fruitflies fall. *Nature*, **342**, 21–2.

Johnson, A. R., Milne, B. T. and Wiens, J. A. (1992) Diffusion in fractal landscapes: simulations and experimental studies of tenebrionid beetle movement. *Ecology*, **73**, 1968–83.

Judson, H. F. (1980) *The Search for Solutions*, Holt, Rinehart and Winston, New York.

Keddy, P. A. (1991) Working with heterogeneity: an operator's guide to environmental gradients, in *Ecological Heterogeneity* (eds J. Kolasa and S. T. A. Pickett), Springer, New York, pp. 181–201.

King, A. W. (1991) Translating models across scales in the landscape, in *Quantitative Methods in Landscape Ecology* (eds M. G. Turner and R. H. Gardner), Springer, New York, pp. 479–517.

Kingsland, S. E. (1985) *Modeling Nature. Episodes in the History of Population Ecology*, University Chicago Press, Chicago.

Kingsolver, J. G., Huey, R. B. and Kareiva, P. M. (1993) An agenda for population and community research on global change, in *Biotic Interactions and Global Change* (eds. P. M. Kareiva, J. G. Kingsolver and R. B. Huey), Sinauer, Sunderland, MA, pp. 480–6.

Kolasa, J. and Pickett, S. T. A. (eds) (1991) *Ecological Heterogeneity*, Springer, New York.

Krummel, J. R., Gardner, R. H. and Sugihara, G. *et al.* (1987) Landscape patterns in a disturbed environment. *Oikos*, **48**, 321–4.

Lande, R. (1987) Extinction thresholds in demographic models of territorial populations. *Am. Nat.*, **130**, 624–35.

Lefkovitch, L. P. and Fahrig, L. (1985) Spatial characteristics of habitat patches and population survival. *Ecol. Modelling*, **30**, 297–308.

Levin, S. A. (1992) The problem of pattern and scale in ecology. *Ecology*, **73**, 1943–67.

Levin, S. A. (1993) Concepts of scale at the local level, in *Scaling Physiologi-*

cal Processes. Leaf to Globe (eds J. R. Ehleringer and C. B. Field), Academic Press, New York, pp. 7–19.

Levins, R. (1970) Extinction, in *Some Mathematical Problems in Biology* (ed. M. Gesternhaber), Amer. Math. Soc., Providence, RI, pp. 77–107.

Loehle, C. (1987) Hypothesis testing in ecology: psychological aspects and the importance of theory maturation. *Q. Rev. Biol.*, **62**, 397–409.

MacArthur, R. H. (1972) *Geographical Ecology*, Harper & Row, New York.

MacArthur, R. H. and MacArthur, J. W. (1961) On bird species diversity. *Ecology*, **42**, 594–8.

May, R. M. and Southwood, T. R. E. (1990) Introduction, in *Living in a Patchy Environment* (eds B. Shorrocks and I. R. Swingland), Oxford, Oxford, pp. 1–22.

McCauley, D. E. (1993) Genetic consequences of extinction and recolonization in fragmented habitats, in *Biotic Interactions and Global Change* (eds P. M. Kareiva, J. G. Kingsolver and R. B. Huey), Sinauer, Sunderland, MA, pp. 217–33.

McCulloch, C. E. and Cain, M. L. (1989) Analyzing discrete movement data as a correlated random walk. *Ecology*, **70**, 383–8.

Meentemeyer, V. and Box, E. O. (1987) Scale effects in landscape studies. in *Landscape Heterogeneity and Disturbance* (ed. M. G. Turner), Springer, New York, pp. 15–34.

Merriam, G. (1988) Landscape dynamics in farmland. *Trends Ecol. Evol.*, **19**, 16–20.

Merriam, G., Henein, K. and Stuart-Smith, K. (1991) Landscape dynamics models, in *Quantitative Methods in Landscape Ecology* (eds M. G. Turner and R. H. Gardner), Springer, New York, pp. 399–416.

Noss, R. F. (1991) Landscape connectivity: different functions at different scales, in *Landscape Linkages and Biodiversity* (ed. W. E. Hudson), Island Press, Washington, pp. 27–39.

Okubo, A. (1980) *Diffusion and Ecological Problems: Mathematical Models*, Springer, New York.

O'Neill, R. V., DeAngelis, D. L., Waide, J. B. and Allen, T. F. H. (1986) *A Hierarchical Concept of Ecosystems*, Princeton University Press, Princeton, NJ.

O'Neill, R. V., Krummel, J. R., Gardner, R. H. *et al.* (1988) Indices of landscape pattern. *Landsc. Ecol.*, **1**, 153–62.

O'Neill, R. V., Gardner, R. H., Turner, M. G. and Romme, W. H. (1992) Epidemiology theory and disturbance spread on landscapes. *Landsc. Ecol.*, **7**, 19–26.

Opdam, P. (1991) Metapopulation theory and habitat fragmentation: a review of holarctic breeding bird studies. *Landsc. Ecol.*, **4**, 93–106.

Opdam, P., van Apeldoorn, R., Schotman, A. and Kalkhoven, J. (1993) Population responses to landscape fragmentation, in *Landscape Ecology of a Stressed Environment* (eds C. C. Vos and P. Opdam), Chapman and Hall, London, pp. 147–71.

Overington, I. and Greenway, P. (1987) Practical first-difference edge detection with subpixel accuracy. *Image Vision Comput.*, **5**, 217–24.

Palmer, M. W. (1988) Fractal geometry: a tool for describing spatial patterns of plant communities. *Vegetatio*, **75**, 91–102.

Peters, R. H. (1991) *A Critique for Ecology*, Cambridge University Press, Cambridge.

Phipps, M. J. (1992) From local to global: the lesson of cellular automata, in *Individual-based Models and Approaches in Ecology* (eds D. L. DeAngelis and L. J. Gross), Chapman and Hall, New York, pp. 165–87.

Pickett, S. T. A. and White, P. S. (1985) *The Ecology of Natural Disturbance and Patch Dynamics*, Academic, New York.

Popper, K. R. (1959) *The Logic of Scientific Discovery*, Hutchinson, London.

Pulliam, H. R. (1988) Sources, sinks and population regulation. *Am. Nat.*, **132**, 652–61.

Reeve, J. D. (1988) Environmental variability, migration and persistence in host-parasitoid systems. *Am. Nat.*, **132**, 810–36.

Rossi, R. E., Mulla, D. J., Journel, A. G. and Franz, E. H. (1992) Geostatistical tools for modeling and interpreting ecological spatial dependence. *Ecol. Monogr.*, **62**, 277–314.

Roth, R. R. (1976) Spatial heterogeneity and bird species diversity. *Ecology*, **57**, 773–82.

Roughgarden, J. (1983) Competition and theory in community ecology. *Am. Nat.*, **122**, 583–601.

Sabelis, M. W., Diekmann, O. and Jansen, V. A. A. (1991) Metapopulation persistence despite local extinction: predator–prey patch models of the Lotka-Volterra type. *Biol. J. Linn. Soc.*, **42**, 267–83.

Saunders, D. A. and Hobbs, R. J. (eds) (1991) *The Role of Corridors*, Surrey Beatty & Sons, Chipping Norton, Australia.

Schimel, D. S., Davis, F. W. and Kittel, T. G. F. (1993) Spatial information for extrapolation of canopy processes: examples from FIFE, in *Scaling Physiological Processes. Leaf to Globe* (eds J. R. Ehleringer and C. B. Field), Academic, New York, pp. 21–38.

Schluter, D. and McPhail, J. D. (1993) Character displacement and replicate adaptive radiation. *Trends Ecol. Evol.*, **8**, 197–200.

Scribner, K. T. and Chesser, R. K. (1993) Environmental and demographic correlates of spatial and seasonal genetic structure in the eastern cottontail (*Sylvilagus floridanus*). *J. Mamm.*, **74**, 1026–44.

Simberloff, D. and Cox, J. (1987) Consequences and costs of conservation corridors. *Conserv. Biol.*, **1**, 63–71.

Slatkin, M. (1974) Competition and regional coexistence. *Ecology*, **55**, 128–34.

Slatkin, M. (1985) Gene flow in natural populations. *Ann. Rev. Ecol. Syst.*, **16**, 393–430.

Soulé, M. E. (1991) Theory and strategy, in *Landscape Linkages and Biodiversity* (ed. W. E. Hudson), Island Press, Washington, pp. 91–104.

Soulé, M. E. and Gilpin, M. E. (1991) The theory of wildlife corridor capability, in *The Role of Corridors* (eds. D. A. Saunders and R. J. Hobbs), Surrey Beatty & Sons, Chipping Norton, Australia, pp. 3–8.

Stauffer, D. (1985) *Introduction to Percolation Theory*, Taylor and Francis, London.

Stenseth, N. C. (1984) Why mathematical models in evolutionary ecology? in *Trends in Ecological Research for the 1980s* (eds J. H. Cooley and F. B. Golley), Plenum, New York, pp. 239–87.

Stenseth, N. C. and Hansson, L. (1981) The importance of population dynamics in heterogeneous landscapes: management of vertebrate pests and some other animals. *Agro-Ecosystems*, **7**, 187–211.

Strong, D. R., Jr (1982) Null hypotheses in ecology, in *Conceptual Issues in Ecology* (ed. E. Saarinen), D. Reidel, Boston, pp. 245–59.

Taylor, A. D. (1988) Large-scale spatial structure and population dynamics in arthropod predator–prey systems. *Ann. Zool. Fennici*, **25**, 63–74.

Taylor, A. D. (1990) Metapopulations, dispersal and predator–prey dynamics: an overview. *Ecology*, **71**, 429–33.

Turner, M. G. (1989) Landscape ecology: the effect of pattern on process. *Ann. Rev. Ecol. Syst.*, **20**, 171–97.

Turner, M. G., Gardner, R. H., Dale, V. H. and O'Neill, R. V. (1989) Predicting the spread of disturbance across heterogeneous landscapes. *Oikos*, **55**, 121–9.

Turner, M. G., Dale, V. H. and Gardner, R. H. (1989) Predicting across scales: theory development and testing. *Landsc. Ecol.*, **3**, 245–52.

Turner, S. J., O'Neill, R. V., Conley, W. *et al.* (1991) Pattern and scale: statistics for landscape ecology, in *Quantitative Methods in Landscape Ecology* (eds M. G. Turner and R. H. Gardner), Springer, New York, pp. 17–49.

Underwood, A. J. (1990) Experiments in ecology and management: their logics, functions and interpretations. *Aust. J. Ecol.*, **15**, 365–89.

Urban, D. L., O'Neill, R. V. and Shugart, H. H. (1987) Landscape ecology. *BioScience*, **37**, 119–27.

Vail, S. G. (1993) Scale-dependent responses to resource spatial pattern in simple models of consumer movement. *Am. Nat.*, **141**, 199–216.

Verboom, J. Metz, J. A. J. and Meelis, E. (1993) Metapopulation models for impact assessment of fragmentation, in *Landscape Ecology of a Stressed Environment* (eds C. C. Vos and P. Opdam), Chapman and Hall, London, pp. 172–91.

Vos, C. C. and Opdam, P. (eds) (1993) *Landscape Ecology of a Stressed Environment*, Chapman and Hall, London.

Ward, D. (1992) The role of satisficing in foraging theory. *Oikos*, **63**, 312–17.

Whittaker, R. H. (1960) Vegetation of the Siskiyou Mountains, Oregon and California. *Ecol. Monogr.*, **23**, 41–78.

Whittaker, R. H. (1965) Dominance and diversity in land plant communities. *Science*, **147**, 250–60.

Wiens, J. A. (1974) Habitat heterogeneity and avian community structure in North American grasslands. *Am. Midl. Natur.*, **91**, 195–213.

Wiens, J. A. (1989a) *The Ecology of Bird Communities. Volume 2. Processes and Variations.* Cambridge University Press, Cambridge.

Wiens, J. A. (1989b) *The Ecology of Bird Communities. Volume 1. Foundations and Patterns.* Cambridge University Press, Cambridge.

Wiens, J. A. (1989c) Spatial scaling in ecology. *Funct. Ecol.*, **3**, 385–97.

Wiens, J. A. (1992a) What is landscape ecology, really? *Landsc. Ecol.*, **7**, 149–50.

Wiens, J. A. (1992b) Ecological flows across landscape boundaries: a conceptual overview, in *Landscape Boundaries* (eds A. J. Hansen and F. di Castri), Springer, New York, pp. 217–35.

Wiens, J. A. and Milne, B. T. (1989) Scaling of 'landscapes' in landscape ecology, or, landscape ecology from a beetle's perspective. *Landsc. Ecol.*, **3**, 87–96.

Wiens, J. A., Crawford, C. S. and Gosz, J. R. (1985) Boundary dynamics: a conceptual framework for studying landscape ecosystems. *Oikos*, **45**, 421–7.

Wiens, J. A., Rotenberry, J. T. and Van Horne, B. (1987) Habitat occupancy patterns of North American shrubsteppe birds: the effects of spatial scale. *Oikos*, **48**, 132–47.

Wiens, J. A., Stenseth, N. C., Van Horne, B. and Ims, R. A. (1993) Ecological mechanisms and landscape ecology. *Oikos*, **66**, 369–80.

Part One

Origin of Landscape Pattern

Landscape pattern arises from two sources: primary physical heterogeneity and secondary biotic heterogeneity which is added to the primary physical pattern by organisms modifying it or adding elements to it. Primary physical patterns elicit responses from organisms, organisms modify and add to physical patterns, and the new, more complex patterns elicit additional responses from organisms. Not only the organisms causing the pattern are affected. The entire community using the landscape may be affected.

Joy Belsky presents examples from African grasslands of the influence of the original physical patterns on landscape use by herbivores. She also investigates how modifications of the secondary biotic heterogeneity can cause higher level repatterning by the combined influence of various organisms, including humans. Carol Johnston focuses on the other extreme where organisms, through effects required for their survival, inadvertently alter the composition and pattern of the landscape by changing the proportions of certain elements or by adding entirely new elements.

Both of the chapters in this section provide strong reminders that organisms do not necessarily accept environments as they find them. They commonly modify and create environments both for themselves and for others. The landscape patterns affecting the ecological processes discussed in other chapters – individual movement, habitat selection, population dynamics, population genetic processes, and species interactions – are therefore potentially also affected by these processes. This section also reminds us that the physical and the biotic patterns in a landscape are not necessarily spatially coincident, and that the boundaries of landscape elements are dynamic responses to, as well as controlling variables on, ecological processes at the landscape scale.

Spatial and temporal landscape patterns in arid and semi-arid African savannas

2

A. Joy Belsky

2.1 INTRODUCTION

> Dense herds of wildebeest, ten-thousand strong, move slowly across the Serengeti plains, creating clouds of dust, consuming every blade, reducing the grasslands to a height of a few centimeters . . .
> Elephants stand in a ghostly woodland, browsing on broken limbs and uprooted trees . . .
> Starving zebra lie on the ground near an overgrazed, trampled, devastated waterhole . . .

These images, derived from television, magazines and newspaper reports, illustrate the ability of large herbivores to alter vegetational landscapes. The images are reinforced by numerous scientific articles and reviews detailing how large mammalian herbivores of arid and semi-arid African savannas control, and sometimes overwhelm, their environments by reducing plant biomass, altering productivity, disturbing soils, transporting nutrients and altering decomposition rates. With vast amounts of empirical evidence showing that large native mammalian herbivores of African savannas exert tremendous influences on their environment, it is easy to conclude that these animals are the dominant

Mosaic Landscapes and Ecological Processes.
Edited by Lennart Hansson, Lenore Fahrig and Gray Merriam.
Published in 1995 by Chapman & Hall, London. ISBN 0 412 45460 2

forces shaping vegetational landscapes and creating savanna pattern. But is this true? A careful reading of the literature suggests not! Although it is evident that native wildlife species have an impact on their environment (Chapter 3), especially when their populations are artificially compressed by human activities, they are much more likely to be influenced by existing vegetation patterns (Chapters 4, 5, 6 and 9) than to influence them.

The above conclusions should not be construed to mean that native wildlife species have only a minor effect on the composition and pattern of savanna landscapes. Most researchers who have erected large-mammal-proof enclosures in African savannas have reported substantial and significant changes in plant community composition and diversity (Lock, 1972; Edroma, 1981; Belsky, 1986a, b, c, 1992), vegetation structure (Belsky, 1984; Smart, Hatton and Spence, 1985), biomass (Strugnell and Pigott, 1978; Guy, 1989), and chemical and physical properties of soils (Lock, 1972; Hatton and Smart, 1984). Researchers have also shown that large herbivores alter fire frequencies in savannas by reducing fuel loads, thereby influencing the ratio of woody to herbaceous biomass (Norton-Griffiths, 1979). In addition, drastic reductions in plant biomass and erosion of soils near rivers and water holes during periods of high herbivore density (Glover, 1963; Sinclair and Fryxell, 1985) illustrate the ability of large mammals to radically alter their physical environments.

In spite of this evidence, a look at the entire picture suggests that large mammals and fire play only secondary roles in structuring the vegetational landscape of pristine African habitats. Of greater importance are geomorphology, soil chemistry and soil moisture (Tinley, 1982; Cole, 1986; Belsky, 1990; Coughenour and Ellis, 1993) (Figure 2.1).

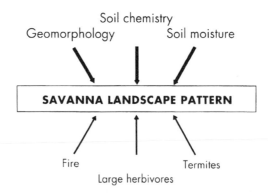

Figure 2.1 Factors influencing landscape pattern in African savannas. The importance of each factor is indicated by arrow width.

2.2 REVIEW OF LANDSCAPE ECOLOGY IN AFRICA

Landscape ecology of sub-Saharan savannas in tropical Africa has seldom been investigated explicitly. (In this paper I use Forman's (1987) definition of landscape as a 'heterogeneous land area composed of a cluster of interacting ecosystems that is repeated in similar form throughout' (p. 217).) In fact, a search of the sub-Saharan savanna literature revealed only six publications with the term 'landscape' in their titles (i.e. Laws, 1970; Pritchard, 1979; Loth and Prins, 1986; Belsky, 1989; Coughenour, 1992; Coughenour and Ellis, 1993); and only four papers (Belsky, 1989; Coughenour, 1991, 1992; Coughenour and Ellis, 1993) discussed spatial and temporal patterns in terms of landscape-scale interactions. Most studies of African savannas were investigated before landscape ecology became widely recognized as a sub-discipline (following publication of books on the topic by Naveh and Lieberman (1984) and Forman and Godron (1986)). Earlier studies focused on other ecological levels, such as ecosystem-level descriptions of productivity and nutrient flux and community-level descriptions of species composition.

Only by carefully and creatively reading between the lines of the existing literature do important landscape-level patterns and processes in African savannas emerge. This creative approach has been encouraged by Risser (1987) and Forman (1987), who exhorted ecologists to think broadly and innovatively about landscape-level questions. In this spirit, I have sifted through a considerable amount of literature on arid and semi-arid sub-Saharan African savannas, searching for patterns that might be related to landscape processes. I will first describe the major landscape patterns that have emerged from this search. Information on several of these patterns is limited to one or a few examples or to one or a few areas, but they most likely represent repeated landscape elements in other savannas also. I will discuss those physical and biological factors that have formed and subsequently maintained landscape pattern. Where information is available, I will also discuss how landscape patterns influence wildlife distributions and behavior, and how wildlife in turn influences the larger landscapes.

In the second and third sections of this chapter, I will discuss two important examples of landscape-level processes involving large herbivores: the impacts of fire and herbivores in creating grassland–woodland mosaics in African savannas; and the role of artificially developed livestock watering points in creating landscape pattern and in altering ecosystem function.

2.3 LANDSCAPES AND LANDSCAPE ECOLOGY IN AFRICAN SAVANNAS

A savanna is most often defined as a tropical community with a continuous herbaceous layer, usually dominated by C_4 grasses, and a discontinuous woody layer of shrubs or trees. Plant productivity in arid and semiarid African savannas is limited first and foremost by rainfall (Deshmukh, 1984; van Wijngaarden, 1985; Scholes, 1991), or to be more exact, by soil moisture. All published regressions of plant productivity versus rainfall show a direct increase in productivity with increasing rainfall. At some sites, this linear response becomes asymptotic at 700–900 mm rainfall, at which point plant growth is controlled by soil nutrient status (Scholes, 1991). Although changes in plant productivity do not necessarily create landscape pattern, in arid and semi-arid regions they appear to be related: changes in plant biomass along soil-moisture gradients are invariably accompanied by changes in species composition and often by changes in vegetational structure. Whenever the transition zone between communities with different compositions is relatively abrupt, visible pattern is created.

2.3.1 DRAINAGES – THE DOMINANT LANDSCAPE PATTERN

At both ends of the rainfall gradient, African savannas are dominated by woody species. At the drier end of the gradient, grasslands are replaced by thorn scrubland; and at the wetter end, they are replaced by woodland or forest (Belsky, 1990; Coughenour and Ellis, 1993). At any point along the soil-moisture continuum, factors that alter moisture conditions – such as changes in frequency and intensity of rainfall, soil water-holding capacity and depth of water penetration – are invariably associated with changes in species composition and productivity, and often with altered stature and density of woody species. These changes in community composition and stature create the most frequently observed repeating landscape pattern in African savannas – ribbons of tall, green, dense vegetation growing along watercourses in otherwise short, dry grasslands (Figure 2.2). The most intense of these patterns are the dense riverine and lacustrine forests growing at the edges of permanent rivers and lakes. Less intense but equally dramatic are the more sparsely wooded communities growing along seasonal drainage lines. A short distance from the lakes and water channels, these green ribbons grade sharply into short herbaceous vegetation. A similar pattern occurs where water drains downslope to the bottom of hills and ridges: grasses, trees and shrubs are denser, taller and greener on lower than on upper slopes. The width of the forests along rivers and the density of trees along ephemeral streams and at the bottom of slopes may be modified by fire and large herbivores; however, the basic pattern – repeating stripes of tall, green grasses, trees and shrubs in short grasslands – dominates the landscape.

Figure 2.2 Riparian vegetation associated with a river meandering through the grassland of the Serengeti National Park, Tanzania. In the background is a wood-land–grassland mosaic created by fire and the activities of large herbivores. Photo by A. R. S. Sinclair.

2.3.2 ESCARPMENTS – PATCH UTILIZATION BY WILDLIFE

In areas of strong relief such as along escarpments associated with the Rift Valley of East Africa, vegetation is strongly patterned. The escarpment within Lake Manyara National Park in northern Tanzania, for example, has a sharp vertical rise of 300–700 m, extending from Lake Manyara at the bottom of the Rift Valley, up a steep slope, to a plateau on top. The vegetation along this gradient consists of a mosaic of small, discrete patches that have an average size of 7 ha (Prins, 1989). The compositions of these patches vary with distance above the water table, water infiltration rate, geological substrate and soil salinity. Plant communities replace one another along the elevational gradient – from water-logged plains containing *Cyperus* swamps, alkaline grasslands, and bushlands near the lake, to better drained woodlands, to wet groundwater forests near the foot of the escarpment, to dry bushlands on the excessively drained slopes, and finally to arid grasslands on the upper plateau (Loth and Prins, 1986).

This high diversity of landscape patches and high productivity associated with year-round water availability near the lake have resulted in Lake Manyara National Park having the highest density of mammalian herbivore biomass in Africa (Coe, Cumming and Phillipson, 1976). But in spite of the known ability of large herbivores to alter the structure and dynamics of plant communities (Cumming, 1982; McNaughton and Georgiadis, 1986; Belsky, 1989; Coughenour, 1991; Skarpe, 1991), landscape pattern in Lake Manyara National Park is determined first and foremost by soil moisture and edaphic conditions (Greenway and Vesey-FitzGerald, 1969; Prins, 1988).

There is little indication that the high biomass of large animals in Lake Manyara National Park is affecting landscape pattern; however, there is ample evidence of the reverse – that landscape pattern is affecting the animals. For example, the distribution, population size and movements of buffalo (*Syncerus caffer*), an intensively studied herbivore in the park, are strongly affected by the spatial relationships of patches within the landscape (Prins and Beekman, 1987; Prins and Iason, 1988). These large ruminants move among vegetation patches both daily and seasonally as they search for better forage and cooler temperatures and as they try to avoid troublesome insects. During the day, buffalo occupy the cool lakeshore habitats where they graze while avoiding biting tsetse flies; but in the evenings they enter woodlands to graze. During most seasons the animals prefer to graze in *Cynodon dactylon*-dominated patches; but in dry seasons they more frequently utilize *Cyperus laevigatus* swamps, *Chloris gayana* patches, and woodlands.

Landscape pattern also determines buffalo mortality patterns. Predation by lion (*Panthera leo*) is the major cause of mortality in buffalo, causing 88% of deaths (Prins and Iason, 1988). Lions attack buffalo most frequently in *Sporobolus spicatus* grasslands within 100 m of the edge of woodlands, which provide cover for the predators. The greatest buffalo mortality, therefore, occurs near the ecotone between grassland and woodland, and it decreases sharply with distance in both directions. However, in spite of the risk associated with this edge, buffalo pass through it twice daily, entering the woodlands every evening and emerging every morning. These daily movements suggest that buffalo in Lake Manyara National Park are utilizing habitat patches in response to food availability, temperature, and escape from insects – not in response to protection from predators (Prins and Iason, 1988). The same conclusion was reached by Sinclair (1977) for buffalo in the Serengeti National Park, Tanzania. There, buffalo prefer riverine habitats over open grasslands, even though lion predation is highest near rivers where lions hide in tall grasses and reeds.

Although edges between grasslands and woodlands are associated with increased risk of predation, buffalo in Lake Manyara National Park are not noticeably more cautious near these ecotones than in other areas (Prins and Iason, 1988). This is in contrast to the Serengeti National Park, where buffalo are noticeably uneasy before passing from grasslands into woodlands (Schaller, 1972). Prins and Iason did note, however, that buffalo in Lake Manyara National Park were more vigilant in woodlands than in other vegetation types, possibly because communication between members of the herd breaks down among the dense trees.

2.3.3 CATENAS – PATCH UTILIZATION BY UNGULATES

The catenary structure of many African savannas contributes significantly to savanna pattern. In areas of undulating topography, the continuing processes of rainfall runoff and soil transport from ridge-tops to valley-bottoms create horizontal soil sequences known as catenas (Milne, 1935; Morison, Hoyle and Hope-Simpson, 1948) (Figure 2.3). The soil sequences are vegetated by different plant communities, creating vegetation zones that vary from the tops of the catenas to the bottoms. The shallow sandy soils that dominate the ridge-tops are vegetated by short, shallowly rooted species, while the heavy clay soils that accumulate in the more mesic valley-bottoms are vegetated by taller and more deeply rooted species (Morison, Hoyle and Hope-Simpson, 1948; Bell, 1970). In between are medium-height grasses.

While often less obvious than vegetation patterns associated with permanent water sources, this sequence of grassland communities along

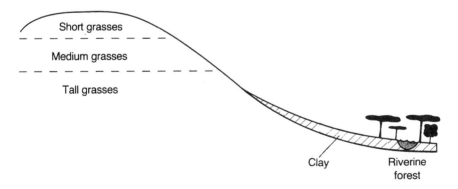

Short grasses

Medium grasses

Tall grasses

Clay

Riverine forest

Figure 2.3 Vegetation and soil patterns along catenas in savannas of sub-Saharan Africa. (Source: Bell, 1970.)

catenas forms a characteristic landscape in many savannas. Catena position, for example, was found to be a strong determinant of species composition in Tsavo National Park, Kenya (Jensen and Belsky, 1989), where upper-catena and lower-catena communities were found to be less similar to adjoining mid-slope communities than to other similarly positioned communities, even those over 30 km distant.

The horizontal vegetation bands associated with catenas have been shown to strongly influence the movement of mammalian herbivores. In the Serengeti National Park, Bell (1970) found that large ungulates prefer the short-grass communities of the upper catenas during the wet season, but that they move downslope as the vegetation senesces during the dry season. He reported that zebra (*Equus burchelli*), the largest ungulate species in the community, moves downslope first, where individuals graze the taller, more fibrous grasses. One to several weeks later, they are followed by the intermediate-sized ungulates, topi (*Damaliscus korrigum*) and wildebeest (*Connochaetes taurinus*). The last group to descend the catena is the smallest species, Thomson gazelle (*Gazella thomsoni*). This sequence of animal movements, which Vesey-FitzGerald (1960) called the 'grazing succession', depends on herbivores having differing tolerances for the tall, fibrous grasses of the lower catena. It may also depend on an improvement in quality of these grasses after they have been grazed and trampled by the earlier-arriving animals (Vesey-FitzGerald, 1960). This grazing succession is reversed in wet seasons, when the smaller species are the first to move upslope.

Because composition and phenology of woody species also vary with catena position (Herlocker, 1976; Jager, 1982), browsers similarly move up- and downslope. In the western Serengeti, Jarman and Jarman (1979) noted that impala (*Aepyceros melampus*), a browser, follows the same pattern of catena utilization as grazers, moving down-catena for the dry season and up-catena for the wet season. This species, interestingly, shows sex-related differences in landscape utilization: females move to the bottom of the catena during the dry season while males remain mid-slope. This difference in catena-use is thought to be related to attempts by male impala to secure and hold territories containing a wider diversity of patch-types than would be possible at lower positions on the catena (Jarman and Jarman, 1979).

2.3.4 LANDSCAPE PATTERN AND ANIMAL MIGRATIONS

The Serengeti National Park consists of > 13 000 km² of rolling grasslands and woodlands in northern Tanzania, with rainfall increasing from less than 450 mm in the south to more than 1100 mm in the north (Belsky, 1983). This gradient in rainfall coincides with a strong edaphic gradient, with soils decreasing in alkalinity, sodicity and nutrient content from south to north (de Wit, 1978). Soil alkalinity and sodicity are highest in the southern Serengeti (the Serengeti plains), where soils are derived from alkaline volcanic ash ejected from nearby volcanoes, and decline northward with distance from the volcanoes and with increasing rainfall, which leaches salts out of upper soil horizons.

This rainfall/soil alkalinity gradient dominates all other factors in organizing plant communities in the Serengeti: short grasslands dominate the southern half of the park, and savannas and woodlands dominate the northern half. This gradient sets in motion the great wildlife migrations of the Serengeti Ecosystem: over 1 300 000 wildebeest (Sinclair, 1979) as well as zebra and Thomson gazelle migrate from the northern savannas and woodlands to the southern shortgrass plains at the beginning of each rainy season; and they return to the northern savannas and woodlands during the dry season (Maddock, 1979). The reason the herbivores leave the northern savannas during the rainy season, when forage is plentiful there, has long been debated. This question has never been completely resolved, but the main hypotheses (discussed by Maddock, 1979) are that:

1. grasses in the southern plains are more nutritious than those of the northern savannas and have more nitrogen and calcium for lactating females (Kreulen, 1975; McNaughton, 1990);
2. predators are more easily avoided in shortgrass plains where cover is lacking (Darling, 1960);
3. animals dislike the wet, muddy soils of the northern savannas in the wet season (Talbot and Talbot, 1963); and
4. migrants more fully utilize total plant productivity in the park by moving between regions (Jarman and Sinclair, 1979).

These reasons are also consistent with those given for migrations of white-eared kob (*Kobus kob leucotis*) in Boma National Park, south-east Sudan (Fryxell and Sinclair, 1988). Large herbivore migrations such as these illustrate that rather than creating the major vegetation patterns, large mammals are influenced by them.

2.3.5 EDAPHIC GRASSLANDS

Most landscapes in semi-arid African savannas are vegetated both by woody species and by grasses. However, treeless grasslands also exist. These grasslands, referred to as 'edaphic grasslands' by Vesey-FitzGerald (1973), are created by conditions such as sodic–saline soils, shallow soils, and waterlogged, seasonally anaerobic conditions (Belsky, 1990), all of which inhibit the growth of woody species. Without experimental investigation, it is sometimes difficult to determine what is preventing the growth of trees and shrubs in a grassland. For example, the absence of trees from one of the largest grasslands in East Africa, the Serengeti plains, has been ascribed both to intensive herbivory (i.e. Bell, 1982; McNaughton, 1983) and low rainfall. However, the shallow, sodic soils of the region (de Wit, 1978), the absence of trees within old enclosures (Belsky, 1986 a, c; Belsky and Amundson, 1986), and the presence of trees in even drier savannas suggest that these grasslands are created by extreme sodic conditions rather than by low soil moisture and herbivory (Belsky, 1990).

2.3.6 MOSAIC GRASSLANDS

Often in sub-Saharan Africa, grasslands and savannas are highly patterned, even on flat, undisturbed landscapes. In the Serengeti, some grasslands form distinct two-phase mosaics, while others contain a variety of discrete patches of variable size and species composition (Anderson and Talbot, 1965; Schmidt, 1975; de Wit, 1978; Belsky, 1983, 1985, 1986a). In spite of nearly ubiquitous grassland pattern, certain areas of the Serengeti are more highly patterned than others. To determine whether certain physical or biological attributes cause or enhance landscape pattern in these areas, Belsky (1988) calculated an index of vegetational heterogeneity for 16 grassland and savanna communities distributed throughout the Serengeti. She then constructed a multiple-regression model to test whether this heterogeneity was related to annual precipitation, recurrent fire, presence of mound-building termites, soil characteristics, herbivore density, or plant growth-patterns. The resulting model indicated that subsurface soil concentrations of sodium had the greatest positive influence on amount and intensity of grassland pattern, followed by presence of mound-building termites. These two factors alone explained 55% of the variability in grassland heterogeneity in the Serengeti National Park.

Both soil sodicity and termite activity emphasize the importance of edaphic factors, as opposed to surficial disturbance and grazing by large mammals, in creating landscape pattern. Since high concentrations of sodium in soils are toxic to most plant species (Epstein, 1972), conditions that improve water infiltration and increase the leaching of salts out of sodic soils improve plant growth (de Wit, 1978; Belsky, 1986a, 1988). Similarly, disturbances that bring alkaline subsurface soils to more highly leached surfaces, as occurs when termites construct mounds in alkaline soils, create patches having higher pH and lower plant cover than the surrounding community. In other cases, termites create highly fertile patches by bringing nutrient-rich subsurface soils aboveground and by adding organic matter to them (Hesse, 1955; Lee and Wood, 1971; de Wit, 1978; Belsky, 1983; Jones, 1990). In many areas, these fertile, termite-derived patches are eventually colonized by woody species and develop into dense woodland clumps (Herlocker, 1976; Jager, 1982), giving a sharply defined, multi-phased pattern to the landscape.

One interesting example of the effects of soil sodicity and salinity on landscape pattern is the two-phase *Andropogon greenwayi* community in the Serengeti plains. This community consists of an 800-km^2 mosaic of two patch-types that range from < 1 m to > 100 m in diameter (Belsky, 1986a). One phase of the community has neutral soils with rapid infiltration rates and is dominated by a dense monoculture (75–100% cover) of the perennial grass *A. greenwayi*. The other phase has more sodic soils and slow infiltration rates and is more sparsely vegetated (10–70% cover) by a mixture of the annual grass *Chloris pycnothrix* and other grasses and herbs. Based on investigations of soils, vegetation, and population dynamics of important grass species in this community, Belsky (1986a) suggested that slight topographic differences in drainage in the highly sodic soils created patches having differing salt concentrations. More highly leached patches were eventually colonized by *A. greenwayi*, and more poorly leached (and sodic) patches were colonized by *C. pycnothrix* and other species. Due to continued differences in water infiltration and drainage between the two phases, differences in soil salinity and species composition between the two phases are maintained.

2.3.7 LANDSCAPE PATTERN CREATED BY LARGE HERBIVORES

Distribution and density of African herbivores are determined primarily by distance from water and by composition, quality and structure of the vegetation (Lamprey, 1963; McNaughton and Georgiadis, 1986; Skarpe, 1991). With only a few important exceptions, discussed in the following sections, the affects of animals on vegetational pattern are localized and often short-lived. This is in contrast to the effects of animals on species composition, decomposition rates and nutrient movement, which are strongly altered by the presence or absence of herbivores (Botkin, Mellilo and Wu, 1981; Belsky, 1992).

Large mammals create vegetational pattern in African savannas predominantly by grazing, browsing, and disturbing the ground with their hooves. These animals create trails, wallows and latrines; they excavate tubers and rhizomes and uproot grasses and herbs; they dig burrows; and they consume aboveground grass shoots and woody foliage and sometimes push over trees. But the activities of mammals that create the longest lasting patterns on the landscape are those that alter soil characteristics, especially in sodic or saline areas (Belsky, 1988). Aardvarks (*Orycteropus afer*) and warthogs (*Phacochoerus aethiopicus*) in Tanzania, for example, excavate dens and tunnels and bring to the surface sodic soils that are only slowly colonized by early successional species (Belsky, 1985); rhizomyid mole rats in Kenya create mounds (mima mounds) that are more favorable for plant growth than intermounds (Cox and Gakahu, 1985); and bat-eared fox (*Otocyon megalotis*) and porcupine (*Hystrix austro-africanae*) in South Africa create microsites in hard-capped soils where detritus and water accumulate and seeds germinate and become established (Dean and Milton, 1991). And in all areas of Africa, large ungulates create trails and compact the ground around waterholes. These trampled and compacted soils inhibit the penetration of seedling roots and therefore retard plant colonization and succession.

Other, less severe animal disturbances produce only short-lived patterns, if any (Belsky, 1985, 1986b). In contrast to more temperate grasslands and forests where disturbances initiate colonization by early successional species, disturbances in pristine African savannas are often revegetated by late-successional herbaceous species that are common in surrounding communities (Belsky, 1986a, b, c, 1987). As a result, patches resulting from these activities have the same composition as the rest of the community and quickly disappear.

Although most landscape elements are not created by large mammals, they may be maintained or intensified by them. For example in the *Andropogon greenwayi* community mentioned above, large herbivores maintain the two-phase mosaic by grazing and compacting the soil with their hooves (Belsky, 1986a). When the community is protected from grazers, the mosaic disappears. Similarly, browsing and brush-beating by large herbivores maintain distinct edges around woodlands by destroying woody invaders into the grassland (Belsky, 1984).

2.4 FIRE, HERBIVORES AND WOODLAND PATCHES

Having argued that animals exert only a minor influence in creating landscape pattern in pristine African savannas, I now discuss an example where large native herbivores play a major role in creating spatial and temporal savanna pattern. The possible contribution of human activities to the creation of this pattern is conjectural.

Until the middle of this century, woodlands and forests dominated much of sub-Saharan Africa. Beginning in the 1950s and 1960s, thousands of square kilometers of woodlands were converted to open savanna by fire, elephants (*Loxodonta africana*) and other forces. Large areas of *Commiphora–Acacia* woodlands and shrublands in Tsavo National Park, Kenya, were converted from prime woodland habitat for elephant and rhinoceros to open savanna, grasslands and bare ground (Glover, 1963; Agnew, 1968); and 50% of woodlands in northern Serengeti National Park were converted into open savannas (Norton-Griffiths, 1979).

Fire and elephants were thought to be the major forces behind this conversion from woodland to savanna. Because fires burn patchily as a result of varying wind speeds, topography and fuel loads (Frost and

Robertson, 1987), dry-season fires create grassland–woodland mosaics, rather than continuous grasslands. Similarly, by utilizing habitats and browsing selectively, elephants contribute to the formation of landscape mosaics. These reductions in woodland area and the creation of grassland–woodland mosaics in Tsavo and the Serengeti paralleled changes occurring at the same time in Uganda (Buechner and Dawkins, 1961; Laws, Parker and Johnstone, 1970), Rwanda (Spinage and Guinness, 1971), Zambia (Caughley, 1976), Zimbabwe (Guy, 1989), and in other parks in Kenya (Western and van Praet, 1973) and Tanzania (Barnes, 1985). Most often this conversion of woodlands to grassland–woodland mosaics was attributed to elephants (the 'elephant problem') (Lamprey *et al.*, 1967; Croze, 1974; Laws, Parker and Johnstone, 1970; among others), but fire (Phillips, 1974; Lock, 1977; Norton-Griffiths, 1979), drought (Greenway and Vesey-FitzGerald, 1969; Norton-Griffiths, 1979), wind and lightning (Spinage and Guinness, 1971), insect eruptions (Walker, 1981), fluctuations in water tables (Western and van Praet, 1973) and senescence of even-aged stands (Young and Lindsay, 1988) were sometimes cited as being equally or more important.

Many of the scientists who worked in East Africa in the 1960s and 1970s viewed the loss of woodlands as catastrophic and permanent (Buechner and Dawkins, 1961; Laws, 1970); however, other investigators proposed that these vegetational changes were short-lived or essentially

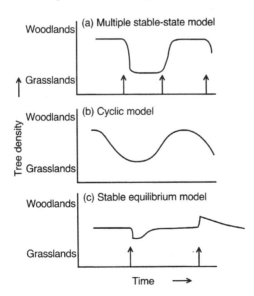

Figure 2.4 Models explaining temporal fluctuation between grasslands and woodlands in sub-Saharan Africa. The arrows within the model indicate the occurrence of a perturbation such as fire, drought or intense herbivory.

cyclic in nature. Several models were proposed to explain the temporal and spatial dynamics of these grassland–woodland mosaics (Figure 2.4):

- **Multiple stable-state models.** The long-dominant view of East African savannas can be described as a multiple stable-state (or multiple climax) model, in which several stable vegetation types exist. A catastrophic event such as the introduction of disease or uncontrolled population growth of a dominant herbivore is thought to cause one vegetation type to be replaced by another, which remains unchanged until the next catastrophic change. This model is illustrated in the Serengeti National Park by the conversion of woodland to grassland, which may have been due to abnormally high concentrations of elephants that over-utilized the woodlands inside the park. Outside the park, elephant habitat had been reduced by human settlement and cultivation and the elephants were therefore compressed into a smaller area (Lamprey et al., 1967; Dublin, Sinclair and McGlade, 1990).

- **Cyclic models.** Many of the recent models have been variations of the stable-limit-cycle model (Holling, 1973), which suggests that populations or communities cycle continuously between various phases. Caughley (1976) suggested that African woodland ecosystems that include elephants could never reach a natural stable equilibrium. In his model, elephants, by browsing and pushing over trees, destroy their woodland habitat so that their own population begins to decline through starvation or dispersal. With a lower elephant density, the woodlands regenerate, creating improved habitat for the elephants. The elephant population subsequently recovers, starting the next cycle of woodland decline. Similarly, Petrides (1974) proposed that stable equilibria between producers and large consumers cannot exist locally and that vegetational mosaics are separate successional stages produced by a cycle of overgrazing, dispersal of grazers, and vegetational recovery. The landscape as a whole, however, may be in equilibrium.

- **Stable-equilibrium models.** Walker and Noy-Meir (1982) proposed a dynamic but non-cyclic model of savanna vegetation. They interpreted the marked temporal changes in savanna vegetation as demonstrating the resilience of savanna communities (i.e. a stable-node model; Holling, 1973) following perturbations such as drought, fire, elephant damage, and disease. Based on a model of interspecific competition between herbaceous and woody species for available soil-water, they concluded that only one stable equilibrium exists for grasses, trees, grazers and browsers. After every deflection from stability, the savanna vegetation eventually returns to the equilibrium position. This model differs from cyclic models in that only one final vegetation type exists, while cyclic models predict that vegetation fluctuates between several types.

2.4.1 MODELS FOR AMBOSELI NATIONAL PARK, KENYA

A cyclic model based on fluctuations in rainfall was suggested by Western and van Praet (1973) to explain woodland losses in Amboseli National Park, Kenya. They proposed that high rainfall causes a rise in the water table in the Amboseli basin and increases salinization of the groundwater. High salinity kills the trees but not the grasses, creating an edaphic grassland. Western and van Praet established that past cycles of expansion and contraction of woodlands in the Amboseli basin were correlated with fluctuations in annual rainfall.

Young and Lindsay (1988) suggested a different cyclic model to explain the fluctuations: rapid die-off of trees in the Amboseli basin, as well as in other African woodlands, may be due to natural, synchronous senescence of even-aged stands. Woodlands in Africa are often even-aged as a result of synchronous seedling establishment after a fire or during a series of favorable years. Consequently, all trees in the stand reach maturity at approximately the same time. They also die at the same time, giving the appearance of responding to an external disturbance.

2.4.2 MODELS FOR TSAVO NATIONAL PARK, KENYA

The multiple stable-state model was originally proposed to explain changes in savanna vegetation in Tsavo National Park, Kenya, where woodlands were thought to have been permanently destroyed by elephants and replaced by grasslands (Glover, 1963; Leuthold, 1977). It was assumed that elimination of poaching from Tsavo in the late 1950s had led to an increase in elephants, which consequently over-utilized and destroyed their woodland resources. This view led to recommendations that 5000 elephants be destroyed (Laws, 1969).

Different models have recently been suggested to explain vegetational changes in the park. A cyclic model developed by Harris and Fowler (1975) differs from models developed for other parks by including indigenous human groups that have traditionally hunted elephants in the region. According to this model, large elephant populations over-browse and suppress woodlands and attract ivory hunters (Spinage, 1973). As hunters reduce the number of elephants, bushlands and woodlands regenerate which, being prime habitat for tsetse fly, brings sleeping sickness to the humans and to their cattle. The hunters retreat from the tsetse-infested woodlands, allowing elephant populations to recover. As the recovering elephant population once again converts woodlands to grasslands, the threat of sleeping sickness is reduced and the hunters return. According to this model, the recent loss of woodlands in Tsavo resulted from the elimination of hunting (now called poaching) from the park in the 1950s.

A second cyclic model proposed that climatic fluctuation was the driving variable behind vegetational changes in Tsavo (Cobb, 1980). Droughts, which re-occur in Tsavo approximately every ten years, reduce grass biomass so that elephants over-utilize the woody vegetation. When the drought is severe, woodlands are damaged and elephants starve. During periods of average to above-average rainfall, both tree and elephant populations recover.

2.4.3 MODELS FOR THE SERENGETI NATIONAL PARK, TANZANIA

Norton-Griffiths (1979) and Pellew (1983) proposed for the Serengeti National Park a stable-limit-cycle model that was based on both grazing and fire. According to this model, during periods of low grazer density, as occurred during a rinderpest epidemic during the first half of this century (Sinclair, 1979), grass biomass in savannas is incompletely consumed and builds up as fuel (Figure 2.5). As a result, dry-season fires are widespread and hot, killing trees at the edge of woodlands, preventing woody regeneration and destroying bush thickets. Woodlands are consequently reduced in size and replaced by grasslands. The rate of woodland reduction is accelerated by elephants, which open up woodland canopy by browsing and by pushing over trees. More grass thus grows within the wooded stand, providing fuel for hotter and more destructive fires.

During the next step in the cycle, the widespread grasslands contribute to the expansion of grazer populations, which then consume the grass that would otherwise fuel fires. Consequently, the frequency and intensity of fires decline. With a reduction in fire frequency, woody species once again establish, grasslands convert to woodlands, and grazer populations are reduced, re-initiating the cycle.

Recently, Dublin, Sinclair and McGlade, (1990) proposed that the conversion of woodlands to grasslands in the Serengeti illustrates the existence of multiple steady-states. Using computer models, they tested whether elephants alone, fire alone, or fire and elephants combined were responsible for the conversion of Serengeti woodlands to grasslands and for maintenance of the new steady-state. They found that fire alone could have converted Serengeti woodlands to grasslands, while elephants alone might be preventing woodland regeneration.

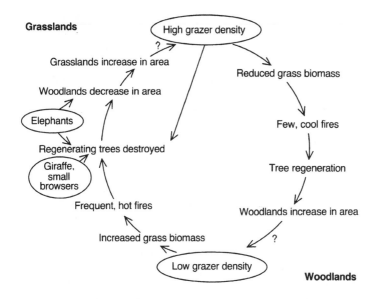

Figure 2.5 Factors causing vegetation cycling between woodlands and grasslands in the Serengeti National Park, Tanzania. (Source: Belsky, 1989.)

2.5 BOREHOLES AND THE IMPACT OF MODERN LIVESTOCK ON SUB-SAHARAN AFRICA

One final example of landscape pattern and ecosystem interactions in African landscapes results from Africa's colonial and post-colonial pastoral economy. Whereas traditional pastoralists in central, eastern and southern Africa were nomadic and migrated seasonally between areas of available water and forage (Lamprey, 1983; Sinclair and Fryxell, 1985; Ellis and Swift, 1988), recent government settlement programs have forced pastoralists to curtail their movements and use fixed and limited areas of rangeland. To provide permanent water resources for livestock during dry seasons, government authorities have developed watering points or boreholes. These boreholes provide the required water, but they also lead to over-utilization of the surrounding vegetation.

Wherever the impacts of borehole development and the resulting heavy concentrations of livestock have been studied, the results have been predictable (Andrew, 1988). Along transects extending from undisturbed savannas and woodlands to boreholes, livestock density and dung concentrations increase, trees and palatable perennial herbs decrease, annuals and unpalatable perennials increase, amount of bare ground, soil compaction and erosion increase, and concentrations of soil nutrients increase (Tolsma, Ernst and Verwey, 1987; Barker, Herlocker and Young,

1989; Georgiadis and McNaughton, 1990). Similar results were also found near boreholes developed for wildlife in Wankie National Park in Zimbabwe, except that native wildlife have fewer negative effects on the vegetation (Weir, 1971).

Barker, Herlocker and Young (1989) measured vegetation and soil changes along 5- and 9-km long transects from savannas to boreholes in Somalia and found that dense livestock concentrations near the boreholes increased soil erosion, the number of livestock trails and the number of unpalatable forage species. Along the transects, herbage cover was reduced from 50% to 6%, and litter was reduced from 2–3% to less than 1%.

In another detailed examination of the effects of livestock around boreholes in Botswana, Tolsma, Ernest and Verwey (1987) analyzed soil nutrient concentrations and plant densities along two 1.5-km long transects. They found that due to the transportation of nutrients from surrounding communities to boreholes via cattle dung, soils within 50 m of the boreholes had significantly higher concentrations of N, Ca, Mg, Mn, P, K, Fe and Na than surrounding communities, and most of these nutrients remained elevated 200 and 400 m away. As a result, phosphorus contents in topsoils of more distant areas (up to 3 km away) were diminished and may have given rise to P deficiencies in livestock (Tolsma, Ernst and Verwey, 1987). Levels of N, K and P in tree tissues near the boreholes were also elevated, but other tissue nutrients were either lower or unchanged. Species composition also varied along the transects: starting with a concentric zone (0–20 m) of nitrophilous, trampling-resistant, unpalatable annuals; followed by a zone (20–50 m) dominated by a dense thicket of woody shrubs; and terminating in a more species-rich, open savanna.

Finally, in a study of the effects of artificial watering points on forage quality in Kenya, Georgiadis and McNaughton (1990) found that tissues of grasses growing 50 m from boreholes had significantly higher concentrations of Ca, K, Mg, Mo and Na, significantly lower concentrations of Al, and significantly higher forage quality (as measured by cell contents, crude protein and cell wall constituents) than grasses growing 800 m away.

These studies demonstrate the degree to which boreholes constructed for livestock or wildlife alter ecosystem function and create landscape pattern by increasing herbivore density and transporting nutrients between ecosystems. The areas near the watering points, often referred to as 'sacrifice areas', are readily visible from satellite photos (Andrew, 1988). What is not visible from the satellites are changes in soil fertility, altered animal-movement patterns, and changed forage quality.

The importance of these boreholes and their increasing concern to ecologists lie in their tendency to raise livestock densities throughout the year, leading to local overgrazing and range deterioration. As livestock impact becomes increasingly severe, denuded and eroded patches develop in a predictable pattern, providing nuclei of degradation and enhancing the process of desertification (Sinclair and Fryxell, 1985). Unless livestock numbers are reduced, adjacent degraded nuclei coalesce and larger areas of the landscape are converted to desert. The movements of animals and nutrients across the landscape, enriching certain zones while depleting others, and the creation of patches that impact human society by impoverishing pastoralists and possibly changing rainfall patterns (Sinclair and Fryxell, 1985), suggest that borehole development may be a critical process that should be addressed by landscape ecologists.

2.6 CONCLUSION

Because of the intense interest ecologists have in herbivore–plant interactions, a large number of empirical studies, reviews and speculative papers have recently been published. Many suggest that animals have the ability to 'control' or 'regulate' their environments, even to the point of creating the major features of the landscapes they inhabit. While most of these studies are to some degree correct, they may be overemphasizing the impacts of large mammals on their environment. A more realistic reading of the data and interpretation of natural history suggest that soil moisture, geomorphology and edaphic properties are the primary creators of landscape pattern in undeveloped African savannas (Cole, 1986; Belsky, 1990), and that factors such as large herbivores and fire only fine-tune these patterns. Only in those areas where animals and plant communities have been strongly manipulated by humans, such as by compression of wildlife into small reserves, by alteration of fire frequency and by borehole construction, does the importance of fire and animals increase. As Monica Cole (1986) stated in an important book on the biogeography and geobotany of tropical savannas, 'the height and spacing of [plant community components] are influenced mainly by soil moisture conditions, while the composition of the vegetation units within each category varies with nutrient status' (p. 42). [All savanna vegetation is] 'subject to continuous change in response to geomorphological evolution of the landscape. . . . Overlying the impacts of these factors are those caused by grazing animals, decomposer fauna, fire, and man's cultural practices, which must be regarded as secondary rather than primary influences' (p. 30).

2.7 SUMMARY

Landscape pattern in arid and semi-arid savannas of sub-Saharan Africa is determined primarily by soil moisture and edaphic properties of soils.

Savannas are patterned by densely wooded riverine and lacustrine forests and by more sparsely wooded seasonal drainages and moist bottoms of catenas, hills and steep escarpments. Any factor that alters soil moisture creates pattern in these strongly water-limited regions. Similarly, alterations in soil chemistry properties create intense vegetation patterns, either by increasing soil fertility or by decreasing soil salinity and alkalinity in infertile or saline environments.

With few exceptions, fire and disturbances by large herbivores are secondary in importance to soil moisture and edaphic properties in creating landscape patterns and regulating landscape dynamics. In relatively pristine grassland communities in national parks and wildlife refuges, animal disturbances and fire scars are often revegetated by species that are already abundant in the surrounding communities, creating short-lived pattern. Only in cases where disturbances seriously alter soil-moisture and edaphic properties or where fire and elephants create grassland–woodland mosaics is the pattern longer-lived.

The effects of large mammals on landscape pattern are intensified when animal concentrations are compressed into small game reserves by human settlement or when boreholes are constructed for wildlife and livestock. Artificially high herbivore densities lead to overgrazing, erosion and grassland deterioration, creating a pattern that is large-scale and long-lived.

ACKNOWLEDGEMENTS

I thank A. R. E. Sinclair and M. B. Coughenour for their thoughtful comments on an earlier draft of this paper.

REFERENCES

Agnew, A. D. O. (1968) Observations on the changing vegetation of Tsavo National Park (East). *E. Afr. Wildl. J.*, **6**, 75–80.

Anderson, G. D. and Talbot, L. M. (1965) Soil factors affecting the distribution of the grassland types and their utilization by wild animals on the Serengeti Plains, Tanganyika. *J. Ecol.*, **53**, 33–56.

Andrew, M. H. (1988) Grazing impact in relation to livestock watering points. *Trends Ecol. Evol.*, **3**, 336–9.

Barker, J. R., Herlocker, D. J. and Young, S. A. (1989) Vegetal dynamics along a grazing gradient within the coastal grassland of central Somalia. *Afr. J. Ecol.*, **27**, 283–9.

Barnes, R. F. W. (1985) Woodland changes in Ruaha National Park (Tanzania) between 1976 and 1982. *Afr. J. Ecol.*, **23**, 215–21.

Bell, R. H. V. (1970) The use of the herb layer by grazing ungulates in the Serengeti, in *Animal Populations in Relation to their Food Resources* (ed. A. Watson), Blackwell, Oxford, pp. 111–24.

Bell, R. H. V. (1982) The effect of soil nutrient availability on community structure in African ecosystems, in *Ecology of Tropical Savannas* (eds B. J. Huntley and B. H. Walker), Springer, Berlin, pp. 193–216.

Belsky, A. J. (1983) Small-scale pattern in four grassland communities in the Serengeti National Park, Tanzania. *Vegetatio*, **55**, 141–51.

Belsky, A. J. (1984) The role of small browsing mammals in preventing woodland regeneration in the Serengeti National Park, Tanzania. *Afr. J. Ecol.*, **22**, 271–9.

Belsky, A. J. (1985) Long-term ecological monitoring in the Serengeti National Park, Tanzania. *J. Appl. Ecol.*, **22**, 449–60.

Belsky, A. J. (1986a) Population and community processes in a mosaic grassland in the Serengeti, Tanzania. *J. Ecol.*, **74**, 841–56.

Belsky, A. J. (1986b) Revegetation of artificial disturbances in grasslands in the Serengeti National Park, Tanzania. I. Colonization of grazed and ungrazed plots. *J. Ecol.*, **74**, 419–37.

Belsky, A. J. (1986c) Revegetation of artificial disturbances in grasslands in the Serengeti National Park, Tanzania. II. Five years of successional change. *J. Ecol.*, **74**, 937–51.

Belsky, A. J. (1987) Revegetation of natural and human-caused disturbances in the Serengeti National Park, Tanzania. *Vegetatio*, **70**, 51–9.

Belsky, A. J. (1988) Regional influences on small-scale vegetational heterogeneity within grasslands in the Serengeti National Park, Tanzania. *Vegetatio*, **74**, 3–10.

Belsky, A. J. (1989) Landscape patterns in a semi-arid ecosystem in East Africa. *J. Arid Env.*, **17**, 265–70.

Belsky, A. J. (1990) Tree/grass ratios in East African savannas: a comparison of existing models. *J. Biogeogr.*, **17**, 483–9.

Belsky, A. J. (1992) Effects of grazing, competition, disturbance, and fire on species composition and diversity of grassland communities. *J. Veg. Sci.*, **3**, 187–200.

Belsky, A. J. and Amundson, R. G. (1986) Sixty years of successional history behind a moving sand dune near Olduvai Gorge, Tanzania. *Biotropica*, **18**, 231–5.

Botkin, D. B., Mellilo, J. M. and Wu, L. S. Y. (1981) How ecosystem processes are linked to large mammal population dynamics, in *Dynamics of Large Mammal Populations* (eds C. W. Fowler and T. D. Smith), Wiley, New York, pp. 373–87.

Buechner, H. K. and Dawkins, H. C. (1961) Vegetation change induced by elephants and fire in Murchison Falls National Park, Uganda. *Ecology*, **42**, 752–66.

Caughley, G. (1976) The elephant problem – an alternative hypothesis. *E. Afr. Wildl. J.*, **14**, 265–83.

Cobb, S. (1980) Tsavo, the first thirty years. *Swara*, **3**, 12–15.

Coe, M. J. Cumming, D. H. M. and Phillipson, J. (1976) Biomass and production of large African herbivores in relation to rainfall and primary production. *Oecologia*, **22**, 341–54.

Cole, M. M. (1986) *The Savannas: Biogeography and Geobotany*, Academic Press, New York.

Coughenour, M. B. (1991) Spatial components of plant–herbivore interactions in pastoral, ranching, and native ungulate ecosystems. *J. Range Manage.*, **44**, 530–42.

Coughenour, M. B. (1992) Spatial modeling and landscape characterization of an African pastoral ecosystem: A prototype model and its potential use for monitoring drought, in *Ecological Indicators* (eds D. H. McKenzie, D. H. Hyatt and J. McDonald), Elsevier Applied Science, London, pp. 787–810.

Coughenour, M. B. and Ellis, J. E. (1993) Landscape and climatic control of woody vegetation in a dry tropical ecosystem: Turkana District, Kenya. *J. Biogeogr.*, **22**, 107–22.

Cox, G. W. and Gakahu, C. G. (1985) Mima mound microtopography and vegetation pattern in Kenyan savannas. *J. Trop. Ecol.*, **1**, 23–36.

Croze, H. (1974) The Seronera bull problem, II. The trees. *E. Afr. Wildl. J.*, **12**, 29–47.

Cumming, D. H. M. (1982) The influence of large herbivores on savanna structure in Africa, in *Ecology of Tropical Savannas* (eds B. J. Huntley and B. H. Walker), Springer, Berlin, pp. 215–45.

Darling, F. F. (1960) *An Ecological Reconnaissance of the Mara Plains in Kenya Colony.* Wildl. Monogr. No. 5, The Wildlife Society.

Dean, W. R. J. and Milton, S. J. (1991) Disturbances in semi-arid shrubland and arid grassland in the Karoo, South Africa: mammal diggings as germination sites. *Afr. J. Ecol.*, **29**, 11–16.

Deshmukh, I. (1984) A common relationship between precipitation and grassland peak biomass for East and southern Africa. *Afr. J. Ecol.*, **22** 181–6.

de Wit, H. A. (1978) Soils and Grassland Types of the Serengeti Plains (Tanzania). PhD Thesis, Agricultural University, Wageningen.

Dublin, H. T., Sinclair, A. R. E. and McGlade, J. (1990) Elephants and fire as causes of multiple stable states in the Serengeti–Mara woodlands. *J. Anim. Ecol.*, **59**, 1147–64.

Edroma, E. L. (1981) The role of grazing in maintaining high species-composition in *Imperata* grassland in Rwenzori National Park, Uganda. *Afr. J. Ecol.*, **19**, 215–33.

Ellis, J. E. and Swift, D. M. (1988) Stability of African pastoral ecosystems: alternate paradigms and implications for development. *J. Range Manage.*, **41**, 450–9.

Epstein, T. (1972) *Mineral Nutrition of Plants: Principles and Perspectives*, John Wiley, New York.

Forman, R. T. T. (1987) The ethics of isolation, the spread of disturbance, and landscape ecology, in *Landscape Heterogeneity and Disturbance* (ed. M. G. Turner), Springer Verlag, New York, pp. 213–29.

Forman, R. T. T. and Godron, M. (1986) *Landscape Ecology*, Wiley & Sons, New York.

Frost, P. G. H. and Robertson, F. (1987) Fire: The ecological effects of fire in savannas, in *Determinants of Tropical Savannas* (ed. B. H. Walker), IRL Press, Oxford, pp. 93–140.

Fryxell, J. M. and Sinclair, A. R. E. (1988) Seasonal migration by white-eared Kob in relation to resources. *Afr. J. Ecol.*, **26**, 17–31.

Georgiadis, N. J. and McNaughton, S. J. (1990) Elemental and fibre contents of savanna grasses: variation with grazing, soil type, season and species. *J. Appl. Ecol.*, **27**, 623–34.

Glover, P. (1963) The elephant problem at Tsavo. *E. Afr. Wildl. J.*, **1**, 30–9.

Greenway, P. J. and Vesey-FitzGerald, D. F. (1969) The vegetation of Lake Manyara National Park. *J. Ecol.*, **57**, 127–49.

Guy, P. R. (1989) The influence of elephants and fire on a *Brachystegia–Julbernardia* woodland in Zimbabwe. *J. Trop. Ecol.*, **5**, 215–26.

Harris, L. D. and Fowler, N. K. (1975) Ecosystem analysis and simulation in Mkomazi Reserve, Tanzania, *E. Afr. Wildl. J.*, **13**, 325–45.

Hatton, J. C. and Smart, N. O. E. (1984) The effect of long-term exclusion of large herbivores on soil nutrient status in Murchison Falls National Park, Uganda. *Afr. J. Ecol.*, **22**, 23–30.

Herlocker, D. (1976) *Woody Vegetation of the Serengeti National Park*, Texas A and M University Press, College Station, Texas.

Hesse, P. R. (1955) A chemical and physical study of the soils of termite mounds in East Africa. *J. Ecol.*, **43**, 449–61.

Holling C. S. (1973) Resilience and stability of ecological systems. *Ann. Rev. Ecol. Syst.*, **4**, 1–24.

Jager, T. (1982) *Soils of the Serengeti Woodlands, Tanzania*, Pudoc, Wageningen.

Jarman, P. J. and Jarman, M. V. (1979) The dynamics of ungulate social organization, in *Serengeti: Dynamics of an Ecosystem* (eds A. R. E. Sinclair and M. Norton-Griffiths), University of Chicago Press, Chicago, pp. 185–220.

Jarman, P. J. and Sinclair A. R. E. (1979) Feeding strategy and the pattern of resource partitioning in ungulates, in *Serengeti: Dynamics of an Ecosystem* (eds A. R. E. Sinclair and M. Norton-Griffiths), University of Chicago Press, Chicago, pp. 130–63.

Jensen, C. L. and Belsky, A. J. (1989) Grassland homogeneity in Tsavo National Park (West), Kenya. *Afr. J. Ecol.*, **27**, 35–44.

Jones, J. A. (1990) Termites, soil fertility and carbon cycling in dry tropical Africa: a hypothesis. *J.Trop. Ecol.*, **6**, 291–305.

Kreulen, D. A. (1975) Wildebeest habitat selection in the Serengeti plains, Tanzania, in relation to calcium and lactation: a preliminary report. *E. Afr. Wildl. J.*, **13**, 297–304.

Lamprey, H. F. (1963) Ecological separation of the large mammal species in the Tarangire Game Reserve, Tanganyika. *E. Afr. Wildl. J.*, **1**, 63–92.

Lamprey, H. F. (1983) Pastoralism yesterday and today: the overgrazing problem, in *Tropical Savannas* (ed. F. Bouliere), Elsevier Press, Amsterdam, pp. 643–66.

Lamprey, H. F., Gover, P. E., Turner, M. I. M. and Bell, R. H. V. (1967) Invasion of the Serengeti National Park by elephants. *E. Afr. Wildl. J.*, **5**, 151–66.

Laws, R. M. (1969) The Tsavo research project. *J. Reprod. Fert., Suppl.*, **6**, 495–531.

Laws, R. M. (1970) Elephants as agents of habitat and landscape change in East Africa. *Oikos*, **21**, 1–15.

Laws, R. M., Parker, I. S. C. and Johnstone, R. C. B. (1970) Elephants and habitats in North Bunyoro, Uganda. *E. Afr. Wildl. J.*, **8**, 163–80.

Lee, K. E. and Wood, T. G. (1971) *Termites and Soils*, Academic Press, London.

Leuthold, W. (1977) Changes in tree populations of Tsavo East National Park, Kenya. *E. Afr. Wildl. J.*, **15**, 61–9.

Lock, J. M. (1972) The effects of hippopotamus grazing on grasslands. *J. Ecol.*, **60**, 445–68.

Lock, J. M. (1977) Preliminary results from fire and elephant exclusion plots in Kabalega National Park, Uganda. *E. Afr. Wildl. J.*, **15**, 229–32.

Loth, P. E. and Prins, H. H. T. (1986) Spatial patterns of the landscape and vegetation of Lake Manyara National Park, Tanzania. *ITC Journal*, **1986**(2), 115–30.

Maddock, L. (1979) The 'migration' and grazing succession, in *Serengeti: Dynamics of an Ecosystem*, (eds A. R. E. Sinclair and M. Norton Griffiths), University of Chicago Press, Chicago, pp. 104–29.

McNaughton, S. J. (1983) Serengeti grassland ecology: the role of composite environmental factors and contingency in community organization. *Ecol. Monogr.*, **53**, 291–320.

McNaughton, S. J. (1990) Mineral nutrition and seasonal movements of African migratory ungulates. *Science*, **345**, 613–15.

McNaughton, S. J. and Georgiadis, N. J. (1986) Ecology and African grazing and browsing mammals. *Ann. Rev. Ecol. Syst.*, **17**, 39–65.

Milne, G. (1935) Some suggested units for classification and mapping, particularly for East African soils. *Soil Res. (Berl.)*, **4**, 183–98.

Morison, C. G. T., Hoyle, A. C. and Hope-Simpson, J. F. (1948) Tropical soil vegetation catenas and mosaics. *J. Ecol.*, **36**, 1–84.

Naveh, Z. and Liberman, A. S. (1984) *Landscape Ecology: Theory and Applications*, Springer-Verlag, New York.

Norton-Griffiths, M. (1979) The influence of grazing, browsing, and fire on vegetation dynamics, in *Serengeti: Dynamics of an Ecosystem* (eds A. R. E. Sinclair and M. Norton-Griffiths), University of Chicago Press, Chicago, pp. 310–52.

Pellew, R. A. P. (1983) The impacts of elephants, giraffe and fire upon the *Acacia tortilis* woodlands of the Serengeti. *Afr. J. Ecol.*, **21**, 41–74.

Petrides, G. A. (1974) The overgrazing cycle as a characteristic of tropical savannas and grasslands in Africa. *Proc. First Inter. Cong. Ecol. (Wageningen)*, pp. 86–91.

Phillips, J. (1974) Effects of fire in forest and savanna ecosystems of sub-Saharan Africa, in *Fire and Ecosystems* (eds T. T. Kozlowski and C. E. Ahlgren), Academic, New York, pp. 435–81.

Prins, H. H. T. (1988) Plant phenology patterns in Lake Manyara National Park. *J. Biogeogr.*, **15**, 465–80.

Prins, H. H. T. (1989) Condition changes and choice of social environment in African buffalo bulls. *Behaviour*, **108**, 297–324.

Prins, H. H. T. and Beekman, J. H. (1987) A balanced diet as a goal of grazing: the food of the Manyara buffalo, in *The Buffalo of Manyara* (ed. H. H. T. Prins), Krips Repro, Meppel. Netherlands, pp. 69–98.

Prins, H. H. T. and Iason, G. R. (1988) Dangerous lions and nonchalant buffalo. *Behaviour*, **108**, 262–97.

Pritchard, J. M. (1979) *Landform and Landscape in Africa*, Edward Arnold, London.

Risser, P. G. (1987) Landscape ecology: state of the art, in *Landscape Heterogeneity and Disturbance* (ed. M. G. Turner), Springer Verlag, New York, pp. 3–14.

Schaller, G. B. (1972) *The Serengeti Lion*, University of Chicago Press, Chicago.

Scholes, R. J. (1991) The influence of soil fertility on the ecology of southern African dry savannas, in *Savanna Ecology and Management* (ed. P. A. Werner), Blackwell, Oxford, pp. 71–5.

Schmidt, W. (1975) Plant communities in permanent plots of the Serengeti Plains. *Vegetatio*, **30**, 133–45.

Sinclair, A. R. E. (1977) *The African Buffalo: A Study of Resource Limitation of Populations*, University of Chicago Press, Chicago.

Sinclair, A. R. E. (1979) The eruption of the ruminants, in *Serengeti: Dynamics of an Ecosystem* (eds A. R. E. Sinclair and M. Norton-Griffiths), University of Chicago Press, Chicago, pp. 82–103.

Sinclair, A. R. E. and Fryxell, J. M. (1985) The Sahel of Africa: ecology of a disaster. *Can. J. Zool.*, **63**, 987–94.

Skarpe, C. (1991) Impact of grazing in savanna ecosystems. *Ambio*, **20**, 351–6.

Smart, N. O. E., Hattton, J. C. and Spence, D. H. N. (1985) The effect of long-term exclusion of large herbivores on vegetation in Murchison Falls National Park, Uganda. *Biol. Conserv*, **33**, 229–45.

Spinage, C. A. (1973) A review of ivory exploitation and elephant population trends in Africa. *E. Afr. Wildl. J.*, **11**, 281–9.

Spinage, C. A. and Guinness, F. E. (1971) Tree survival in the absence of elephants in the Akagera National Park, Rwanda. *J. Appl. Ecol.*, **8**, 723–8.

Strugnell, R. G. and Pigott, R. G. (1978) Biomass, shoot-production and grazing of two grasslands in the Rwenzori National Park, Uganda. *J. Ecol.*, **66**, 73–96.

Talbot, L. M. and Talbot, M. H. (1963) The wildebeest in western Masailand, East Africa. Wildlife Monograph No. 12, The Wildlife Society.

Tinley, K. L. (1982) The influence of soil moisture balance in ecosystem pattern in southern Africa, in *Ecology of Tropical Savannas* (eds B. J. Huntley and B. H. Walker), Springer, Berlin, pp. 175–92.

Tolsma, D. J., Ernst, W. H. O. and Verwey, R. A. (1987) Nutrients in soil and vegetation around two artificial waterpoints in eastern Botswana. *J. Appl. Ecol.*, **24**, 991–1000.

van Wijngaarden, W. (1985) *Elephants–Grass–Trees–Grazers*, ITC Publication No. 4, Enschede, The Netherlands.

Vesey-FitzGerald, D. (1960) Grazing succession among East African game animals. *J. Mammal.*, **41**, 161–72.

Vesey-FitzGerald, D. (1973) *East African Grasslands*, East African Publishing House, Nairobi.

Walker, B. H. (1981) Is succession a viable concept in African savanna ecosystems? in *Forest Succession* (eds D. C. West, H. H. Shugart and D. B. Botkin), Springer Verlag, New York, pp. 431–47.

Walker, B. H. and Noy-Meir, I. (1982) Aspects of the stability and resilience of savanna ecosystems, in *Ecology of Tropical Savannas* (eds B. J. Huntley and B. H. Walker), Springer, Berlin, pp. 556–90.

Weir, J. S. (1971) The effect of creating additional water supplies in a Central African national park, in *The Scientific Management of Animal and Plant Communities for Conservation* (eds E. Duffey and A. S. Watt), Blackwell Science, Oxford, pp. 367–85.

Western, D. and van Praet, C. (1973) Cyclical changes in the habitat and climate of an East African ecosystem. *Nature*, **241**, 104–6.

Young, T. P. and Lindsay, W. K. (1988) Disappearance of acacia woodlands: the effect of size structure. *Afr. J. Ecol.*, **26**, 69–72.

Effects of animals on landscape pattern

3

Carol A. Johnston

3.1 INTRODUCTION

Studies of animal–patch interactions have generally focused on how animals are affected by patchy habitats (Wiens, 1985; Johnson, Milne and Wiens, 1992; Levin, 1992) rather than how they create them (but see Belsky, this volume). Scientists who observe animal activities that may alter habitat usually quantify and interpret those behaviors in terms of the life history of the animal, rather than consequences to the landscape, hence it is often assumed that their influence is minimal. However, there is growing evidence of the ability of animals to influence landscape pattern and process (Botkin, Melillo and Wu, 1981; Naiman, 1988; Johnston, Pastor and Naiman, 1993).

Animal disturbances create two-dimensional patches on the land surface (Pickett and White, 1985) and three-dimensional 'patch bodies', volumetric landscape units that have boundaries with upper and lower strata in addition to boundaries with adjacent surface patches (Johnston and Naiman, 1987). Patch bodies may be subaqueous (e.g. sediments underlying beaver ponds), or subterranean (e.g. rodent burrows), but they have considerable ramifications to ecosystem processes and landscape dynamics. This three-dimensional view of the environment has been well accepted for decades in limnology (e.g. epilimnion versus hypolimnion: Hutchinson, 1957).

The concept of the patch body is particularly relevant to animal influences on the physical landscape, because the majority of those influences occur underground. Only a few animals other than man can directly alter the landscape surface, but many animals can burrow within the soil

Mosaic Landscapes and Ecological Processes.
Edited by Lennart Hansson, Lenore Fahrig and Gray Merriam.
Published in 1995 by Chapman & Hall, London. ISBN O 412 45460 2

(Meadows and Meadows, 1991). Below-ground animals are also capable of consuming a larger proportion of net primary productivity than can their above-ground counterparts (Hairston, Smith and Slobodkin, 1960), causing organic matter and associated nutrients to become incorporated into the physical landscape (Coleman, Reid and Cole, 1983; Brown and Gange, 1990; Martin et al., 1992).

Rates of patch and patch body formation, whether biotically or abiotically controlled, are related to the rate and distribution of disturbance (Pickett and White, 1985). While these rates are well known for small patches (e.g. on a scale of square meters) with short-lived plant species (Paine and Levin, 1981; Connell and Keough, 1985; Sousa, 1985), the creation and ontogeny of disturbance-induced patches at the landscape scale (e.g. on a scale of square kilometers) is poorly understood because of the large spatial and long temporal scales over which they occur.

Animal-induced disturbances are often so diffusely distributed in space or time as to make patch definition difficult at the landscape scale. Those animals that can disturb their environment sufficiently to create patches either have exceptional behavioral traits that allow them to physically manipulate their environment (e.g. beavers, fossorial rodents), or have large populations that can exploit a resource over an extensive area before moving to another location (e.g. gypsy moth). Animals that do not create discrete patches can still have landscape-level effects (e.g. alteration of plant communities and nutrient cycling via selective herbivory), but the study of those effects requires experimental manipulation such as exclosures to distinguish them from other landscape processes (Tiedemann and Berndt, 1972; Cargill and Jefferies, 1984; Hatton and Smart, 1984; Pastor et al., 1988).

The magnitude of animal-induced disturbances is related to the number of animals present, which changes over time in response to fairly predictable patterns of birth, death and mortality (Botkin, Melillo and Wu, 1981). Less well known are the food and habitat selection criteria that govern where an individual or colony will establish (Senft et al., 1987), but progress is being made in this area with the development and integration of remote sensing and geographic information systems (GIS) (Dahlsted et al., 1981; Johnston, 1993). If the population dynamics and environmental effects of an animal population are known, an expert system can be developed to predict future disturbances (Coulson et al., 1991).

The relative impact of animal influences on the physical landscape is a function of both animal and environmental attributes:

- the severity of the perturbation,
- the susceptibility of landscape elements to disturbance,
- the animal's population density,

- the spatial distribution of the population and its effects, and
- the mobility of the population (i.e. changes in its spatial distribution over time).

For example, moose (*Alces*) consume large quantities of plant biomass per individual, allowing them to alter soil organic matter content at much lower population densities than smaller herbivores such as prairie dogs (*Cynomys ludovicianus*) (Pastor *et al.*, 1988; Whicker and Detling, 1988). However, the colonial nature of prairie dog populations concentrates their cumulative effect within a small area, promoting a patchier landscape than would occur under the relatively diffuse disturbance caused by the dispersed and mobile moose population. All of these factors need to be considered when evaluating the influence of animals on landscapes.

Mechanisms by which animals create landscape patches include herbivory (Blais, 1954; Lock, 1972; McNaughton, 1985; Turner and Bratton, 1987; Johnston and Naiman, 1990a), flooding (Johnston and Naiman, 1990b), and physical perturbation of the soil (Laycock and Conrad, 1967; Bratton, 1975; Huntly and Inouye, 1988; Whicker and Detling, 1988). Other authors have reviewed the ecosystem and landscape-level effects of herbivory (Andersen, 1987a; Brown and Gange, 1990; Huntly, 1991; Pacala and Crawley, 1992; Chapter 2), so the focus of this chapter is on animal alteration of the physical landscape by flooding or digging.

3.2 PATCH FORMATION BY FLOODING

Beavers are unique among animals other than man in their ability to construct dams that alter the flow of water through the landscape. Beaver ponds create spot disturbance patches (*sensu* Forman and Godron, 1986) in a matrix of upland forest; where before there was only a narrow stream corridor among the forest matrix, a beaver dam creates a patch of water with very different properties to the stream and forest it replaced. The stream is converted from a lotic to a lentic environment, and the terrestrial system is converted to an aquatic one. The flooding of soils limits the flora to species that are tolerant of inundation (e.g. submerged aquatics), or can import oxygen from overlying aerobic strata (e.g. wetland emergents). As beaver ponds go through their life cycle of flooding, aging and abandonment, the vegetation within them changes from forest to open water to wetland. A beaver-impounded landscape consists of hundreds of beaver ponds in various stages of development, a spatial mosaic of active and abandoned impoundments.

My colleagues and I have studied the effects of pond construction by an expanding beaver population on the extent and functioning of wetlands at Voyageurs National Park (VNP) in northern Minnesota, USA

(Naiman, Johnston and Kelley, 1988; Broschart, Johnston and Naiman, 1989; Johnston and Naiman 1990b, c; Johnston, Pastor and Pinay, 1992). We used historical aerial photos and a GIS to map and measure beaver ponds built over the past 50 years, and field studies to determine how their construction affects landscape pattern and biogeochemistry.

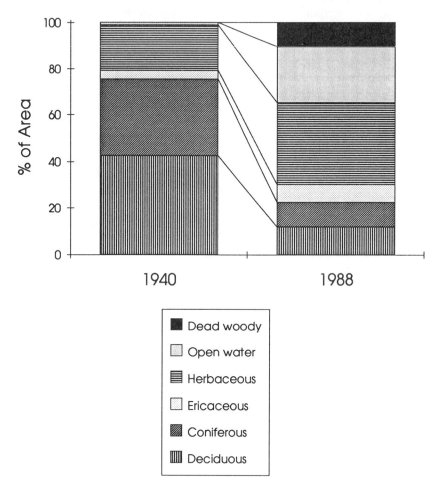

Figure 3.1 Vegetation types in areas before and after impoundment by beaver dams. Cumulative area = 3687 ha.

Beaver impoundments greatly influenced vegetation composition (Figure 3.1). In 1940, 80% of the areas that would be flooded by beaver had some type of woody vegetation, and open water was negligible. In 1988, after flooding by beavers, only 30% of the area had live woody vegetation remaining, and deciduous wet forest was almost completely eliminated. The area of herbaceous wetland vegetation was twice what it was in 1940, largely due to the conversion of forest lands to marshes and wet meadows. Therefore, there was a change in vegetation structure (woody versus herbaceous) as well as community type as a result of beaver impoundment.

Once a forest is disturbed by beaver impoundment, it does not return to its original vegetation state, at least within time-scales of 40 to 60 years (Remillard, Gruendling and Bogucki, 1987; Johnston and Naiman, 1990c). This lack of patch extinction may be due in part to recurring disturbance by beaver, but also to the dramatic alteration of environmental conditions caused by flooding and consequent soil changes that persist long after beaver have abandoned the site (Wilde, Youngberg and Hovind, 1950). The patch would thus be considered a semistable spot disturbance patch, remaining distinct in structure and composition from the surrounding forest matrix (Forman and Godron, 1986).

The establishment of new ponds during recovery of beaver population levels from historic overtrapping provides a good example of the interaction between animal density and patch formation (Chapter 4). The beaver population of VNP was low during the first half of this century, and beaver ponds covered only about 1% of the landscape as of 1940. By 1958, about 100 beaver colonies had impounded 10% of the landscape, and by 1986, some 300 colonies had impounded 13% of the landscape (Broschart, Johnson and Naiman, 1989; Johnston and Naiman, 1990b). Between 1940 and 1961, beavers created new ponds at the rate of 0.42% of the landscape per year (i.e. each 10 000 ha area of landscape would have an additional 42 ha of beaver ponds each year). By comparison, the urbanization of land around the Milwaukee metropolitan area was 0.64% per year (Sharpe *et al.*, 1981), and the abandonment of cropland in a county in Georgia was 0.8% per year (Turner, 1987). Therefore, beavers are capable of altering landscapes at rates comparable with those for human activities.

These data demonstrate that beaver can quickly create new patches through their dam-building activities, rapidly spreading into suitable habitats as their population increases. This is due in part to beaver biology and behavior (Bergerud and Miller 1977). Young beaver leave their parent colony at about two years of age in search of new habitat (Bradt, 1938; Townsend, 1953; Aleksiuk, 1968; Svendsen, 1980), usually dispersing within a 16 km radius of their natal pond (Beer, 1955; Libby, 1957; Hibbard, 1958; Leege, 1968). Given this dispersal radius and an average annual litter size of 3–4 kits per year (Hodgdon and Hunt, 1953; Osborn, 1953; Brenner, 1964), it would be possible for beaver to colonize areas as far as 736 km from an initial nucleus over a 46-year period.

The rate of patch formation by ponding was not linear over this period of population increase: as the beaver population density increased

and habitats susceptible to flooding decreased, the rate of patch formation and the average area per patch decreased. In the area studied there
were 71 ponds in 1940, with an average area of 3.7 ha. Between 1940 and
1961, approximately 25 new ponds were created each year, with no
significant change in the average area of newly created ponds. Between
1961 and 1986, however, only 10 new ponds per year were created, and
the average area per new pond was significantly lower than those created
during the initial time period (Johnston and Naiman, 1990c). Furthermore, ponds created before 1961 tended to be spatially isolated, whereas
after 1961 new ponds were clustered between or attached to the larger
old ponds. The smaller new ponds filled in the valley bottoms, forming
a percolated network (*sensu* Gardner *et al.*, 1991) across the landscape.

The pattern and magnitude of pond growth varied according to the
decade in which the pond was initiated (i.e. its cohort). There were
significant differences across cohorts in average area of like-aged pond
sites (i.e. 10, 20, 30 or 40 years of age) (Figure 3.2). The regression of the
growth curve for the 1940 pond cohort was also significantly different

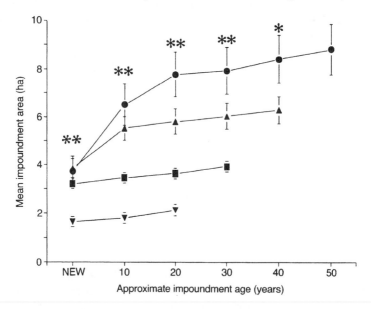

Figure 3.2 Average impoundment area, by pond cohort. • = 1940 pond cohort
($y = 1.27 \ln(x) + 3.72$); ▲ = 1948 pond cohort ($y = 0.63 \ln(x) + 3.85$); ■ = 1961
pond cohort ($y = 0.18 \ln(x) + 3.16$); ▼ = 1972 pond cohort ($y = 0.14 \ln(x) +
1.62$). * and ** indicate significant effects of impoundment age on impoundment
area (Kruskal-Wallis one-way ANOVA) at 0.05 and 0.01 significance levels.
(Reprinted, with permission, from *Ecology*. Copyright 1990 by the Ecological
Society of America. 'Aquatic Patch Creation in Relation to Beaver Population
Trends'; C. A. Johnston and R. J. Naiman, v. 71, n. 4, pp. 1617–21, Fig. 1.)

than those for all subsequent cohorts. Therefore, not only were the 1940 cohort ponds significantly larger to begin with than ponds initiated in 1972 or later, they also grew at a faster rate once established, doubling in average area after only two decades (Figure 3.2). The 1948 cohort ponds also grew rapidly during the first decade after establishment (43% increase in average pond site area), but there were no significant differences among the growth curves for the 1948, 1961 and 1972 age classes. Since 1961, average pond site area increases in all age classes have been linear and small, ranging from 0.1 to 0.8 ha per decade.

The combination of high rates of pond creation, larger initial pond area and rapid growth made the 1940, 1948 and 1961 pond cohorts the most spatially influential (Figure 3.3). As of 1986, these cohorts constituted 75% of the total number of pond sites and 90% of the total area impounded. The establishment of new ponds was the primary cause of increased cumulative pond site area prior to 1961 (70% of the total increase), the rest being due to the enlargement of existing ponds. The proportions were reversed after 1972. Therefore, ponds constructed by beaver during their first few decades of occupancy have the greatest impact on the landscape.

The abrupt change in new pond characteristics after 1961 is believed to have been due to the attainment of a geomorphic threshold controlling

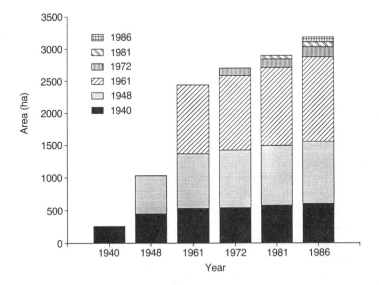

Figure 3.3 Cumulative pond area, by age class. (Reprinted, with permission, from *Ecology*. Copyright 1990 by the Ecological Society of America. 'Aquatic Patch Creation in Relation to Beaver Population Trends'; C. A. Johnston and R. J. Naiman, v. 71, n. 4, pp. 1617–21, Fig. 2.)

landscape susceptibility to this type of disturbance. Beaver ponds are restricted to locations where a relatively low dam (1–2 m) can impound a large area of water, so there is a topographic limit to the sites that can potentially be affected. The beaver at VNP flooded first those areas that would make the largest ponds with the greatest potential for expansion. As more and more of the potentially floodable sites were occupied, new pond creation decreased and was limited to sites where only small ponds could be established (Johnston and Naiman, 1990c). This is consistent with the finding of other researchers that stream gradient and/or dimensions are significantly correlated with beaver colony density (Slough and Sadleir, 1977; Howard and Larson, 1985; Beier and Barrett, 1987). Habitat factors also found to be important are substrate type (Retzer *et al.*, 1956; Howard and Larson, 1985) and food supply (Slough and Sadleir, 1977).

In addition to altering the surface of the landscape, the establishment of beaver ponds caused both short-term (e.g. anaerobic conditions) and long-term (e.g. nutrient and carbon accumulation) changes in soils underlying the ponds (Naiman, Johnston and Kelley, 1988; Johnston, Pastor and Pinay, 1992). Although some of these subterranean patch bodies coincided spatially with visible surface patch features, others did not. An anaerobic patch body below a beaver-abandoned wet meadow coincided spatially with the overlying disturbance patch, but soil carbon storage was related to past rather than present vegetation (i.e. pre-impoundment wetland versus disturbance patch vegetation). Zones of high nitrate concentration in soil water sampled at 25 cm were ephemeral in location and unrelated to changes in vegetation, nor did they always coincide with zones of high nitrate concentration at the 50 cm depth (Johnston, Pastor and Pinay, 1992). The spatial displacement among the different types of patch bodies was dependent on lag times in their response to disturbance. The subterranean zones of high nitrate concentrations and anaerobic conditions, both of which are related to microbial activity, were most dynamic, followed by changes in surface vegetation and soil carbon.

Whereas the above discussion has focused on ecosystem and landscape effects of beaver pond construction, this activity may have global significance to atmospheric trace gas concentrations. Methane fluxes were measured from a hydrosequence consisting of four different zones (upland forest, wet meadow, submergent marsh and deep open water) associated with a beaver pond at VNP (Naiman, Manning and Johnston, 1991). Annual methane fluxes from the deep water (8.73 g C m^{-2} year^{-1}) and submergent zones (10.78 g C m^{-2} year^{-1}) were significantly higher than those from the wet meadow (0.36 g C m^{-2} year^{-1}) and forested zones (0.25 g C m^{-2} year^{-1}) (Naiman, Manning and Johnston, 1991). Other researchers of boreal wetlands (Roulet, Ash and Moore, 1992) have found similarly high fluxes of methane from beaver ponds (7.6 g CH$_4$ m^{-2} year^{-1}), much higher than the rate from marshes or forested swamps

$(\sim 0.1$ g CH_4 m^{-2} year^{-1}). If our data from VNP are representative of beaver activity throughout North America, then about 1% of the recent rise in atmospheric methane may be attributable to pond creation by beavers (Naiman, Manning and Johnston, 1991).

3.3 PATCH FORMATION BY DIGGING

A variety of animals ranging in size from insects (Rusek, 1992) to elephants (Laws, 1970) influence the physical landscape by digging or wallowing. Although animals such as the wild boar (*Sus scrofa*) dig from aboveground (Bratton, 1975), most animals that perturb the soil do so by burrowing below-ground. On a dry weight basis, animals consititute only 0.01% of the soil mass, but the density of individuals (Table 3.1) and their cumulative effects can be substantial. For example, it has been

Table 3.1 Estimates of proportions, dry weight and number of living organisms in a hectare of soil in a humid temperate region, to a depth of 15 cm. (Source: Buol, Hole and McCracken, 1973; nomenclature follows Brusca and Brusca, 1990)

Organism Phylum–Subphylum–Class–Subclass–Order (examples)	Dry weight		Estimated number of individuals
	%	kg/ha	
Nematoda	0.001	20	2.5×10^9
Mollusca–none–Gastropoda (snails, slugs)	0.001	20	2×10^3
Annelida (earthworms, potworms)	0.005	100	7×10^3
Arthropoda–Crustacea (isopods, crayfish)	0.000 5	10	4×10^{17}
Arthropoda–Cheliceriformes–Chelicerata–Arachnida–			
Acari (mites)	0.000 1	2	4×10^5
Araneae (spiders)	0.000 1	2	5×10^5
Opiliones (harvestmen)	0.000 05	1	2.5×10^4
Arthropoda–Uniramia–Myriapoda–Diplopoda and Chilopoda (millepedes and centipedes)	0.001	20	1×10^3
Arthropoda–Uniramia–Myriapoda–Symphyla	0.0001	25	3.8×10^7
Arthropoda–Uniramia–Insecta–Oligoen-tomata–			
Collembola (springtails)	0.000 1	2	4×10^5
Arthropoda–Uniramia–Insecta–Pterygota–			
Hymenoptera (ants)	0.0002	5	5×10^6
Diptera, Coleoptera, Lepidoptera	0.0001	35	5×10^7
Chordata–Vertebrata			
(Mice, voles, moles)	0.000 5	10	4×10^5
(Rabbits, squirrels, gophers)	0.000 6	12	10
(Foxes, badgers, bear, deer)	0.000 5	10	<1
(Birds)	0.000 5	10	100
Total	0.013 75	284	–

estimated that the upper 60 cm of a Wisconsin prairie soil is turned over once each century by the activity of ants, worms and rodents (Curtis, 1959).

Soil perturbation and augmentation by animals may occur in ways that are not intuitive. Cicada nymphs produce a special kind of soil structure in medium-textured soils in Idaho, Utah and Nevada (Hugie and Passey, 1963). A New Zealand petrel (*Pachyptela turtur*) nests in extensive burrows that it makes in the soil. Roosting crows cough up enough fine gravel from their gizzards to change the texture of the silt loam beneath them to a gravelly silt loam over a thousand years (Buol, Hole and McCracken, 1973). Other animals (crustacea, reptiles, amphibia, badgers, foxes and birds) may have influences that are locally important. Unfortunately, many of these influences are reported anecdotally, and there are few studies quantifying their ecosystem or landscape-level significance.

3.3.1 VERTEBRATES

Many families in the rodent order exhibit fossorial behavior, excavating often extensive systems of below-ground tunnels associated with foraging and/or den building: moles (Talpidae), shrews (Soricidae), pocket gophers (Geomyidae), ground squirrels and prairie dogs (Sciuridae), pocket mice and kangaroo rats (Heteromyidae), and mountain beavers (Aplodontiidae). Pre-settlement rodent populations had a large influence on North American prairie soil development, annually bringing 10–90 Mg ha^{-1} of subsoil material to the surface (Thorp, 1949; Anderson, 1987b). Mielke (1977; cf. Huntly and Inouye, 1988) argued that, '. . . [T]he activities of fossorial rodents may provide an explanation for the genesis of North American Prairie soils.'

(a) Pocket gophers

The landscape-level significance of pocket gopher digging was reported as early as 1923 by Grinnell (cf. Huntly and Inouye, 1988), who asserted that '. . . our native plant life, on hill and mountainside, in canyon and mountain meadow, would soon begin to depreciate were the gopher population completely destroyed.' The range of pocket gophers in the Geomyidae family (*Thomomys, Cratogeomys* and *Geomys* spp.) covers most of northern Mexico, central and western USA, and the central plains of Canada, including most grasslands and arid shrublands from sea level to alpine areas (Burt and Grossenheider, 1964; Mielke, 1977). Although the Geomyidae family is exclusively North American, fossorial rodents that are morphologically and behaviorally similar occur in savanna, grassland, and arid scrub habitats in South America, Asia, Africa and Europe (Andersen, 1987a; Nevo, 1979).

Densities of pocket gophers are frequently in the range of 50–100 individuals ha⁻¹, ranging from 12 individuals ha⁻¹ in long-term exclosures (Mohr and Mohr, 1936) to >130 individuals ha⁻¹ in heavily populated areas (Howard, 1961; Nevo, 1979). Gophers are solitary for much of their lives, and have a home range of about 120 m² for females and 200 m² for males of *Thomomys bottae* (Burt and Grossenheider, 1964). Pocket gophers are seldom seen aboveground, spending most of their time in extensive tunnel systems built to access subsurface plant parts. Although they prefer sites with high soil nitrogen (Inouye, Huntly and Tilman, 1987; Hobbs *et al.*, 1988) and high primary productivity (Tilman, 1983; Reichman and Smith, 1985; Inouye *et al.*, 1987), the only areas reported to preclude gopher activity are those with very shallow soils where bedrock is at or close to the surface (Cantor and Whitham, 1989; Hobbs and Mooney, 1991).

As they construct tunnels, pocket gophers move subsoil to the ground surface and deposit it in fan-shaped mounds of soil 20–50 cm in diameter. In areas with seasonal snow cover, gophers burrow in snow and fill the tunnels with soil, which is left on the surface of the ground in rope-like cores after snowmelt (Laycock, 1958). The patches of material brought to the surface by gophers may cover 10–30% of the ground (Hooven, 1971; Turner *et al.*, 1973; Foster and Stubbendieck, 1980; Hobbs and Mooney, 1985; Spencer *et al.*, 1985), although there may be strong clumping of gopher disturbances in any particular year due to the connectivity of tunnels (Hobbs and Mooney, 1991). A 6-year study of gopher soil disturbance found that the area of new disturbance varied annually from 8% to 43%, and that sites were disturbed multiple times at a return rate of 1.6 ± 0.1 year (Hobbs and Mooney, 1991).

This reworking of the soil by gophers greatly alters the properties of the soil surface. Gopher mounds are composed of subsoil brought to the surface that frequently differs in texture, water-holding capacity, and nutrient content relative to undisturbed surface soils (Grant, French and Folse, 1980; Anderson, 1987a, b). Nutrient concentrations may be higher (Grant and McBrayer, 1981; Andersen and MacMahon, 1985; Koide, Huenneke and Mooney, 1987) or lower (McDonough, 1974; Spencer *et al.*, 1985; Koide, Huenneke and Mooney, 1987; Inouye, Huntly and Tilman, 1987; Inouye *et al.*, 1987) in gopher mounds than in undisturbed soil, depending on the fertility of the subsoil brought to the surface. For example, gopher mounds in a Minnesota old field had much lower nitrogen concentrations than did undisturbed surface soils because the subsoil contains only one-fifth the total nitrogen of surface soils (110 ppm versus 580 ppm: Huntly and Inouye, 1988). Conversely, soil brought to the surface by gophers on Mount St Helens was higher in nutrients than the volcanic ash that covered it (Andersen and MacMahon, 1985). Gopher activity thus increases the heterogeneity of the soil surface by creating patches of high and low fertility soil. Associated with these changes in surface soil characteristics are changes in flora (e.g. increase in forbs and annuals), fauna (e.g. grasshopper abundance) and light penetration (Huntly and Inouye, 1988).

While the below-ground perturbances caused by pocket gopher tunnels are less visible than their aboveground counterparts, they may be equally significant to the landscape. Andersen (1987b) found that 41–87% of soil excavated by pocket gophers was packed into old tunnels as backfill. Effects of backfilling on above and below-ground ecosystem processes remain unknown, but backfilled soil has been found to differ in bulk density and nutrient content from undisturbed soil (Andersen, 1987b; Huntly and Inouye, 1988). Abandoned gopher burrows are used by amphibians, reptiles and other mammals (Vaughan, 1961).

Over periods of decades to centuries, above- and below-ground soil mixing by gophers influences soil development and increases fertility (Grinell, 1923; Mielke, 1977; Hole, 1981). For example, a subalpine grassland subject to a 31-year gopher removal experiment had significantly lower levels of several soil nutrients than did a control area containing gophers (Laycock and Richardson, 1975).

Over longer time periods, the activities of gophers may interact with physical landscape forces to produce special geomorphic features: mima mounds and stone stripes. Mima mounds are roughly circular soil lenses up to ~ 2m high and 25–50 m in diameter that occur at densities of 50 to >100 ha^{-1} (Cox and Gakahu, 1986). They are thought to result from the long-term accumulations of soil by pocket gophers in the US (Scheffer, 1958; Mielke, 1977; Cox and Gakahu, 1986), and by similar fossorial rodents on other continents (Cox and Roig, 1986; Cox, Lovegrove and Siegfried, 1987). Mima mounds have deeper, more fertile soils (Mielke, 1977; Cox and Gakahu, 1985), and higher net primary production than intermound areas (Ross, Tester and Breckenridge, 1968; Cox and Zedler, 1986).

Stone stripes are characteristic features of the Columbia Plateau in the western US (Cox, 1990). They are narrow beds of sorted, bare stones oriented perpendicular to the topographic contour, sharply demarcated from adjacent well-vegetated soils (Cox and Hunt, 1990). After testing several geogenic and biogenic hypotheses regarding their origin, Cox and Hunt (1990) concluded that stripe formation and maintenance involves the dominant forces of pocket gopher tunneling on gentle upper slopes and outcrop weathering processes on lower slopes, with erosion playing a secondary role at both locations. Thus, the susceptibility of the landscape to stripe formation by pocket gopher activity was moderated by topography.

(b) Praire Dogs

Unlike the solitary pocket gopher, prairie dogs (*Cynomys* spp.) are greg-arious animals, living in large 'towns' tens to hundreds of hectares in area (Dahlsted *et al.*, 1981; Knowles, 1986) at densities of 10–55 individuals ha^{-1} (O'Meilia, Knopf and Lewis, 1982; Knowles, 1986; Archer, Garrett and Detling, 1987). Colonies are generally located on sites with deep, pro-ductive, gently sloping (<7%) soils where flooding is unlikely (Dahlsted *et al.*, 1981). Viewed as competitors with cattle for rangeland resources, eradication programs have reduced prairie dog populations by more than 98% over only a few decades (Summers and Linder, 1978), and studies of their effects are restricted to relatively pristine preserves such as Wind Cave National Park and Devils Tower National Monument (Chapter 13).

A typical prairie dog burrow has two entrances, a depth of 1–3 m, a total length of 15 m, and a diameter of 10–13 cm (Sheets, Linder and Dahlgren, 1971), requiring the excavation of 200–225 kg of soil (Whicker and Detling, 1988). Much of this soil is deposited in mounds 0.3–0.6 m high and 1–2 m in diameter at the burrow entrances (Burt and Gros-senheider, 1964). These soil mounds may number 50–300 ha^{-1} (O'Meilia, Knopf and Lewis, 1982; White and Carlson, 1984; Archer, Garrett and Detling, 1987) and have physical and chemical properties that may remain unaltered for hundreds or thousands of years (Carlson and White, 1987).

In addition to changes in vegetation induced by the physical alteration of the soil surface, prairie dogs actively maintain the area around burrow entrances to keep them free from vegetation (Whicker and Detling, 1988). This vegetation clipping, in combination with herbivory by prairie dogs and associated grassland herbivores (e.g. bison), causes rapid shifts in canopy height (Archer, Garrett and Detling, 1987), plant morphology (Detling and Painter, 1983; Whicker and Detling, 1988), above- and below-ground plant biomass (Coppock *et al.*, 1983), and plant species composition within the colony patch (Osborn and Allan, 1949; Koford, 1958; Bonham and Lerwick, 1976; Coppock *et al.*, 1983; Agnew, Uresk and Hansen, 1986).

The reduction in plant cover and increase of bare soil associated with prairie dog colonies has effects on microclimate. Soil moisture could either be lower due to increased evaporation from bare soil, or higher due to decreased interception of precipitation by vegetation (Knapp and Seastedt, 1986) and decreased transpiration (Archer and Detling, 1986). Soil temperatures are 2.5 °C higher on prairie dog colonies than in more densely vegetated uncolonized areas nearby (Archer and Detling, 1986), which could lead to increased rates of nitrogen mineralization in the subsurface soil (Whicker and Detling, 1988).

3.3.2 INVERTEBRATES

Invertebrates are abundant in soils (Table 3.1) and cause substantial soil mixing. Based on published leaf litter consumption rates (Knollenberg, Merritt and Lawson, 1985) and annelidae densities commonly found in soil (Table 3.1), I calculated that earthworms annually consume 1825 kg ha^{-1} of above-ground biomass, an amount much higher than that consumed by most vertebrates (Table 3.2). Given annual litterfall rates in deciduous forests of about 3000 kg ha^{-1} (Gosz, Likens and Bormann, 1972; Merriam, Dwyer and Wegner, 1982) this represents a much larger proportion of net annual production than that consumed by most herbivores (Table 3.2).

Table 3.2 Annual biomass consumed by herbivores per unit area and as a proportion of production. (Source: McInnes et al., 1992)

Herbivore	Biomass consumed (kg ha^{-1} year^{-1})	Biomass consumed/ annual production	Reference
Earthworms	1825	0.61	See text
Moose	0.1–21	0.00–0.03	McInnes et al. (1992)
Bison	16–240	0.01–0.16	Coppock et al. (1983)
Snow geese	780–900	0.80	Cargill and Jefferies (1984)
Ungulates	996–6242	0.17–0.95	McNaughton (1985)

Termites are impressive mound-builders, building nests as large as 3 m high and 15 m in diameter at densities of up to 75 ha^{-1} (Buol, Hole and McCracken, 1973). The surface and below-ground influence of termites is also substantial: bald spots made by termite nests occupied 20% of the surface area of a Russian pasture, and termite tunnels in semi-arid regions can be as deep as 8 m (Buol, Hole and McCracken, 1973). Termite mounds are particularly prevalent in central and western Africa (Nye, 1955; Watson, 1962; Carroll, 1969). In addition to their landscape-level effects, termites are a significant source of atmospheric methane due to anaerobic fermentation in their digestive tracts (Zimmerman et al., 1982; Seiler, Conrad and Scharffe, 1984; Fraser et al., 1986; Khalil et al., 1990). Termites contribute 4.0×10^{13} g CH_4 year^{-1}, i.e. 7.6% of all global sources (Schlesinger, 1991).

Mound-building ants carry subsoil material to the surface, creating mounds ~0.3 m in cross-section in undisturbed prairies (*Formica cinera*), and as large as 1 m high and 2 m wide in forest gaps (*Formica exsectoides*) (Baxter and Hole, 1967; Salem and Hole, 1968; Denning, 1973). Prairie mound-building ants were abundant prior to European settlement of the US, but cannot survive in cultivated land and are presently confined to areas such as cemeteries, railroad rights-of-way, and wetlands. However, the relatively uniform clay content of prairie soil profiles is thought to be a relict of the ants' former soil mixing activities, indicative of the upward translocation of clay from the B to the A horizon (Milfred, 1966; Baxter and Hole, 1967).

Ant species that are not mound builders also move large volumes of soil, although their influence is not as patchy. At Harvard Forest in the eastern US, ants that make small (1 cm high and 7 cm wide) surficial deposits have completely mixed the upper 35 cm of soil (Lyford, 1963). On the Great Plains, ants create graveled bare areas as large as 6 m in diameter (Thorp, 1949). Soil scientists estimate that surface clearing and accumulation of soil by ants affects 1–4% of landscape area (Buol, Hole and McCracken, 1973).

3.4 SUMMARY

Animals can greatly influence surface and subsurface features of the landscape through direct and indirect physical manipulation of the environment. Fossorial rodents, earthworms and other soil fauna directly manipulate the environment via tunneling, whereas beavers do so indirectly via dam building. The magnitude of physical alteration is apparently not proportional to animal body size, but to behavioral adaptations that govern the animal's manipulative capabilities as individuals (e.g. pocket gophers) and colonies (e.g. ants, termites, prairie dogs).

The physical influence of animals may be diffuse or patchy, with patch sizes ranging from square centimeters to square kilometers. Patch creation can occur at multiple scales, even when caused by the same species; pocket gophers create landscape patches at spatial scales of decimeters and temporal scales of years (i.e. burrow entrance mounds), as well as larger patches at temporal scales of decades to centuries (i.e. mima mounds). Animal-induced alteration of the physical landscape has secondary effects on ecosystem processes that can be significant even at global scales (e.g. methane production from termite mounds and beaver ponds).

Human activities such as agriculture and trapping have reduced the abundance of many landscape-altering animals to only a fraction of their pre-settlement populations, so that much of the evidence of their influence comes from preserve areas. Given the historical significance of animals to soil formation in landscapes such as the North American prairies, their near eradiction may have future impacts difficult to remedy using artificial means. Fortunately, if the recovery of beaver populations in North America is any indication, these animals may be able to recover from centuries of decimation in a relatively short time. We need to seek ways in which human and animal alteration of the physical environment can occur within the same landscape.

ACKNOWLEDGEMENTS

Research support from the National Science Foundation (DEB–9119614) is gratefully acknowledged.

REFERENCES

Agnew, W., Uresk, D. W. and Hansen, R. M. (1986) Flora and fauna associated with prairie dog colonies and adjacent ungrazed mixed-grass prairie in western South Dakota. *J. Range Manage.* **39**, 135–9.

Aleksiuk, M. (1968) Scent-mound communication, territoriality and population regulation in beaver (*Castor canadensis* Kuhl). *J. Mamm.*, **49**, 759–62.

Andersen, D. C. (1987a) Below-ground herbivory in natural communities: a review emphasizing fossorial animals. *Q. Rev. Biol.*, **62**, 261–86.

Andersen, D. C. (1987b) *Geomys bursarius* burrowing patterns: influence of season and food patch structure. *Ecology,* **68**, 1306–18.

Andersen, D. C. and MacMahon, J. A. (1985) Plant succession following the Mount St. Helens volcanic eruption: facilitation by a burrowing rodent, *Thomomys talpoides*. *Am. Midl. Nat.*, **114**, 62–9.

Archer, S. and Detling, J. K. (1986) Evaluation of potential herbivore mediation of plant water status in a North American mixed-grass prairie. *Oikos*, **47**, 287–91.

Archer, S., Garrett, M. G. and Detling, J. K. (1987) Rates of vegetation change associated with prairie dog (*Cynomys ludovicianus*) grazing in North American mixed-grass prairie. *Vegetatio*, **72**, 159–66.

Baxter, F. P. and Hole, F. D. (1967) Ant (*Formica cinerea*) pedoturbation in a prairie soil. *Soil Sci. Soc. Am. Proc.*, **31**, 425–8.

Beer, J. R. (1955) Movements of tagged beaver. *J. Wildl. Manage.*, **19**, 492–3.

Beier, P. and Barrett, R. H. (1987) Beaver habitat use and impact in Truckee River Basin, California. *J. Wildl. Manage.*, **51**, 794–9.

Bergerud, A. T. and Miller, D. R. (1977) Population dynamics of Newfoundland beaver. *Can. J. Zool.*, **55**, 1480–92.

Blais, J. R. (1954) The recurrence of spruce budworm infestations in the past century in the Lac Seul area of northwestern Ontario. *Ecology,* **35**, 62–71.

Bonham, C. D. and Lerwick, A. (1976) Vegetation changes induced by prairie dogs on shortgrass range. *J. Range Manage.*, **29**, 221–5.

Botkin, D. B., Melillo, J. M. and Wu, L. S.-Y. (1981) How ecosystem processes are linked to large mammal population dynamics, in *Dynamics of Large Mammal Populations* (eds C. F. Fowler and T. D. Smith), New York, John Wiley & Sons, pp. 373–87.

Bradt, G. W. (1938) A study of beaver colonies in Michigan. *J. Mamm.*, **19**, 139–62.

Bratton, S. P. (1975) The effect of the European wild boar (*Sus scrofa*) on gray beech forest in the Great Smokey Mountains. *Ecology*, **69**, 1356–66.

Brenner, F. J. (1964) Reproduction of the beaver in Crawford County, Pennsylvania. *J. Wildl. Manage.*, **28**, 743–7.

Broschart, M. R., Johnston, C. A. and Naiman, R. J. (1989) Prediction of beaver colony density using impounded habitat variables. *J. Wildl. Manage.*, **53**, 929–34.

Brown, V. K. and Gange, A. C. (1990) Insect herbivory below ground. *Adv. Ecol. Res.*, **201**, 1–58.

Brusca, R. C. and Brusca, G. J. (1990) *Invertebrates*, Sinauer Associates, Sunderland, Massachusetts.

Buol, S. W., Hole, F. D. and McCracken, R. J. (1973) *Soil Genesis and Classification*, Iowa State University Press, Ames.

Burt, W. H. and Grossenheider, R. P. (1964) *Field Guide to the Mammals*, Houghton Mifflin Company, Boston.

Cantor, L. F. and Whitham, T. G. (1989) Importance of belowground herbivory: pocket gophers may limit aspen to rock outcrop refugia. *Ecology*, **70**, 962–70.

Cargill, S. M. and Jefferies, R. L. (1984) The effects of grazing by lesser snow geese on the vegetation of a subarctic salt marsh. *J. Appl. Ecol.*, **21**, 669–86.

Carlson, D. C. and White, E. M. (1987) Effects of prairie dogs on mound soils. *Soil Sci. Soc. Am. J.*, **51**, 389–93.

Carroll, P. H. (1969) Soil-dwelling termites in the southwest region of the Ivory Coasts. *Soil Survey Horizons*, **10**, 3–16.

Coleman, D. C., Reid, C. P and Cole, C. V. (1983) Biological strategies of nutrient cycling in soil systems. *Adv. Ecol. Res.*, **13**, 1–55.

Connell, J. H. and Keough, M. J. (1985) Disturbance and patch dynamics of subtidal marine animals on hard substrata, in *The Ecology of Natural Disturbance and Patch Dynamics* (eds S.T.A. Pickett and P.S. White), Academic Press, Orlando, pp. 125–51.

Coppock, D. L., Ellis, J. E., Detling, J. K. and Dyer, M. I. (1983) Plant-herbivore interactions in a North American mixed-grass prairie II. Responses of bison to modification of vegetation by prairie dogs. *Oecologia*, **56**, 10–15.

Coulson, R. N., Lovelady, C. N., Flamm, R. O. *et al.* (1991) Intelligent geographic information systems for natural resource management, in *Quantitative Methods in Landscape Ecology* (eds M. G. Turner and R. H. Gardner), Springer-Verlag, New York, pp. 153–72.

Cox, G. W. (1990) Form and dispersion of Mima mounds in relation to slope steepness and aspect on the Columbia Plateau. *Great Basin Nat.*, **50**, 21–31.

Cox, G. W. and Gakahu, C. G. (1985) Mima mound microtopography and vegetation pattern on Kenyan savannas. *J. Trop. Ecol.*, **1**, 23–36.

Cox, G. W. and Gakahu, C. G. (1986) A latitudinal test of the fossorial rodent hypothesis of Mima mound origin in western North America. *Zeitschrift für Geomorphologie*, **30**, 485–501.

Cox, G. W. and Hunt, J. (1990) Nature and origin of stone stripes on the Columbia Plateau. *Landsc. Ecol.*, **5**, 53–64.

Cox, G. W. and Roig, V. G. (1986) Argentinian Mima mounds occupied by ctenomyid rodents. *J. Mamm.*, **67**, 428–32.

Cox, G. W. and Zedler, J. B. (1986) The influence of mima mounds on vegetation patterns in the Tijuana estuary salt marsh, San Diego County, California. *Bull. S. Calif. Acad. Sci.*, **83**, 158–72.

Cox, G. W., Lovegrove, B. G. and Siegfried, W. R. (1987) The small stone content of mima-like mounds in the South African Cape region. *Catena*, **14**, 165–76.

Curtis, J. T. (1959) *The Vegetation of Wisconsin: An Ordination of Plant Communities*, University of Wisconsin Press, Madison.

Dahlsted, K. J., Sather-Blair, S., Worcester, B. K. and Klukas, R. (1981) Application of remote sensing to prairie dog management. *J. Range Manage.*, **34**, 218–23.

Denning, J. L. (1973) I. Measurement and calculation of hydraulic conductivity using physical and morphometric techniques. II. Ant pedoturbation in a poorly drained Calamine silty clay loam (Typic Haplaquoll). MS Thesis, University of Wisconsin Press, Madison.

Detling, J. K. and Painter, E. L. (1983) Defoliation responses of western wheatgrass populations with diverse histories of prairie dog grazing. *Oecologia*, **57**, 65–71.

Forman, R. T. T. and Godron, M. (1986) *Landscape Ecology*, John Wiley & Sons, New York.

Foster, M. A. and Stubbendieck, J. (1980) Effects of the plains pocket gopher (*Geomys busarius*) on rangeland. *J. Range Manage.*, **33**, 75–8.

Fraser, P. J., Rasmussen, R. A., Creffield, J. W. *et al.* (1986) Termites and global methane – another assessment. *J. Atmos. Chem.*, **4**, 295–310.

Gardner, R. H., Turner, M. G., O'Neill, R. V. and Lavorel, S. (1991) Simulation of the scale-dependent effects of landscape boundaries on species persistence and dispersal, in *Role of Landscape Boundaries in the Management and Restoration of Changing Environments* (eds. M. M.

Holland, P. G. Risser and R. J. Naiman), Chapman & Hall, New York, pp. 76–89.

Gosz, J. R., Likens, G. E. and Bormann, F. H. (1972) Nutrient content of litter fall on the Hubbard Brook Experimental Forest, New Hampshire. *Ecology*, **53**, 769–84.

Grant, W. E. and McBrayer, J. F. (1981) Effects of mound formation by pocket gophers (*Geomys bursarius*). *Pedobiologia*, **22**, 21–8.

Grant, W. E., French, N. R. and Folse, L. J. Jr (1980) Effects of pocket gopher mounds on plant production in a shortgrass prairie ecosystem. *Southwest Nat.*, **2**, 215–24.

Grinnell, J. (1923) The burrowing rodents of California as agents in soil formation. *J. Mamm.*, **4**, 137–49.

Hairston, N. G., Smith, F. E. and Slobodkin, L. B. (1960) Community structure, population control, and competition. *Am. Nat.*, **94**, 421–5.

Hatton, J. C. and Smart, N. O. E. (1984) The effect of long-term exclusion of large herbivores on soil nutrient status in Murchison Falls National Park, Uganda. *Afr. J. Ecol.*, **22**, 23–30.

Hibbard, E. A. (1958) Movements of beaver transplanted in North Dakota. *J. Wildl. Manage.*, **22**, 209–11.

Hobbs, R. J. and Mooney, H. A. (1985) Community and population dynamics of serpentine grassland annuals in relation to gopher disturbance. *Oecologia*, **67**, 342–51.

Hobbs, R. J. and Mooney, H. A. (1991) Effects of rainfall variability and gopher disturbance on serpentine annual grassland dynamics. *Ecology*, **72**, 59–68.

Hobbs, R. J., Gulmon, S. L., Hobbs, V. J. and Mooney, H. A. (1988) Effects of fertiliser addition and subsequent gopher disturbance on a serpentine annual grassland community. *Oecologia*, **75**, 291–5.

Hodgdon, K. W. and Hunt, J. H. (1953) *Beaver management in Maine*, Maine Dept. Inland Fisheries and Game, Game Div. Bull. 3.

Hole, F. D. (1981) Effects of animals on soil. *Geoderma*, **25**, 75–112.

Hooven, E. F. (1971) Pocket gopher damage on ponderosa pine plantations in southwestern Oregon. *J. Wildl. Manage.*, **35**, 346–53.

Howard, R. J. and Larson, J. S. (1985) A stream habitat classification system for beaver. *J. Wildl. Manage.*, **49**, 19–25.

Howard, W. E. (1961) A pocket gopher population crash. *J. Mamm.*, **42**, 258–60.

Hugie, V. K. and Passey, H. B. (1963) Cicadas and their effect upon soil genesis in certain soils in southern Idaho, northern Utah and northeastern Nevada. *Soil Sci. Am. Proc.*, **27**, 78–82.

Huntly, N. J. (1991) Herbivores and the dynamics of communities and ecosystems. *Ann. Rev. Ecol. Syst.*, **22**, 477–503.

Huntly, N. J. and Innouye, R. S. (1988) Pocket gophers in ecosystems: patterns and mechanisms. *BioScience*, **38**, 786–93.

Hutchinson, G. E. (1957) *A Treatise on Limnology. Vol. 1, Part 1.* John Wiley & Sons, New York.

Inouye, R. S., Huntly, N. J. and Tilman, D. (1987) Pocket gophers, vegetation and soil nitrogen along a successional sere in east central Minnesota. *Oecologia*, **72**, 178–84.

Inouye, R. S., Huntly, N. J., Tilman, D. and Tester, J. R. (1987) Old-field succession on a Minnesota sand plain. *Ecology*, **68**, 12–26.

Johnson, A. R., Milne, B. T. and Weins, J. A. (1992) Diffusion in fractal landscapes: simulations and experimental studies of tenebrionid beetle movements. *Ecology*, **73**, 1968–83.

Johnston, C. A. (1993) Introduction to quantitative methods and modeling in community, population, and landscape ecology, in *Environmental Modeling with GIS* (eds M. R. Goodchild, B. O. Parks and L. T. Steyaert), Oxford University Press, New York, pp. 276–83.

Johnston, C. A. and Naiman, R. J. (1987) Boundary dynamics at the aquatic–terrestrial interface: The influence of beaver and geomorphology. *Landsc. Ecol.*, **1**, 47–57.

Johnston, C. A. and Naiman, R. J. (1990a) Browse selection by beaver: effects on riparian forest composition. *Can. J. For. Res.*, **20**, 1036–43.

Johnston, C. A. and Naiman, R. J. (1990b) The use of a geographic information system to analyze long-term landscape alteration by beaver. *Landsc. Ecol.*, **4**, 5–19.

Johnston, C. A. and Naiman, R. J. (1990c) Aquatic patch creation in relation to beaver population trends. *Ecology*, **71**, 1617–21.

Johnston, C. A., Pastor, J. and Pinay, G. (1992) Quantitative methods for studying landscape boundaries, in *Landscape Boundaries: Consequences for Biotic Diversity and Ecological Flows*, (eds F. di Castri and A. Hansen), Springer-Verlag, New York, pp. 107–25.

Johnston, C. A., Pastor, J. and Naiman, R. J. (1993) Effects of beaver and moose on boreal forest landscapes, in *Landscape Ecology and Geographical Information Systems* (eds S. H. Cousins, R. Haines-Young and D. Green), Taylor & Francis, London, pp. 237–54.

Khalil, M. A. K., Rasmussen, R. A., French, J. R. J. and Holt, J. A. (1990) The influence of termites on atmospheric trace gases: CH_4, CO_2, CHl_3, N_2O, CO, H_2, and light hydrocarbons. *J. Geophys. Res.*, **95**, 3619–34.

Knapp, A. K. and Seastedt, T. R. (1986) Detritus accumulation limits productivity of tallgrass prairie. *BioScience*, **36**, 662–8.

Knollenberg, W. G., Merritt, R. W. and Lawson, D. L. (1985) Consumption of leaf litter by *Lumbricus terrestris* (Oligochaeta) on a Michigan woodland floodplain. *Am. Midl. Nat.*, **113**, 1–6.

Knowles, C. J. (1986) Some relationships of black-tailed prairie dogs to livestock grazing. *Great Basin Nat.*, **46**, 198–203.

Koford, C. B. (1958) *Prairie Dogs, Whitefaces, and Blue Grama.* Wildlife Monograph No. 3.

Koide, R. T., Huenneke, L. F. and Mooney, H. A. (1987) Gopher mound soil reduces growth and affects ion uptake of two annual grassland species. *Oecologia*, **72**, 284–90.

Laws, R. M. (1970) Elephants as agents of habitat and landscape change in East Africa. *Oikos*, **21**, 1–15.

Laycock, W. A. (1958) The initial pattern of revegetation of pocket gopher mounds. *Ecology*, **39**, 346–51.

Laycock, W. A. and Conrad, P. W. (1967) Effect of grazing on soil compaction as measured by bulk density on a high elevation cattle range. *J. Range Manage.*, **20**, 136–40

Laycock, W. A. and Richardson, B. Z. (1975) The long-term effects of pocket gopher control on vegetation and soils of a subalpine grassland. *J. Range Manage.*, **28**, 458–62.

Leege, T. A. (1968) Natural movements of beavers in southeastern Idaho. *J. Wildl. Manage.*, **32**, 973–6.

Levin, S. A. (1992) The problem of pattern and scale in ecology. *Ecology*, **73**, 1943–67.

Libby, W. L. (1957) Observations on beaver movements in Alaska. *J. Mamm.*, **38**, 269.

Lock, J. M. (1972) The effects of hippopotamus grazing on grasslands. *J. Ecol.*, **60**, 445–67.

Lyford, W. H. (1963) Importance of ants to Brown Podzolic soil genesis in new England. Harvard Forum Paper 7, Petersham, Massachusetts.

Martin, A., Mariotti, A., Balesdent, J. and Lavelle, P. (1992) Soil organic matter assimilation by a geophagous tropical earthworm based on δ^{13}C measurements. *Ecology*, **73**, 118–28.

McDonough, W. T. (1974) Revegetation of gopher mounds on aspen range in Utah. *Great Basin Nat.*, **34**, 267–74.

McInnes, P. F., Naiman, R. J., Pastor, J. and Cohen, Y. (1992) Effects of moose browsing on vegetation and litter of the boreal forest, Isle Royale, Michigan, USA. *Ecology*, **73**, 2059–75.

McNaughton, S. J. (1985) Ecology of a grazing ecosystem: The Serengeti. *Ecol. Monogr.*, **55**, 259–94.

Meadows, P. S. and Meadows. A. (1991) *The Environmental Impact of Burrowing Animals and Animal Burrows*, Clarendon Press, Oxford.

Merriam, G. L., Dwyer, L and Wegner, J. (1982) Litterfall in two Canadian deciduous woods: quality, quantity, and timing. *Holarctic Ecology*, **51**, 1–9.

Mielke, H. W. (1977) Mound building by pocket gophers (*Geomyidae*): their impact on soils and vegetation in North America. *J. Biogeogr.*, **4**, 171–80.

Milfred, C. J. (1966) Pedography of three soil profiles of Wisconsin representing Fayette, Tama, and Underhill series. PhD Thesis, University of Wisconsin, Madison.

Mohr, C. O. and Mohr, W. P. (1936) Abundance and digging rate of pocket gophers (*Geomys busarius*). *Ecology,* **17**, 325–7.

Naiman, R. J. (1988) Animal influences on ecosystem dynamics. *BioScience,* **38**, 750–2.

Naiman, R. J., Johnston, C. A. and Kelley, J. C. (1988) Alteration of North American streams by beaver. *BioScience,* **38**, 753–62.

Naiman, R. J., Manning, T. and Johnston, C. A. (1991) Beaver population fluctuations and tropospheric methane emissions in boreal wetlands. *Biogeochemistry,* **12**, 1–15.

Nevo, E. (1979) Adaptive convergence and divergence of subterranean mammals. *Ann. Rev. Ecol. Syst.,* **10**, 269–308.

Nye, P. H. (1955) Some soil-forming processes in the humid tropics. IV. Action of the soil fauna. *J. Soil Sci.,* **6**, 73–83.

O'Meilia, M. E., Knopf, F. L. and Lewis, J. C. (1982) Some consequences of competition between prairie dogs and beef cattle. *J. Range Manage.,* **35**, 580–5.

Osborn, B. and Allan, P. F. (1949) Vegetation of an abandoned prairie-dog town in tall grass prairie. *Ecology,* **30**, 322–32.

Osborn, D. J. (1953) Age classes, reproduction, and sex ratios of Wyoming beaver. *J. Mamm.,* **34**, 27–44.

Pacala, S. W. and Crawley, M. J. (1992) Herbivores and plant diversity. *Am. Nat.,* **140**, 243–60.

Paine, R. T. and Levin, S. A. (1981) Intertidal landscapes: disturbance and the dynamics of pattern. *Ecol. Monogr.,* **51**, 145–78.

Pastor, J., Naiman, R. J., Dewey, B. and McInnes, P. (1988) Moose, microbes, and the boreal forest. *BioScience,* **38**, 770–7.

Pickett, S. T. A. and White, P. S. (eds) (1985) *The Ecology of Natural Disturbance and Patch Dynamics,* Academic, Orlando.

Reichman, O. J. and Smith, S. (1985) Impact of pocket gopher burrows on overlying vegetation. *J. Mamm.,* **66**, 720–5.

Remillard, M. M., Gruendling, G. K. and Bogucki, D. J. (1987) Disturbance by beaver (*Castor canadensis* Kuhl) and increased landscape heterogeneity, in *Landscape Heterogeneity and Disturbance* (ed. M. G. Turner), Springer-Verlag, New York, pp. 103–22.

Retzer, J. L., Swope, H. M., Remington, J. D. and Rutherford, W. H. (1956) Suitability of physical factors for beaver management. *Colorado Dept. Game Fish Tech. Bull. No. 2,* pp. 1–33.

Ross, B. A., Tester, J. R. and Breckenridge, W. J. (1968) Ecology of mima-type mounds in northwestern Minnesota. *Ecology,* **49**, 172–7.

Roulet, N. T., Ash, R and Moore, T. R. (1992) Low boreal wetlands as a source of atmospheric methane. *J. Geophys. Res.,* **97**, 3937–49.

Rusek, J. (1992) Distribution and dynamics of soil organisms across eco-tones, in *Landscape Boundaries: Consequences for Biotic Diversity and Eco-*

logical Flows (eds A. J. Hansen and F. di Castri), Springer-Verlag, New York, pp. 196–214.

Salem, M. Z. and Hole, F. D. (1968) Ant (*Formica exsectoides*) pedoturbation in a forest soil. *Soil Sci. Soc. Am. Proc.*, **32**, 563–7.

Scheffer, V. B. (1958) Do fossorial rodents originate Mima-type micro-relief? *Am. Midl. Nat.*, **59**, 505–10.

Schlesinger, W. H. (1991) *Biogeochemistry: An Analysis of Global Change*, Academic Press, San Diego.

Seiler, W., Conrad, R. and Scharffe, D. (1984) Field studies of methane emission from termite nests into the atmosphere and measurements of methane uptake by tropical soils. *J. Atmos. Chem.*, **1**, 171–86.

Senft, R. L., Coughenour, M. B., Bailey, D. W. *et al.* (1987) Large herbivore foraging and ecological hierarchies. *BioScience*, **37**, 789–99.

Sharpe, D. M., Stearns, F. W., Burgess, R. L. and Johnson, W. C. (1981) Spatio-temporal patterns of forest ecosystems in man-dominated landscapes of the eastern United States, in *Perspectives in Landscape Ecology* (eds S. P. Tjallingii and A. A. Veer), Centre for Agricultural Publication and Documentation, Wageningen, pp. 109–16.

Sheets, R. G., Linder, R. L. and Dahlgren, R. B. (1971) Burrow systems of prairie dogs in South Dakota. *J. Mamm.*, **52**, 451–3.

Slough, B. G. and Sadleir, R. M. F. S. (1977) A land capability classification system for beaver (*Castor canadensis*). *Can. J. Zool.*, **55**, 1324–35.

Sousa, W. P. (1985) Disturbance and patch dynamics on rocky intertidal shores, in *The Ecology of Natural Disturbance and Patch Dynamics* (eds S.T.A. Pickett and P. S. White), Academic Press, Orlando, pp. 101–24.

Spencer, S. R., Cameron, G. N. and Eshelman, B. D. *et al.* (1985) Influence of pocket gopher mounds on a Texas coastal prairie. *Oecologia*, **66**, 111–15.

Summers, C. A. and Linder, R. L. (1978) Food habits of the black-tailed prairie dog in western South Dakota. *J. Range Manage.*, **31**, 134–6.

Svendson, G. E. (1980) Population parameters and colony composition of beaver (*Castor canadensis*) in southeast Ohio. *Am. Midl. Nat.*, **104**, 47–56.

Thorp, J. (1949) Effects of certain animals that live in soils. *Sci. Monthly*, **68**, 180–91.

Tiedemann, A. R. and Berndt, W. H. (1972) Vegetation and soils of a 30-year deer and elk exclosure in central Washington. *Northwest Sci.*, **46**, 59–66.

Tilman, D. (1983) Plant succession and gopher disturbance along an experimental gradient. *Oecologia*, **60**, 285–92.

Townsend, J. E. (1953) Beaver ecology in western Montana with special reference to movements. *J. Mamm.*, **34**, 459–79.

Turner, G. T., Hansen, R. M., Reid, V. H. *et al.* (1973) Pocket gophers and Colorado mountain rangeland. *Colorado State University Experiment Station Bull.* 554S, Fort Collins.

Turner, M. G. (1987) Spatial simulation of landscape changes in Georgia: a comparison of 3 transition models. *Landsc. Ecol.*, **1**, 29–36.

Turner, M. G. and Bratton, S. P. (1987) Fire, grazing, and the landscape heterogeneity of a Georgia barrier island, in *Landscape Heterogeneity and Disturbance* (ed. M. G. Turner), Springer-Verlag, New York, pp. 86–101.

Vaughan, T. A. (1961) Vertebrates inhabiting pocket gopher burrows in Colorado. *J. Mamm.*, **42**, 171–4.

Watson, J. P. (1962) The soil below a termite mound. *J. Soil. Sci.*, **13**, 46–51.

Whicker, A. D. and Detling, J. K. (1988) Ecological consequences of prairie dog disturbance. *BioScience*, **37**, 778–85.

White, E. M. and Carlson, D. C. (1984) Estimating soil mixing by rodents. *Proc. S. Dak. Acad. Sci.*, **63**, 34–7.

Wiens, J. A. (1985) Vertebrate responses to environmental patchiness in arid and semiarid ecosystems, *The Ecology of Natural Disturbance and Patch Dynamics* (eds S. T. A. Pickett and P. S. White), Academic Press, Orlando, pp. 169–93.

Wilde, S. A., Youngberg, C. T. and Hovind, J. T. (1950) Changes in composition of ground water, soil fertility, and forest growth produced by the construction and removal of beaver dams. *J. Wildl. Manage.*, **14**, 123–8.

Zimmerman, P. R., Greenberg, J. P., Wandiga, S. O. and Crutzen, P. J. (1982) Termites: A potentially large source of atmospheric methane, carbon dioxide, and molecular hydrogen. *Science*, **218**, 563–5.

Part Two

Response of Individuals and Populations to Landscape Pattern

Patterns of movements are central to understanding landscape dynamics. Organisms affect, and are affected by, different elements in a landscape by movements between patches. However, movements can be very complex and occur at various spatial scales for various processes such as foraging, territorial defence, mate-finding, between-patch dispersal and migration between regions. These topics are covered in the chapter by Rolf Ims.

Movements between habitat patches of varying quality will, at least in part, depend on the species densities in the more preferred patches, either by direct reaction to resources per capita or by subordinate individuals being forced to move to lower quality habitat. A theoretical framework based on such premises is presented by Douglas Morris. Based on evolutionary arguments, he presents formal models for density-dependent habitat selection. He also provides tests of predictions from these models by comparison with field data on the distribution of individuals of one species between neighboring habitats. This analysis is thus related to two habitats within a fairly simple landscape.

Michal Kozakiewicz develops ideas and provides field data for more diverse landscapes, where habitats will change in quality both in space and over time. This final chapter of this section also provides insight into individual adaptations for survival in heterogeneous areas and provides an introduction to the following section on dynamics and genetics of fragmented or subdivided populations.

Movement patterns related to spatial structures

4

Rolf A. Ims

4.1 INTRODUCTION

The ability of individuals to move in space, although highly variable between species, is a general characteristic of all organisms. In fact the very persistence of species on a micro-evolutionary time-scale requires the ability to escape from natural, long-term environmental changes (e.g. climate changes) (Holt, 1990). At shorter (ecological) time-scales, movement capacities of organisms are universally important to ecological phenemona (Wiens *et al.*, 1993). For this reason, the study of movement patterns deserves a key position within all disciplines of ecology.

In landscape ecology, which focuses on the effect of spatial landscape features on ecological processes (Forman and Godron, 1986; Urban, O'Neill and Shugart, 1987; Turner, 1989; Wiens *et al.*, 1993), movement patterns are central to fundamental topics such as connectivity, patch and boundary dynamics, spread of disturbances, and source–sink and metapopulation dynamics. Hence, one of the most basic and important questions in landscape ecology is: How is the movement behavior of individual organisms affected by landscape mosaic structures? In this chapter I review current knowledge related to this question. My emphasis will be on proximate relations. Considerations about ultimate determinants of movement strategies and how movement behavior of individual animals translates to processes at higher organizational levels (e.g. popu-

Mosaic Landscapes and Ecological Processes.
Edited by Lennart Hansson, Lenore Fahrig and Gray Merriam.
Published in 1995 by Chapman & Hall, London. ISBN 0 412 45460 2

lations and communities) are treated in other chapters of this volume (Chapters 5, 6, 10 and 11).

Information about animal movements and their relations to spatial patterns may be found widely scattered in the ecological and ethological literature. It is not my aim to provide a comprehensive review of this information. My intention is threefold:

1. to identify the main approaches to the study of movement patterns;
2. to evaluate the kind of insight various approaches have provided to our understanding of how movement patterns are affected by spatial structures; and finally,
3. to suggest what kind of studies will be most important for enhancing our knowledge on this topic.

4.2 COMPONENTS OF MOVEMENTS AND SPATIAL PATTERNS: DEFINING THE CONCEPTS

4.2.1 MOVEMENT EVENTS AND PATTERNS

A **movement event** represents a spatial displacement of an individual. Causally, it may have both a passive (stochastic) and an active (behavioral) component. For example, an aphid accidentally displaced by strong wind may be considered as an entirely **passive displacement**. On the other hand, a honey bee's straight-lined flight toward a nectar source is a movement event based on an active, **behavioral decision**.

A **movement pattern** of an individual that emerges as the sum of movement events (often termed 'steps') over some time period may thus result from a series of both behavioral decisions and passive displacements. This chapter is concerned with how various spatial structures interact with movement events (either resulting from passive displacements or behavioral decisions) to shape the movement patterns.

4.2.2 SPATIAL PATTERNS

A multitude of environmental factors may influence movement patterns. Whether they represent resources, biotic enemies or abiotic factors, they are often spatially heterogeneous and, when depicted on a map (Figure 4.1), may form complex spatial mosaics. Much work within landscape ecology has been devoted to describing and quantifying spatial mosaics (e.g. Brunt and Conley, 1990; Turner and Gardner, 1991). Much less effort has gone into the study of how organisms respond to such mosaics.

Some environmental factors form spatial structures that are rather static (e.g. substrate type; Brown *et al.*, 1992), while others are highly dynamic (e.g. thermal patches; Huey, 1991). The spatial dynamics of

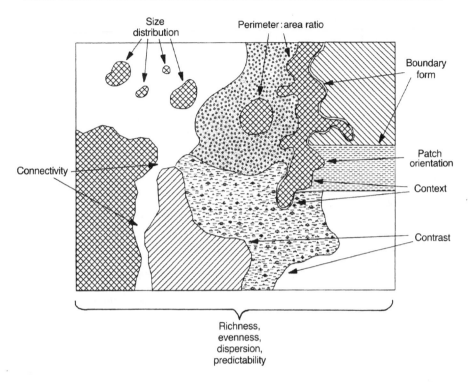

Figure 4.1 Spatial structures of a hypothetical habitat patch mosaic and examples of descriptors often used by landscape ecologists. (From Wiens *et al.*, 1993, reproduced with permission.)

predators, competitors and resources interact strongly with the movement pattern of the organism itself, but this chapter focuses on spatial structures which usually have a slower spatial dynamic than the movement rate of the focal organism. In particular, the kinetics of highly dynamic biological interactions such as predation and competition, although being spatially structured, will not be dealt with here.

4.2.3 THE RELATIONSHIP BETWEEN MOVEMENT BEHAVIORS AND SPATIAL SCALE

An animal's movement decisions relative to spatial structures depend on the kind of life process conducted. For instance, an individual's response to a specific spatial structure may depend on whether it is searching for mates or food items. Different life processes and their associated movement behaviors are manifested at different spatial scales, from food searching in a local patch to migratory movements on regional or continental scales. The relationship between spatial scale and movement behaviors associated with various life processes can be conceptualized in a hierarchical manner (Table 4.1; see also Senft *et al.*, 1987; Urban, O'Neill and Shugart, 1987; Wiens, 1989; Merriam, Henein and Stuart-Smith, 1991). The set of spatial structures most influential to the resultant movement pattern at a given level in such a hierarchy may also be suggested (Table 4.1). Note, however, that movement decisions at any level may be simultaneously affected by spatial structures at levels below and above in the hierarchy (Senft *et al.*, 1987; Wiens, 1989; Kotliar and Wiens, 1990; Gautestad and Mysterud, 1993).

Table 4.1 Animal movement modes and influential spatial structures classified according to a spatial scale hierarchy (see also Senft *et al.*, 1987; Kotliar and Wiens, 1990)

Spatial scale	Movement type (life process)	Spatial structures
Resource patch	Food item searching (foraging)	Food item distribution Food patch shape and size Small-scale obstructions
Habitat patch	Patch searching, traplining, territory patrolling	Food patch configuration Shelter Abiotic factors and topography
Patch mosaic (landscape)	Dispersal	Patch parameters (e.g. size, shape, isolation) Landscape parameters (e.g. connectivity, dispersion)
Region	Migration	Large-scale topography Large-scale barriers

Both quantitative (i.e. absolute spatial scale) and qualitative aspects (i.e what kind of spatial features are important) of movement scale hierarchies, exemplified by Table 4.1, will depend on species-specific spatial constraints (Wiens *et al.*, 1993; Chapter 6). For example, the home-range scale for a moose may exceed the spatial scale of several mouse landscapes. Although classically being associated with broad scale phenomena, landscape ecology should not be a priori restricted to certain spatial scales. The effect of spatial patterns on ecological processes, irrespective of absolute spatial scale, ought to be the common denominator of landscape ecological studies (Turner, 1989; Wiens *et al.*, 1993).

It is also important to bear in mind that spatial structure-movement pattern relations at any spatial scale may be affected by 'non-spatial factors' such as the internal state of the individual (e.g. nutritional condition), its age and its life cycle stage (see below). Indeed, landscape ecology, although often being considered as a relatively holistic science (e.g. Naveh and Lieberman, 1990), cannot escape from the important biological details from which the more broad scale processes and patterns emerge (Huston, DeAngelis and Post, 1988; DeAngelis and Gross, 1992; Wiens *et al.*, 1993).

4.3 THEORY AND MODELS OF MOVEMENT PATTERNS

Although often not explicitly stated, models of movement behavior address themselves to specific spatial scales (Table 4.1). Besides the spatial scale to which the various models pertain, I will pay particular attention to how spatial structure enters the models, to what extent other biological variables affecting movement patterns are taken into account and whether a link has been established between movement patterns and population or community processes. A summary of this evaluation is given in Table 4.2.

Table 4.2 Theories of movement patterns according to relevant spatial scales, the kind of spatial structures they pertain to (see Table 4.1), whether they take into account interactions with biological factors (e.g. internal state of the individuals, competition, predation) and whether attempts have been made to link movement patterns, spatial structures and population/community processes

Theory	Scale range	Spatial structures	Biological factors	Link to population or community processes
Optimal searching	Resource patch – Home range	Food resources Shelter/ landmarks	Internal state	None
Optimal foraging	Resource patch – Home range	Food resources Shelter	Internal state Competition, Predation	Demography
Habitat selection and dispersal	Habitat patch – Mosaic	Patch shape and edges Connectivity and patch configuration	Social structure Population density	Demography and population dynamics
Diffusion and random walk	Within patch – Region	Mosaics of suitable and non-suitable areas	Usually none	Population dynamics Biological invasions

4.3.1 SEARCH THEORY

What kind of movement pattern should a naive animal adopt to explore space more efficiently in the search for resources (either within or among food patches)? Optimal search theory (e.g. Haley and Stone, 1980; Bell, 1991) suggests answers to this question.

A random walk movement pattern generated by movement events based on random turning angles and step lengths, will never be the most efficient movement mode, simply because the probability of crossing the previous track and of entering patches already searched through will be high (Bell, 1991). One way of avoiding the problem of crossing paths is to adopt a systematic movement pattern. Spiralling and parallel sweeping are examples of hypothetical, systematic search patterns that may be efficient (Haley and Stone, 1980). The most efficient systematic search pattern will depend on the animal's detection radius to a given resource as well as the spatial arrangement of that resource (Bell, 1991). For instance, the distance between the successive loops in a spiral, or sweeps in a parallel sweep, should optimally be twice the maximum detection distance to a resource patch (or food item) (Bell, 1991). Since the maximum detection distance to a patch may be orders of magnitude greater than distances to individual resource items within a patch, animals should change the scale of their movements when entering a patch. The fine-scale intra-patch search pattern, characterized by more frequent turns and shorter sweeps/loops compared to patch search movements, are equivalent to the well-known area-restricted search concept (Tinbergen, Impekoven and Franck, 1967; Croze, 1970). A scale-specific systematic search pattern in a fractal 'resource landscape' (i.e. self-similar resource distribution over a range of spatial scales; Sugihara and May, 1990) will give rise to fractal movement patterns (Wiens and Milne, 1989; Johnson et al., 1992).

Some degree of interference of stochastic factors (e.g. resulting in passive displacements from an optimal search trajectory) will always be present. Moderate stochasticity may, however, not be damaging to search efficiency. Some elements of randomness superimposed on a predominantly systematic search pattern actually may increase search efficiency (Hoffman, 1983; Bell, 1991).

Philopatric animals are equipped with spatial memory and thus have some experience with the area in which they carry out their regular life processes (i.e. their home range). On well-known ground a stereotypic, systematic movement pattern (e.g. parallel sweeping) may be an inefficient way of searching for resources when their positioning can be memorized. Other physical structures (e.g. vertical features) may serve as landmarks facilitating orientation. Patchily distributed, renewable resources may be most efficiently exploited on a rotational basis by **traplining** (Jansen, 1971; Gill and Wolf, 1977; Gill, 1988) A trapline is a foraging circuit which is learned and followed during consecutive searching bouts. In addition to minimizing the distance between resource patches visited, an optimal trapline should take advantages of other spatial features such as shelter (cover) to minimize exposure to various risk factors (e.g. predators) (McNamara and Houston, 1987; Brown *et al.*, 1988, 1992).

4.3.2 OPTIMAL FORAGING THEORY

Optimal foraging theory (MacArthur and Pianka, 1966; Schoener, 1971; Charnov, 1976; Stephens and Krebs, 1986) also has some bearing on animal movement patterns in relation to spatial structures. Optimal foraging theory predicts an animal's allocation of time to foraging and searching mainly at the two lowest levels of the hierarchy given in Table 4.1. Spatial structures enter optimal foraging models as the size and quality of discrete food patches and sometimes as the dispersion of food patches in space (e.g. random or uniform). Although the size and spatial dispersion of food patches implicitly affect movement behavior through determining patch residency and search times, searching behavior and the resultant movement pattern does generally not appear to be a dynamic attribute in foraging models. Linking foraging theory to optimal search theory may be a way to advance the understanding of how foraging and movement patterns are related.

Optimal foraging theory, like most ecological theory, is not spatially explicit (Wiens *et al.*, 1993). For instance, in cases where the dispersion of resource patches has been modeled, it is typically approximated by general probabilistic distributions with well-defined properties (Iwasa, Higashi and Yamahura, 1981). Recent theoretical developments, however, incorporate some irregularities typical of real-life resource mosaics. Arditi and D'Acoragne (1988) (see also Kacelnik and Bernstein, 1988) found that relaxing the common assumption of discreteness of resource patches did not qualitatively change the prediction of the marginal value theorem by Charnov (1976). Further theoretical investigations are needed to check whether predictions made by other classical models are robust in the same way.

Optimal foraging theory has also been criticized by landscape ecologists (e.g. Senft *et al.*, 1987) for being relatively limited with respect to spatial scale; the decisions made by individuals during foraging are governed by patterns and processes at small spatial scales (patch or home-range scale; Table 4.1) and ignore the fact that animals may have to integrate foraging decisions over a range of spatial scales (see e.g. Gautestad and Mysterud, 1993; Chapter 6).

Optimal foraging theory often explicitly links behavioral decisions to 'non-spatial factors' such as the internal state of the individual (e.g. Mangel and Clark, 1986) and to risk factors (e.g. McNamara and Houston, 1987). The links between foraging decisions, population dynamics and species interactions have also been explored theoretically to some extent (e.g. Engen and Stenseth, 1989; Bernstein, Kacelnik and Krebs, 1991). But again, movement responses to spatially explicit resource distributions have not been a part of such explorations.

4.3.3 THEORY OF HABITAT SELECTION AND DISPERSAL

Relatively few attempts have been made to predict how explicit spatial structures at the habitat mosaic scale (Table 4.1) affect movement patterns (dispersal). Classical habitat selection theories (Fretwell and Lucas, 1970; Rosenzweig, 1981; Morris, 1987; Chapter 5) and dispersal theories (Hamilton and May, 1977; Southwood, 1977; Stenseth, 1983; Hansson, 1991; Stenseth and Lidicker, 1992) are not spatially explicit. Recent patch emigration models, however, have done much to remedy this deficiency. Stamps, Buechner and Krishnan (1987a, b) built patch **size**, patch **shape** and patch **edge permeability** into simulation models to predict patch-specific emigration rates for territorial animals. Emigration rate was predicted to increase with the edge-to-size ratio of a habitat patch (a measure of patch shape). The increase was most pronounced from patches with 'soft' edges (high edge permeability). Relaxing the territoriality assumption of Stamps, Buechner and Krishnan does not change this prediction (Turchin, 1986). Regarding patch edge **curvature**, concave edges have been predicted to act as funnels channeling emigration from tips or peninsulas of habitat islands (Forman and Godron, 1986; Hanski and Peltonen, 1988; Hardt and Forman, 1989).

Non-spatial factors such as population density and social organization may interact with boundary properties to affect emigration rates. Wiens, Crawford and Gosz (1985) predicted that emigration rate should increase as a function of the density in the patch, but most rapidly from patches with permeable edges and/or a high edge-to-size ratio. While emigration rate should increase gradually with density for a social species, emigration may not occur before a certain threshold density (e.g. maximum group size) is reached for colonial or group forming species (Wiens *et al.*, 1993). Wiens *et al.* (1993) also showed how such interactions can be built into models of population dynamics.

In contrast to patch emigration models for which it may suffice to consider patch-specific parameters such as density, size, shape and permeability, models for predicting immigration rates onto patches also need to take into account characteristics of the entire patch mosaic such as connectivity, context and isolation (Figure 4.1). The complexity of patch immigration models (in terms of number of parameters to be included) approaches patch dynamics and landscape models (for reviews of such models see Fahrig, 1991; Merriam, Henein and Stuart-Smith, 1991).

Few patch immigration models have been analyzed. Fahrig and Paloheimo (1988) modeled cabbage butterfly immigration onto habitat patches with a spatially explicit configuration. The movement behavior was modeled as rather inflexible, long-distance migration with a directional bias. In this case the spatial arrangements of patches were relatively unimportant to patch-specific immigration rates. This prediction is likely to be valid for organisms with the capacity for long-distance, unconstrained movement paths (Fahrig and Paloheimo, 1988) such as insects with an aerial dispersal (migration) mode.

The dispersal pathways of other organisms, however, may be severely constrained by the spatial features of landscapes and should be modeled as a dynamic variable. In particular, certain habitat specialists may respond behaviorally to a boundary between their preferred habitat and the surrounding matrix as if it was a 'hard edge' (*sensu* Wiens, Crawford and Gosz, 1985; Stamps, Buechner and Krishnan, 1987a). Such organisms (sometimes called matrix-sensitive species) need continuous tracts of habitat for dispersal, and mosaic parameters such as connectivity (Figure 4.1) determine to a large extent movement patterns. Matrix-sensitive species are particularly vulnerable to habitat fragmentation (Soule, 1986; Simberloff, 1988; Fahrig and Merriam, 1993).

There has been considerable debate about whether the establishment of **movement corridors** may remedy the negative effects of habitat fragmentation (Simberloff and Cox, 1987; Noss, 1987; Saunders and Hobbs, 1991; Hobbs, 1992; Simberloff *et al.*, 1992; Merriam and Saunders, 1993; Mann and Plummer, 1993). Movement corridors are narrow strips of habitat connecting otherwise isolated habitat patches. The debate has centered around the question of whether habitat corridors functionally connect habitats in terms of movements of individuals and to what extent this enhances the persistence of populations.

Although there are theoretical suggestions that connectivity as such may enhance population persistence (e.g. Fahrig and Merriam, 1985; Lefkovitch and Fahrig, 1985; Merriam, Henein and Stuart-Smith, 1991; but see Simberloff and Cox, 1987; Heinein and Merriam, 1990; Simberloff *et al.*, 1992, for opposite views), the literature contains relatively few predictions about which structural characteristics of corridors enhance connectivity. A potentially important characteristic is **corridor width** (Harrison, 1992). Although the notion that wider corridors are more efficient corridors (in terms of connectivity) has been accepted as a general recommendation for corridor design (Noss, 1987; Harrison, 1992; Merriam and Saunders, 1993), it has not been verified either theoretically or empirically.

4.3.4 DIFFUSION AND RANDOM WALK MODELS

The diffusion approach to modeling movement patterns has a long tradition in ecology (see Okubo, 1980; Turchin, 1991, for reviews). Such models have been developed along two lines.

Traditionally, mathematical diffusion models have been used to predict the spread of organisms in space as a population or a community process, e.g. as an expansion of a species' range (Skellam, 1951). Although based on a framework adopted from physics, in which movements of individuals are assumed to be random (analogues to particles) in a spatially featureless environment, modeling movements as a simple diffusion process has proved to be rather successful, e.g. for predicting population expansion rates, on some spatial scales (Levin, 1992). A strength of this approach is that the movement patterns of individuals can be directly linked to processes at higher organizational levels, for example by incorporating dispersal as a diffusion term in models of population dynamics.

The other line of development, which is a more recent one, involves computer simulations of movement patterns in spatially explicit mosaics (e.g. Gardner et al., 1989; Wiens and Milne, 1989; Johnson, Milne and Wiens, 1992; Johnson et al., 1992). Typically random walk algorithms (simple or first-order correlated random walks) are applied in spatial mosaics modeled as lattices of suitable and non-suitable (impermeable) cells (Figure 4.2). The cell size corresponds to the grain size (the smallest spatial scale at which an organism recognizes spatial heterogeneity; Wiens, 1989; Kotliar and Wiens, 1990). Spatial structures at larger scales emerge as clusters of cells. The clusters might or might not be connected (percolating networks; Gardner et al., 1989; Johnson et al., 1992) depending on the fraction of the cells that has been designated as suitable area. Such spatial mosaics can be modeled to resemble the spatial structures of real landscapes (e.g. Johnson, Milne and Wiens, 1992) or may be made by assigning a given proportion of suitable and non-suitable cells at random (Gardner et al., 1989).

The computer simulation approach to modeling individual movements in spatial mosaics may serve two related purposes. First, it may serve as a null model against which real movement patterns of individuals may be compared (e.g. Turchin, 1986). The second purpose is to check whether a random walk is a reasonable approximation of real movement behaviors and hence, whether they can be modeled as simple diffusion terms in models of population dynamics.

The fact that random walk simulations have been most commonly used as null models in relation to insect movement patterns, may indicate that random walks are considered more realistic for 'hardwired' arthropods than for behaviorally more 'sophisticated' animals such as ver-

Figure 4.2 The principles of simulation modeling of beetle movements in a lattice landscape. The size of the pixels in the lattice usually corresponds to the grain size (*sensu* Wiens, 1989) and the step length of the beetle. Black pixels represent patches which are inaccessible (impermeable) for the beetle, whereas the white area represents suitable habitat through which the beetle may percolate. See text for further explanation.

tebrates. Although there are some general differences across taxa, the widespread capacity for spatial memory and orientation in insects (Johnson, 1969; Danthanarayana, 1986; Bell, 1991) brings into question the value of this null model approach. Indeed, including more biologically plausible movement decisions of the animal modeled, for example by including elements of optimal search theory in the algorithms, can make the simulation models more useful vehicles for improving our understanding of movement pattern in relation to spatial structures (e.g. Root and Kareiva, 1984; Turchin, 1987; Odendaal, Turchin and Stermitz, 1988; Crist *et al.*, 1992).

Recently, more complexity (e.g. non-random movements and abstractions of spatial heterogeneity) has also been incorporated into mathematical diffusions models (Kareiva and Odell, 1987; Turchin, 1991; Morris, 1993; Vail, 1993). Further development along such lines may be very valuable because the inclusion of spatial structure–movement pattern interactions as terms in partial differential equation models may be a way to directly link spatial structures to demographic processes.

4.4 EMPIRICAL STUDIES OF MOVEMENT PATTERNS

The empirical literature on movement patterns as they relate to spatial structures is very heterogeneous with respect to both the approach and the quality and quantity of the data. Generally, the literature on invertebrates (especially insects) is far more voluminous and of better quality than the equivalent literature on vertebrates. For example, numerous insect studies have tested theoretical predictions using replicated, manipulative experiments. In contrast, studies on vertebrate movement patterns are typically descriptive and often based on anecdotal observations. The paucity of solid empirical information on space use by vertebrates in relation to landscape patterns has been pointed out by several authors (e.g. Stamps, Buechner and Krishnan, 1987a; Opdam, 1990, 1991; Rolstad, 1991).

Below I review empirical studies of movement pattern–spatial structure interactions, not with the aim of providing a complete synthesis; the literature is still too fragmentary to provide any basis for such a synthesis. My intention is rather to highlight the insight provided by the different approaches to empirical investigations of animal movements with emphasis on the most recent developments. In the same manner as for the theory reviewed above, the empirical approaches will be related to the movement scale hierarchy of Table 4.1.

4.4.1 COMPARING MOVEMENT PATTERNS TO RANDOM WALK SIMULATIONS

As pointed out above, random walk simulations have mainly been aimed at predicting movement patterns of insects, in particular beetles. Experimental studies (Wiens and Milne, 1989; Johnson, Milne and Wiens, 1992; Crist *et al.*, 1992) comparing observed movement patterns of ground beetles and ants with the expectations from simulations of random walks have shown that real beetle pathways usually differ from first-order correlated random walks. This is also the case when the input parameters (turning angles and step lengths) (Crist *et al.*, 1992) and the spatial structure of the lattice (Johnson, Milne and Wiens, 1992) are empirically based. Generally, real movements have a higher displacement rate and are less complex (lower fractal dimension) than simulated movements (Crist *et al.*, 1992). This is likely to be caused by the fact that most movement patterns have a directional bias (e.g. Kennedy, 1951; Johnson, 1969).

Not surprisingly, displacement rates of ground insects are dependent on the spatial structure of the vegetation as well as the particular species studied (Wiens and Milne, 1989; Crist *et al.*, 1992). Furthermore, they apparently respond to scale-dependent changes in the vegetation structure (Johnson *et al.*, 1992). However, the most intriguing and potentially important result from these studies is that the structure of the pathways (measured by their fractal dimension) appears to be similar for different species, in different vegetation types and over the range of spatial scales being studied (Crist *et al.*, 1992). This gives hope for establishing certain generalizations about movement patterns that are valid across species, spatial and temporal scales (section 4.4.5).

4.4.2 MOVEMENTS WITHIN HOME RANGES

The very mechanistic approach to the analysis of insect movement pattern reviewed above, is clearly not favored by students of vertebrate space use patterns. Instead the home-range concept (Burt, 1943) generally holds a much stronger position in studies of vertebrates (perhaps particularly in mammals) than in studies of invertebrates. Generally, home ranges are described in terms of delineated areas or utilization distributions estimated from spatial point patterns (e.g. radio-telemetry positions) (Andreassen *et al.*, 1993). Since an animal's positions in time and space have been discretized by the sampling procedure and, furthermore, since the sequence of positions in time is not taken into account by home-range analysis methods, home-range descriptors may not reveal important details about the actual movement pattern (Andreassen *et al.*, 1993).

Although some relevant information about movement patterns can be extracted from the vast literature on home-range use, there are very few studies explicitly addressing how the size of home ranges (the most widely used home-range descriptor) increases or decreases in response to experimental alteration of spatial arrangement of resources and shelter (Rolstad, 1991). Examining space use responses of voles (*Microtus*) and forest grouses (Capercaillie) to experimental fragmentation of resource patches at the home-range scale (Table 4.1), Ims, Rolstad and Wegge (1993) found that both species expanded their home ranges to include several resource patches. In terms of movement pattern descriptors this may imply an increased step length variance (i.e. shorter steps within and longer steps between resource patches).

4.4.3 EMIGRATION RATES FROM HABITAT PATCHES

Many insect studies have examined emigration rate from habitat patches as a function of patch structure (Table 4.1) and biological factors such as patch-specific population density and presence of predators. Although the lattice models (Figure 4.2) and their associated empirical tests, described in the previous section, also include the patch scale (since clusters of suitable cells form habitat patches), the 'lattice approach' usually assumes that movements are confined within networks of hard-edged patches (i.e. no edge crossings).

Experiments manipulating patch size for phytophagous insects have shown that there often is an inverse relationship between patch size and emigration rates (Back, 1980, 1984; Kareiva, 1983, 1985; Turchin, 1986; Lawrence, 1988) as predicted by simulation models (Turchin, 1986; Stamps, Buechner and Krishnan, 1987a). This is, however, not always the case (Back, 1988a, b). Similarly, immigration rates onto patches have been found to be either independent of patch size or to increase with patch size (Kareiva, 1985; Back, 1988a, b; Lawrence, 1988). Furthermore, patch-specific population density may either be neutral (Kareiva, 1985; Midtgaard, F., unpublished) or influential (Back, 1988b; Lawrence, 1988) to the propensity of insects to cross patch edges.

Although habitat patch size has been manipulated in some vertebrate studies (Foster and Gaines, 1991; Robinson et al., 1992), emigration rates from patches of different sizes have not been explicitly analyzed. The higher turnover rates in subpopulations of small mammals on small habitat patches than on large habitat patches (Foster and Gaines, 1991) may, however, indicate that emigration rate decreases with patch size.

The conflicting results from the many patch emigration studies of insects may simply reflect species-specific differences. Indeed, different species may perceive the same patch boundary (e.g. measured as an abrupt change in the vegetation structure) differently (Wiens, Crawford and Gosz, 1985; Stamps, Buechner and Krishnan, 1987a). There may be a gradient from species with a 'hard-edged' response (no crossings of a boundary) to very 'soft-edged' species (no response to a given boundary) (Durelli et al., 1990; Figure 4.3). Edge responses may be difficult to predict. For example, highly mobile animals such as small rodents (Steen, 1994) and carabids (Stork, 1990) may show surprisingly hard-edged responses to rather small-scale spatial structures such as road clearances (Mader, 1984; Swihart and Slade, 1984).

Even though species differences may be expected, there is a danger that the quite heterogeneous outcome of patch emigration studies and other insect movement pattern studies, may be confounded to a large extent by factors not controlled by the experimenters. Indeed, many of the most recent movement studies on insects cited above have ignored potentially important biological characteristics of their study animals. For example, often the sex of the individuals is not noted, although substantial sex differences in movement strategies may be expected (Thornhill and Alcock, 1983). Similarly, an individual's particular life cycle stage (e.g. Graf and Sokolowski, 1989) and internal condition (e.g.

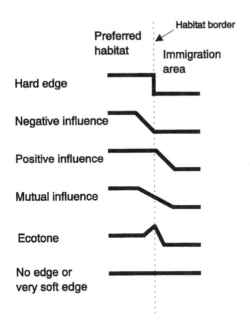

Figure 4.3 Potential edge effects expressed as changes in distribution of individuals across a patch border. (Source: Durelli *et al.*, 1990.)

whether fed or not; Carter and Dixon, 1982) are factors that may introduce considerable unexplained heterogeneity or biases. Certainly, the proximate and ultimate factors underlying an individual's movement decisions (many of which are known; Johnson, 1969; Southwood, 1977; Dingle, 1978; Swingland and Greenwood, 1983; Stenseth and Lidicker, 1992) certainly may have a pervasive influence on movement pattern–spatial pattern interactions. Not including essential biological mechanisms/causations, the 'high-tech slant' of recent movement studies (e.g. Wiens and Milne, 1989; Johnson, Milne and Wiens, 1992) may become a rather sterile enterprise.

4.4.4 MOVEMENTS IN LANDSCAPES (HABITAT PATCH MOSAICS)

The capability of long-distance movements become important at the landscape scale. For example, the configuration of habitat patches appeared to be relatively unimportant to patch immigration rates for cabbage butterflies which dispersed over large distances in a unidirectional fashion (Fahrig and Paloheimo, 1988; section 4.3.3), whereas Lawrence (1988) found that both immigration and emigration rates were affected by patch isolation for milk weed beetles with a more restricted dispersal ability. Some insects apparently are totally lost and die if displaced from a habitat patch (Kareiva, 1986), while others are excellent long-distance migrators with sophisticated orientation mechanisms taking advantage of spatial landscape features in combination with other cues (Johnson, 1969).

The role of habitat corridors in connecting landscape elements is highly controversial due to the lack of solid empirical evidence (Hobbs, 1992; Simberloff et al., 1992; Mann and Plummer, 1993). Despite the fact that many questions about corridor features–animal movement interactions are highly suitable for experimental explorations (e.g. using small mammals; Bennett, 1990; Lorenz and Barrett, 1990; Merriam and Lanoue, 1990; La Polla and Barrett, 1993), few studies have tested the recommendations about corridor design now being implemented as large-scale, expensive conservation measures (Simberloff et al., 1992; Mann and Plummer, 1993).

Two recent experimental studies manipulating the widths of movement corridors for *Microtus* voles (Andreassen, Halle and Ims, 1993; La Polla and Barrett, 1993) have challenged the widely accepted notion that wider corridors necessarily are more efficient corridors in terms of transferring dispersing animals (e.g. Harrison, 1992). Andreassen, Halle and Ims tested dispersal movement rates in *Microtus oeconomus* in 310-m long corridors of three widths (3 m, 1 m and 0.4 m). In this experiment, the intermediate width (1 m) corridor provided the best connection between habitat patches. The mechanisms underlying this result were that voles avoided entering the narrowest corridor, while many zig-zag movements reduced the progress in the widest corridor. The much shorter corridor (10 m long) in La Polla and Barrett's study of *Microtus pennsylvanicus* probably functioned mainly as a connector (Mann and Plummer, 1993) for within home-range movements. However, in this case a 1-m wide corridor also stimulated more movements than a wider (5 m) corridor. Hence, these two studies suggest that corridor width–connectivity relations may be qualitatively the same for different species performing different movement modes on different spatial scales.

The question about how habitat corridors affect movement patterns have so far not had any substantial effect on the focus of insect movements studies, although a few modeling and experimental studies have been carried out in linear habitat patches (e.g. Kareiva, 1986, 1987; Morris, 1993; Vail, 1993).

4.4.5 INTERACTIONS BETWEEN MOVEMENT PATTERNS AND SPATIAL SCALE

A critical question of movement pattern analysis in landscape ecology is the problem of extrapolation across spatial scales (Wiens, 1989; Johnson et al., 1992; Wiens et al., 1993). For instance, can one draw inferences about dispersal/migration patterns on the habitat mosaic scale (Table 4.1) from empirical movement pattern analysis at smaller spatial scales? The general problem of extrapolations across scales has attracted considerable interest from theoreticians (e.g. Krummel *et al.*, 1987; O'Neill, 1989; King, 1991), and concepts such as transmutation effects have been introduced to highlight possible nonlinear changes in process rates that may occur across scale domains (Huston, DeAngelis and Post, 1988; O'Neill, 1989; Johnson *et al.*, 1992). Although such abstract concepts, of which quite a few are adopted from physics, certainly may have a heuristic value, the problem is primarily an empirical and biological one. The difference between home-range movements and dispersal movements may serve as a good example.

Movements within home ranges are conceptually different from dispersal movements (i.e. movements between home ranges) for two reasons. First, movement decisions within home ranges are made on familiar ground based on spatial memory, whereas spatial structures encountered during dispersal movements typically are novel to an individual. Second, the different motivations underlying dispersal movements and the movements associated with the daily activities within an animal's home range, may lead to very different movement responses to equivalent spatial structures. For example, concave patch boundaries seem to facilitate boundary crossings in dispersing animals (Hanski and Peltonen, 1988), while animals may be more likely to cross boundaries with a convex curvature during regular foraging movements (Hardt and Forman, 1989). Similarly, the high sensitivity to small-scale spatial structures during an insect's trivial flight (e.g. mate- or food-searching flights) is fundamentally different from the ignorance of such structures by the same insect when migrating across a landscape (Johnson, 1969). Whether certain spatial structures elicit the same kind of responses by animals in different movement modes and over different spatial scales (cf. the example of vole movement responses to corridor width above; see also Crist *et al.*, 1992; Ims, Rolstad and Wegge, 1993), needs to be addressed by empirical studies actually comparing movement responses across scales.

4.5 SUMMARY

It has been argued that a mechanistic understanding of spatial processes is necessary for landscape ecology to become a tool for predicting ecological consequences of human disturbances rapidly changing the environment (Wiens *et al.*, 1993). A crucial element of such a mechanistic understanding is knowledge about how movement patterns of individual organisms are affected by the spatial structures of their environment. Ultimately, such knowledge is necessary for tackling urgent conservation

issues such as genetic and demographic consequences of habitat fragmentation and the spread of disturbances in space.

Unfortunately, neither theory nor empirical knowledge on animal
movement patterns are currently sufficiently developed to serve as a
mechanistic basis for landscape ecology. Elements of relevant theory on
the subject are scattered over several approaches, each favored by different scientists with very different backgrounds; from ethology (optimal
search theory), through behavioral ecology (optimal foraging) and population ecology (habitat selection and dispersal theory) to mathematics
and computer modeling (e.g. diffusion models). The aim of this review
has been to identify the various approaches and their strengths and
weaknesses. Throughout I have suggested avenues for further developments that many improve the different approaches. In particular, combining elements from different approaches (e.g. incorporating optimal search
algorithms into simulation models to predict dispersal movements), may
allow the development of theoretical models that more accurately predict
movement responses to spatial heterogeneity.

But, most importantly, more solid empirical data are needed both for
suggesting more biologically realistic models and for testing several
fundamental questions about spatial structure–movement relations that
are still basically untested. The movement corridor controversy, now
blooming in the literature, does not imply that less is known about
corridors than any other spatial aspect potentially affecting movement
patterns. On the positive side is the fact that many of the questions are
testable for a range of organisms.

Although the predominance of experimental studies on insects published so far may suggest that insects are more experimentally tractable
than many other taxa, many fundamental questions in movement ecology
are within the reach of experimental treatment for vertebrates too.
Indeed, experimentally testing the same questions on a wide taxonomic
spectrum is probably the best way of establishing general principles in
landscape ecology.

ACKNOWLEDGEMENTS

The constructive comments by Arild Gautestad, Karine Hertzberg, Chris-
ter Solbreck, Nils Chr. Stenseth, Nigel G. Yoccoz and two anonymous
referees to an earlier draft of this chapter, improved it greatly. Harry P.
Andreassen is thanked for sharing many of his ideas with me. My own
research on movement pattern–spatial structure interactions have been
generously supported by research grants to the projects: 'Habitat fragmentation: Implications for the dynamics of populations' and 'The effect
of habitat corridors on the biodiversity of agricultural landscapes' from
the Science Foundation of Norway (NFR).

REFERENCES

Andreassen H., Halle, S. and Ims, R. A. (1994) Optimal design of movement corridors in root voles – not too wide and not too narrow (submitted).

Andreassen, H., Ims, R. A., Stenseth, N. C. and Yoccoz, N. G. (1993) Investigating space use by radiotelemetry and other methods: An updated methodological guide, in *The Biology of Lemmings* (eds N. C. Stenseth and R. A. Ims), Academic Press, London, pp. 689–716.

Arditi, R. and D'Acoragne, B. (1988) Optimal foraging on arbitrary food distributions and the definition of habitat patches. *Am. Nat.*, **131**, 837–46.

Back, C. E. (1980) Effects of plant density and diversity on the population dynamics of a specialist herbivore, the striped cucumber beetle *Acalymma vittata* (Fab.). *Ecology*, **61**, 1515–30.

Back, C. E. (1984) Plant spatial pattern and herbivore population dynamics: plant factors affecting the movement patterns of a tropical cucumber specialist (*Acalymma innunum*). *Ecology*, **64**, 175–90.

Back, C. E. (1988a) Effects of host plant size on herbivory density: patterns. *Ecology*, **69**, 1090–102.

Back, C. E. (1988b) Effects of host plant patch size on herbivore density: underlying mechanisms. *Ecology*, **69**, 1103–17.

Bell, W. J. (1991) *Searching Behaviour*, Chapman & Hall, London.

Bennett, A. (1990) Habitat corridors and the conservation of small mammals in fragmented forest environments. *Landsc. Ecol.*, **4**, 109–22.

Bernstein, C., Kacelnik, A. and Krebs, J. R. (1991) Individual decisions and the distribution of predators in a patchy environment. II. The influence of travel costs and structure of the environment. *J. Anim. Ecol.*, **60**, 205–25.

Brown, J. S., Kotler, R. J., Smith, R. J. and Wirtz, W. O. (1988) The effect of owl predation on the foraging behaviour of heteromyid rodents. *Oecologia*, **76**, 408–15.

Brown, J. S., Arel, Y., Abramsky, Z. and Kotler, B. (1992) Patch use by gerbils (*Geribillus allenbyi*) in sandy and rocky habitats. *J. Mamm.*, **73**, 821–9.

Brunt, J. W. and Conley, W. (1990) Behavior of a multivariate algorithm for ecological edge detection. *Ecol. Modelling*, **49**, 179–203.

Burt, W. H. (1943) Territoriality and home range concepts as applied to mammals. *J. Mamm.*, **24**, 346–52.

Carter, M. C. and Dixon, A. F. G. (1982) Habitat quality and the foraging behaviour of coccinellid larvae. *J. Anim. Ecol.*, **51**, 865–78.

Charnov, E. L. (1976) Optimal foraging: the marginal value theorem. *Theor. Pop. Biol.*, **9**, 129–36.

Crist, T. O., Guertin, D. S., Wiens, J. A. and Milne, B. T. (1992) Animal movement in heterogeneous landscapes: an experiment with *Eleodes* beetles in short grass prairie. *Funct. Ecol.*, **6**, 536–44.

Croze, H. (1970) Search image in carrion crows. *Z. Tierpsychol. Suppl.*, **5**, 1–85.

Danthanarayana, W. (ed.) (1986) *Insect flight. Dispersal and migration*, Springer Verlag, Berlin.

DeAngelis, D. L. and Gross, L. J. (eds) (1992) *Individual Based Models and Approaches in Ecology*, Chapman & Hall, London.

Dingle, H. (ed.) (1978) *Evolution of Insect Migration and Diapause*, Springer Verlag, New York.

Durelli, P., Studer, M., Marchand, I. and Jacob, S. (1990) Population movements of arthropods between natural and cultivated areas. *Biol. Conserv.*, **54**, 193–207.

Engen, S. and Stenseth, N. C. (1989) Age specific optimal diets and optimal foraging tactics: a life history approach. *Theor. Pop. Biol.*, **36**, 281–95.

Fahrig, L. (1991) Simulation methods for developing general landscape-level hypotheses of single-species dynamics, in *Quantitative Methods in Landscape Ecology* (eds M. G. Turner and R. H. Gardner), Springer-Verlag, Berlin.

Fahrig, L. and Merriam, G. (1985) Habitat patch connectivity and population survival. *Ecology*, **66**, 1762–8.

Fahrig, L. and Merriam, G. (1994) Conservation of fragmented populations. *Conserv. Biol*, **8**, 50–9.

Fahrig, L. and Paloheimo, J. (1988) Effect of spatial arrangement of habitat patches on local population size. *Ecology*, **69**, 468–75.

Forman, R. T. T. and Gordon, M. (1986) *Landscape Ecology*, John Wiley & Sons, New York.

Foster, J. and Gaines, M. S. (1991) The effect of a successional habitat mosaic on a small mammal community. *Ecology*, **72**, 1358–73.

Fretwell, S. D. and Lucas, H. L. (1970) On territorial behaviour and other factors influencing habitat distribution in birds. *Acta Biotheor.*, **19**, 16–36.

Gardner, R. H., O'Neill, R. V., Turner, M. G. and Dale, V. H. (1989) Quantifying scale-dependent effects of animal movements with simple percolation models. *Landsc. Ecol.*, **3**, 217–27.

Gautestad, A. and Mysterud, I. (1993) Physical and biological mechanisms in animal movement processes. *J. Appl. Ecol.*, **30**, 523–35.

Gill, F. B. (1988) Trapline foraging by hermit hummingbirds: competition for undefended renewable resource. *Ecology*, **69** 1933–42.

Gill, F. B. and Wolf, L. L. (1977) Non-random foraging by sunbirds in a patchy environment. *Ecology*, **58**, 1284–96.

Graf, S. A. and Sokolowski, M. B. (1989) Rover/sitter *Drosophila melanog-*

aster larval foraging polymorphism as a function of larval development, food, patch quality and starvation. *J. Insect Behav.*, **2**, 301–13.

Haley, K. B. and Stone, L. D. (1980) (eds) *Search Theory and Applications*, Plenum Press, New York.

Hamilton, W. D. and May, R. M. (1977) Dispersal in stable habitats. *Nature*, **269**, 578–81.

Hanski, I. and Peltonen, A. (1988) Island colonization and peninsulas. *Oikos*, **51**, 105–6.

Hansson, L. (1991) Dispersal and connectivity in metapopulations. *Biol. J. Linn. Soc.*, **42**, 89–103.

Hardt, R. A. and Forman, R. T. T. (1989) Boundary form effects on woody colonization of reclaimed mines. *Ecology*, **70**, 1252–60.

Harrison, R. L. (1992) Towards a theory of inter-refuge corridor design. *Conserv. Biol.*, **6**, 293–5.

Heinein, K. and Merriam, G. (1990) The element of connectivity where corridor quality is variable. *Landsc. Ecol.*, **4**, 157–70.

Hobbs, R. J. (1992) The role of corridors in conservation: solution or bandwagon? *Trends Ecol. Evol.*, **7**, 389–92.

Hoffman, G. (1983) The random elements in the systematic search behaviour of the desert isopod *Hemilepistus reaumuri*. *Behav. Ecol. Sociobiol.*, **13**, 93–106.

Holt, R. D. (1990) The micro-evolutionary consequences of climate change. *Trends Ecol. Evol.*, **5**, 311–15.

Huey, R. B. (1991) Physiological consequences of habitat selection. *Am. Nat.*, **137**, S91–115.

Huston, M., DeAngelis, D. and Post, W. (1988) New computer models unify ecological theory. *BioScience*, **38**, 682–91.

Ims, R. A., Rolstad, J. and Wegge, P. (1993) Predicting space use responses to habitat fragmentation: Can voles *Microtus oeconomus* serve as an experimental model system (EMS) for capercaillie grouse in boreal forest. *Biol. Conserv.*, **63**, 261–8.

Iwasa, Y., Higashi, M. and Yamahura, N. (1981) Prey distribution as a factor determining the choice of optimal foraging strategy. *Am. Nat.*, **117**, 710–23.

Jansen, D. H. (1971) Euglossine bees as long distance pollinators of tropical plants. *Science*, **171**, 203–5.

Johnson, A. R., Milne, B. T. and Wiens, J. A. (1992) Diffusion in fractal landscapes: Simulations and experimental studies of tenebrionid beetle movements. *Ecology*, **73**, 1968–83.

Johnson, A. R., Wiens, J. A., Milne, B. T. and Crist, T. O. (1992) Animal movements and population dynamics in heterogeneous landscapes. *Landsc. Ecol.*, **71**, 63–75.

Johnson, C. G. (1969) *Insect Migration and Dispersal by Flight*, Methuen, London.

Kacelnik, A. and Bernstein, C. (1988) Optimal foraging and arbitrary food distributions: Patch models gain a lease of life. *Trends Ecol. Evol.*, **3**, 251–3.

Kareiva, P. (1983) Local movements in herbivorous insects: applying a passive diffusion model to mark–recapture field experiments. *Oecologia*, **57**, 322–7.

Kareiva, P. (1985) Finding and loosing plants by *Phyllotreta*: patch size and surrounding habitat. *Ecology*, **66**, 1810–17.

Kareiva, P. (1986) Patchiness, dispersal, and species interactions: Consequences for communities of herbivorous insects, in *Community Ecology*, (eds. J. Diamond and T. J. Case), Harper & Row, New York.

Kareiva, P. (1987) Habitat fragmentation and the stability of predator–prey interactions. *Nature*, **326**, 388–90.

Kareiva, P. and Odell, G. M. (1987) Swarms of predators exhibit 'prey-taxis' if individual predators use area restricted search. *Am. Nat.*, **130**, 233–70.

Kennedy, J. S. (1951) The migration of desert locus (*Shistocerca gregaria*, Forsk.). I. The behaviour of swarms. II. A theory of long range migration. *Phil. Trans. R. Soc.* London, B, **235**, 163–290.

King, A. W. (1991) Translating models across scales in the landscape, in *Quantitative methods in Landscape Ecology* (eds M. G. Turner and R. H. Gardner), Springer, New York, pp. 479–517.

Kotliar, N. B. and Wiens, J. A. (1990) Multiple scales of patchiness and patch structure: a hierarchial framework for the study of heterogeneity. *Oikos*, **59**, 253–60.

Krummel, J. R., Gardner, R. H., Sugihara, G. *et al.* (1987) Landscape patterns in disturbed environments. *Oikos*, **48** 321–4.

La Polla, V. N. and Barrett, G. W. (1993) Effect of corridor width and presence on the dynamics of the meadow vole (*Microtus pennsylvanicus*). *Landsc. Ecol.*, **8**, 25–37.

Lawrence, W. S. (1988) Movement ecology of the red milkweed beetle in relation to population size and structure. *J. Anim. Ecol.*, **57**, 21–35.

Lefkovitch, L. P. and Fahrig, L. (1985) Spatial characteristics of habitat patches and population survival. *Ecol. Modelling*, **30**, 297–308.

Levin, S. (1992) The problem of pattern and scale in ecology. *Ecology*, **73**, 1943–67.

Lorenz, G. and Barrett, G. W. (1990) Influence of simulated landscape corridors on house mice (*Mus musculus*) populations. *Am. Midl. Nat.*, **123**, 348–56.

MacArthur, R. and Pianka, E. R. (1966) On optimal use of a patchy environment. *Am. Nat.*, **100**, 609–10.

Mader, H. J. (1984) Animal habitat isolation by agricultural fields. *Biol. Conserv.*, **29**, 81–96.

Mangel, M. and Clark, C. W. (1986) Towards an unified foraging theory. *Ecology*, **67**, 1127–38.

Mann, C. C. and Plummer, M. L. (1993) The high costs of biodiversity. *Science*, **160**, 1868–71.

McNamara, J. M. and Houston, A. I. (1987) Starvation and predation as factors limiting population size. *Ecology*, **68**, 1515–19.

Merriam, G. and Lanoue, A. (1990) Corridor use by small mammals: field measurements of three experimental types of *Peromyscus leucopus*. *Landsc. Ecol.*, **4**, 123–31.

Merriam, G. and Saunders, D. (1993) Corridors in restoration of fragmented landscapes, in *Nature Conservation 3: Reconstruction of Fragmented Ecosystems. Global and Regional Perspectives* (eds R. Hobbs, P. Erlich and D. Saunders), Surrey Beatty & Sons, Australia, pp. 71–87.

Merriam, G., Henein, K. and Stuart-Smith, K. (1991) Landscape dynamics models, in *Quantitative methods in landscape ecology* (eds M. G. Turner and R. H. Gardner), Springer-Verlag, Berlin, pp. 400–16.

Morris, D. W. (1987) Tests of density dependent habitat selection in a patchy environment. *Ecol. Monogr.*, **57**, 269–81.

Morris, W. F. (1993) Predicting the consequences of plant spacing and biased movement for pollen dispersal by honey bees. *Ecology*, **74**, 493–500.

Naveh, Z. and Lieberman, A. S. (1990) *Landscape Ecology. Theory and Applications*, Springer Verlag, New York.

Noss, R. F. (1987) Corridors in real landscapes: A reply to Simberloff and Cox. *Conserv. Biol.*, **1**, 159–64.

Odendaal, F. J., Turchin, P. and Stermitz, F. R. (1988) An incidental-effect hypothesis explaining aggregation of males in a population of *Euphydryas anacia*. *Am. Nat.*, **132**, 735–49.

Okubo, A. (1980) *Diffusion and Ecological Problems: Mathematical Models*, Spinger-Verlag, Heidelberg.

O'Neill. R. V. (1989) Perspectives in hierarchy and scales, in *Perspectives in Ecological Theory* (eds J. Roughgarden, R. M. May and S. A. Levin), Princeton University Press, Princeton, New York, pp. 140–56.

Opdam, P. (1990) Dispersal in fragmented populations: the key to survival, in *Species Dispersal in Agricultural Habitats* (eds R. G. H. Bunche and D. C. Howard), Belhaven Press, New York.

Opdam, P. (1991) Metapopulation theory and habitat fragmentation: a review of holarctic breeding bird studies. *Landsc. Ecol.*, **5**, 93–106.

Robinson, G. R., Holt, R. D., Gaines, M. S. *et al.* (1992) Diverse and conflicting effects of habitat fragmentation. *Science*, **257**, 524–6.

Rolstad, J. (1991) Consequences of forest fragmentation for the dynamics of bird populations: conceptual issues and the evidence. *Biol. J. Linn. Soc.*, **42**, 149–63.

Root, R. B. and Kareiva, P. (1984) The search for resources by cabbage butterflies in simple and diverse habitats: Ecological consequences and the adaptive significance of Markovian movements in a patchy environment. *Ecology,* **65**, 147–65.

Rosenzweig, M. L. (1981) A theory of habitat selection. *Ecology,* **62**, 327–35.

Saunders, D. and Hobbs, R. J. (eds) (1991) *Nature Conservation 2: The Role of Corridors,* Surrey Beatty, Chipping Norton, Australia.

Schoener, T. W. (1971) Theory of feeding strategies. *Ann. Rev. Ecol. Syst.,* **11**, 369–404.

Senft, R. L., Coughenour, M. B. and Bailey, D. W. *et al.* (1987) Large herbivore foraging and ecological hierarchies. *BioScience,* **37**, 789–99.

Simberloff, D. (1988) The contribution of population and community biology to conservation science. *Ann. Rev. Ecol. Syst.,* **19**, 473–511.

Simberloff, D. and Cox, D. (1987) Consequences and costs of conservation corridors. *Conserv. Biol.,* **1**, 63–71.

Simberloff, D., Farr, J. A., Cox, J. and Mehlman, D. W. (1992) Movement corridors: conservation bargains or poor investement. *Conserv. Biol.,* **6**, 493–504.

Skellam, J. G. (1951) Random dispersal in theoretical populations. *Biometrica,* **38**, 196–218.

Soule, M. E. (1986) (ed.) *Conservation Biology: The Science of Scarcity and Diversity,* Sinauer, Sunderland, Massachusetts.

Southwood, T. R. E. (1977) Habitat, the template of ecological strategies? *J. Anim. Ecol.,* **50**, 337–65.

Stamps, J. A., Buechner, M. and Krishnan, V. V. (1987a) The effects of edge permeability and habitat geometry on emigration from patches of habitat. *Am. Nat.,* **129**, 533–52.

Stamps, J. A., Buechner, M. and Krishnan, V. V. (1987b) The effect of habitat geometry on territorial defense costs: intruder pressure in bounded populations. *Am. Zool.,* **27**, 307–25.

Steen, H. (1994) Long distance dispersal in *Microtus oeconomus. Ann. Zool. Fennici,* **31**, 271–9.

Stenseth, N. C. (1983) Causes and consequences of dispersal in small mammals, in *The Ecology of Animal Movements* (eds I. Swingland and P. Greenwood), Oxford University Press, Oxford, pp. 63–101.

Stenseth, N. C. and Lidicker, W. Z. (eds) (1992) *Animal Dispersal: Small Mammals as a Model,* Chapman & Hall, London.

Stephens, D. W. and Krebs, J. R. (1986) *Foraging Theory,* Princeton University Press, New Jersey.

Stork, N. (ed.) (1990) *The Role of Ground Beetles in Ecological and Environmental Studies,* Intercept, Hampshire.

Sugihara, G. and May, R. M. (1990) Application of fractals in ecology. *Trends Ecol. Evol.,* **5**, 79–86.

Swihart, R. K. and Slade, N. A. (1984) Road crossings in *Sigmodon hispidus* and *Microtus ochrogaster. J. Mamm.,* **65**, 357–60.

Swingland, I. and Greenwood, P. (1983) *The Ecology of Animal Movements*, Oxford University Press, Oxford.

Thornhill, R. and Alcock, J. (1983) *The Evolution of Insect Mating Systems*, Harvard University Press, Cambridge, Massachusetts.

Tinbergen, N., Impekoven, M. and Franck, D. (1967) An experiment on spacing out as a defence against predation. *Behaviour*, **28**, 307–21.

Turchin, P. (1986) Modelling the effect of patch host size on Mexican bean beetle emigration. *Ecology*, **67**, 124–32.

Turchin, P. (1987) The role of aggregation in the response of Mexican bean beetles to host-plant density. *Oecologia*, **71**, 577–82.

Turchin, P. (1991) Translating foraging movements in heterogeneous environments into spatial distribution of foragers. *Ecology*, **72**, 1253–66.

Turner, M. (1989) Landscape ecology: the effect of patterns on processes. *Ann. Rev. Ecol. Syst.*, **20**, 171–97.

Turner, M. G. and Gardner, R. H. (eds) (1991) *Quantitative Methods in Landscape Ecology*, Springer-Verlag, Berlin.

Urban, D. L., O'Neill, R. V. and Shugart, H. H. (1987) Landscape ecology: a hierarchical perspective can help scientists to understand spatial problems. *BioScience*, **37**, 119–27.

Vail, S. (1993) Scale dependent responses to resource spatial pattern in simple models of consumer movements. *Am. Nat.*, **141**, 199–216.

Wiens, J. A. (1989) Spatial scaling in landscape ecology. *Funct. Ecol.*, **3**, 386–97.

Wiens, J. A. and Milne, B. (1989) Scaling of landscapes in landscape ecology, or, landscape ecology from a beetles perspective. *Landsc. Ecol.*, **3**, 87–96.

Wiens, J. A., Crawford, C. S. and Gosz, J. R. (1985) Boundary dynamics: a conceptual framework for studying landscape ecosystems. *Oikos*, **45**, 421–7.

Wiens, J. A., Stenseth, N. C., Van Horne, B. and Ims, R. A. (1993) Ecological mechanisms and landscape ecology. *Oikos*, **66**, 369–80.

Habitat selection in mosaic landscapes 5

Douglas W. Morris

5.1 INTRODUCTION

Landscape ecology, with its emphasis on spatial patterns and processes, articulates a pressing need to consider spatial heterogeneity and spatial dynamics in studies of population dynamics, species interactions and evolution (e.g. Turner and Gardner, 1991). A parallel perspective argues that predictive landscape ecology must incorporate evolutionary principles developed in the more traditional ecological disciplines (Morris and Brown, 1992). This chapter attempts to meet both objectives by integrating evolutionary theories of habitat selection with an empirical and applied framework for landscape ecology.

I begin by reviewing single-species models of habitat selection. I demonstrate how density-dependent models can be tested with data frequently available to landscape ecologists, how the models can be used to infer spatial scale as well as temporal dynamics in habitat quality, and how they can be extended to multiple-species communities. I contrast the utility of new methods with traditional approaches and conclude by posing a series of questions that should be solved as we develop studies of landscape ecology from a habitat selection perspective (Chapters 2, 6 and 9). My intent is not just to discuss the theory, but to demonstrate how it can be applied to solving problems in landscape ecology (Chapter 1).

Most of my examples are drawn from the population and community dynamics of mammals. The bias is more than one of familiarity. Many of the models have not yet been tested with, or applied to, other groups of organisms.

Mosaic Landscapes and Ecological Processes.
Edited by Lennart Hansson, Lenore Fahrig and Gray Merriam.
Published in 1995 by Chapman & Hall, London. ISBN 0 412 45460 2

5.2 DENSITY-DEPENDENT HABITAT SELECTION

5.2.1 SINGLE-SPECIES MODELS

Theories of density-dependent habitat selection assume that, over some range of population densities, reproductive success should decline with increasing population density (Figure 5.1). Increased density can be expected, among other things, to place higher demands on resources in short supply and on the availability of breeding sites, to magnify risks of predation, and to increase susceptibility to pathogens. These effects vary among habitats and each habitat can, for a given population, be represented by a characteristic fitness-density function (Figure 5.1). At low population size, individuals should congregate in the habitat yielding maximum fitness. As density increases individuals should occupy alternative habitats whenever their expected reproductive success in those habitats equals or exceeds that in already occupied habitat (Fretwell and Lucas, 1970).

Several models address different assumptions about how individuals should distribute themselves among habitats (Fretwell and Lucas, 1970; Lomnicki, 1988; Pulliam, 1988; Milinski and Parker, 1991; Kacelnik, Krebs and Bernstein, 1992; Oksanen, Oksanen and Fretwell, 1992). The most familiar of these, the ideal free distribution (Fretwell and Lucas, 1970), predicts that densities should be adjusted such that an individual's average reproductive success is equivalent in each habitat (Figure 5.1). Thus, the population size of a species in any given landscape, and its average density over the landscape, are going to be functions of the quality and distribution of habitats (Holt, 1985).

Experimental tests of the ideal free model have concentrated on the behavioral decisions of individual 'foragers' (references in Milinski and Parker, 1991). Tests at the landscape scale have been elusive because it is frequently impossible to obtain the necessary replicated simultaneous data on reproductive success and population density across a variety of habitats (for exceptions, see Krebs, 1971; Whitham, 1978, 1980; Morris, 1989a, 1991). Yet landscape tests are essential if theories of density-dependent habitat selection are to contribute to the development of landscape ecology, and vice versa. Landscape pattern represents the geographical and evolutionary context within which habitat selection modifies local population densities and community composition. The resulting patterns of relative abundance and species diversity alter the landscape, and highlight the dynamic linkage between landscape and theories of habitat selection. Neither can be understood in ignorance of the other. Can we modify the original theory to enable tests at the landscape scale?

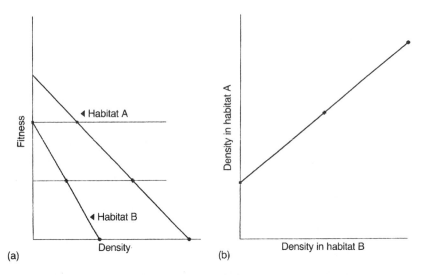

Figure 5.1 (a) A simple representation of the ideal free model of density-dependent habitat selection. Two habitats are shown, each with a characteristic shape and decline in reproductive success with increasing density. At low density, individuals should choose habitat A because their expected fitness is greater than in habitat B. The expected fitness in habitat A will be reduced with increases in density. Individuals should begin to occupy habitat B when the average fitness there is equivalent to that in A. The densities should be adjusted by movement between habitats such that the average reproductive success is equivalent in both (horizontal lines, the pairs of points represented by symbols are replotted in (b)). The pair of habitats depicted here are perceived to differ from one another qualitatively (different slopes) and quantitatively (different intercepts). Discussion of more complicated shapes for fitness-density curves can be found in Fretwell and Lucas (1970), Fretwell (1972), Milinski and Parker (1991), Kacelnik, Krebs and Bernstein (1992), and Morris (1992, 1994).

(b) An isodar generated from the fitness-density curves depicted in (a). The isodar plots the set of densities in habitat A versus those in habitat B such that the expected reproductive success of an individual is the same in both (the intersections of all possible horizontal lines with the fitness–density curves). The fitness–density curves, in this case, diverge from one another, yielding an isodar with slope > 1.0.

5.2.2 ISODAR THEORY

Imagine a density-dependent habitat-selecting species occupying a landscape composed of two habitats as depicted in Figure 5.1(a). According to the ideal free assumption, the respective densities of individuals in the two habitats will be given by the intersection of each habitat's fitness–density function with a set of horizontal lines corresponding to equal reproductive success in both habitats. These densities can be replotted as an isodar (Figure 5.1(b)), a line along which the expected reproductive success of individuals is the same in each habitat (Morris, 1987a, b, 1988). To draw an isodar for the two habitats represented in Figure 5.1(a), plot the density in habitat A against the corresponding density in habitat B such that the fitness is the same in each (examples of these densities are indicated by symbols). The isodar represents the solution to an evolutionarily stable strategy of ideal density-dependent habitat selection. The intercept corresponds to the how far apart the fitness–density curves lie from one another. The slope specifies the relative slopes of the respective fitness–density curves (Morris, 1988). Empirical isodars can be easily generated from estimates of population density in different habitats across any landscape.

Two kinds of habitat differences are likely to have dramatic effects on the slopes and intercepts of fitness–density curves, and on the isodars generated by them. First, imagine a quantitative difference whereby the two habitats differ from one another only in the amount of resource available for consumption. Because the habitats are assumed equivalent in every other respect, individuals should be equally efficient at garnering resources from each. Nevertheless, an individual exploiting the rich habitat at any given population density can expect to have more resources available to convert into reproduction and survival than it can expect by exploiting the poor habitat. The fitness–density curve of the rich habitat will lie above that of the poor one. The isodar will have a non-zero intercept (Figure 5.1).

Now imagine that the two habitats have the same resource renewal, but that they differ in some qualitative respect (e.g. habitat structure or the identity of resources). Individuals can expect to be more efficient at harvesting resources and converting them into descendants in one habitat than in the other. This qualitatively superior habitat can support a greater density if the resource is harvested to the same level in both habitats, but less is spent on non-foraging activities (e.g. foraging costs) than in the inferior habitat. The per capita impact on average fitness will be less than in the inefficiently exploited habitat. Alternatively, efficient consumers may reduce the renewal rate of resource (Holt, 1984). Each individual living in the efficiently exploited habitat would have a larger effect on competing individuals than would those living in the other. The fitness–density curves for each scenario will have different slopes, as will the resulting isodars. Isodars can thus detect not only density-dependent habitat selection, they can also infer the kind of habitat differences involved in habitat choice (Figure 5.1; Morris, 1988).

Preliminary isodars have yielded encouraging results. Studies on insular rodents in the Gulf of Maine (Crowell, 1983), on forest rodents in Ontario (Morris, 1988, 1989b), on prairie rodents in Alberta (Morris, 1992; Figure 5.2) and on desert rodents in Israel (Abramsky, Rosenzweig and Pinshow, 1991; Rosenzweig, 1991) produced significant isodars consistent with the theory's predictions. It appears that estimates of population density can be used to infer relative qualities of habitats in natural landscapes.

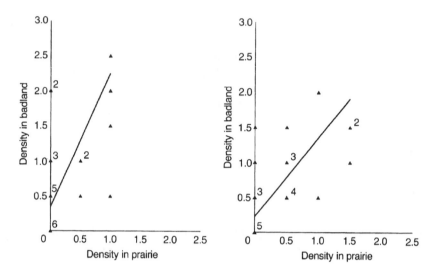

Figure 5.2 Isodars contrasting the density of deer mice (*Peromyscus maniculatus*) occupying prairie and badland habitats in southern Alberta. Badland habitat supports a greater density of deer mice than does prairie. (Geometric mean regression; source: Morris, 1992.)

5.3 PUTTING ISODARS TO WORK IN LANDSCAPE ECOLOGY

5.3.1 INFERENCES OF SPATIAL SCALE

The potential of landscape ecologists to test and apply spatial theories depends upon the investigators' ability to correctly identify the scale(s) at which crucial processes, such as dispersal, occur (Kareiva, 1990; Kotliar and Wiens, 1990; Levin, 1992; Chapter 1). For some species or sets of species, and types of interactions, this may be sufficient to provide insights into patterns of spatial distribution. For many other species and their interactions it will be necessary to integrate purely spatial models with models that specify the quality of patches in heterogeneous landscapes. This is the principal domain of habitat selection theory.

Current theories have identified three scale-dependent processes likely to dominate decisions on habitat choice (Figure 5.3). At some small spatial scale, individuals will be unable to discriminate between habitats and will exploit each equally. This scale should vary with the size and

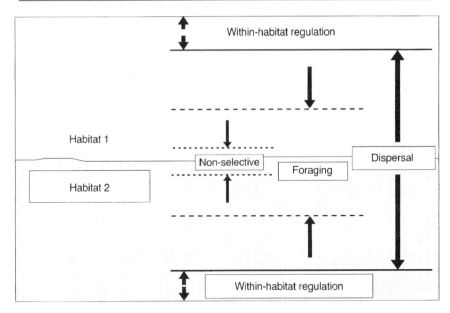

Figure 5.3 The scales of habitat selection between two homogeneous habitats sharing a common border. At some small scale near the boundary, individuals will be unable to discriminate between the habitats and will thus be non-selective in habitat use. At a somewhat larger scale, individuals whose home range spans the boundary will preferentially allocate 'foraging' in one habitat or the other. At a still larger scale, habitat selection can occur only by moving the home range from one habitat to the other (dispersal). Beyond the dispersal scale individuals are incapable of habitat selection, and population dynamics in the two habitats will occur independently of one another. (Source: Morris, 1992.)

perception of the organism, and with the nature of the boundary between habitats.

At the scale of a single home range, individuals can differentially allocate exploitation activities among alternative patches. But differential exploitation carries a cost. The gains that an individual achieves by selecting one patch over another must compensate for the time and energy spent traveling through or around the non-selected patch. An individual that encounters patches in the proportions in which they occur in the environment (a so-called fine-grained forager; MacArthur and Levins, 1964) should become non-selective in habitat use even though the average fitness to be gained in the better habitat exceeds that in the alternative (Rosenzweig, 1974, 1981; Brown and Rosenzweig, 1986).

At a larger scale, individuals can select one habitat over another (or the mix of the habitats in the home range) only by dispersal. Dispersal also carries a cost, but one that is fundamentally different from that of foraging. Individuals attempting to maximize their reproductive success by dispersal should move from one habitat to another only when the increased fitness to be gained there compensates for the lost reproductive potential during dispersal and establishment of the new home range (Morris, 1987a, 1992). Individuals should change habitats only when the expected fitness in the alternative exceeds that of the currently occupied habitat.

The different scales of habitat selection have profound influences on isodars, and on the 'connectedness' of population dynamics among habitats. At the non-selective scale, the two habitats are used indiscriminately. The isodar should pass through the origin with a slope of 1.0. At the foraging scale, exploitation of the 'rich' patch subsidizes exploitation of the 'poor' one. The subsidy, in a home range including both habitats, devalues the apparent quality of the rich habitat, and inflates that of the poor one. Exploitation in a mixed-habitat home range will be reduced relative to what it would be in a home range located entirely within a rich habitat. The opposite occurs in the poor habitat. Home-range size in mixed habitats should thereby be larger (and average density less) than would occur among sets of home ranges located only within the rich habitat. Home-range size in mixed habitats would be smaller (and average density higher) than in only poor habitat. It can thus be seen that foraging cost reduces the difference between the fitness–density curves of the two habitats. The isodar intercept is similarly reduced (Figure 5.4).

The opposite effect occurs at the larger dispersal scale. The 'quality' of the newly colonized habitat must exceed that in the immigrant's previous habitat if dispersal is to result in no loss in reproductive success. This means that the apparent quality of the new habitat must be greater than would occur if there was no dispersal cost. A habitat will be colonized only if its density is lower (and its expected reproductive success thereby greater) than that required for cost-free habitat selection. The isodar intercept will be increased (Figure 5.4). Density-dependent risks of dispersal will increase the isodar slope (Morris, 1992). Beyond the dispersal scale, habitats may be effectively disconnected from the effects of density-dependent habitat selection.

Each scale of habitat selection, as well as the effective limits on the ability of habitat selection to regulate population size, can be evaluated with isodars (Morris, 1992). The basic protocol involves establishing belt transects capable of assessing population density across distinct boundaries between habitats. Transect segments of different lengths, and at varying distances from the habitat boundary, are contrasted with one another by regression to look for the tell-tale differences in isodar intercepts and slopes that identify the shift from non-selective, through foraging, to dispersal scales of density-dependent habitat selection (Figure 5.5).

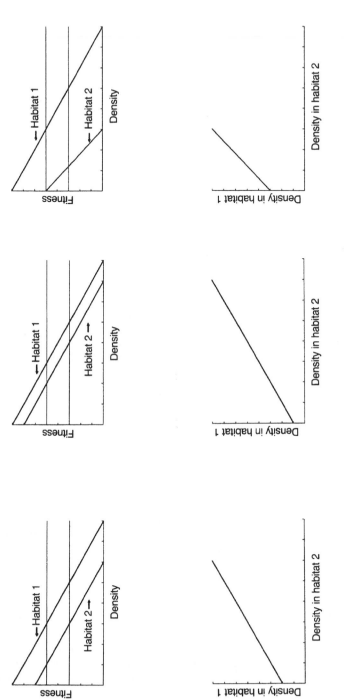

Figure 5.4 The effects of foraging and dispersal costs on the slopes and intercepts of isodars. Fitness–density curves are plotted, with the resulting isodars given below. (Source: Morris, 1992.)

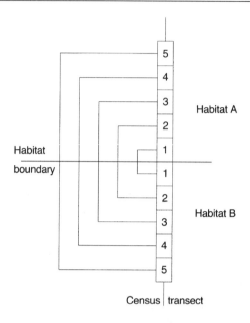

Figure 5.5 A protocol to assess foraging and dispersal scales of density-dependent habitat selection. Replicated census transects bisect two habitats occupied by the species of interest. Each set of 'connecting lines' represents a separate regression comparing densities in segments located at different distances from the boundary. The regressions assess predicted shifts in isodar intercepts and slopes that document the limits of foraging and dispersal scales of habitat selection (thorough analyses would evaluate densities in segments of increasing length). (Source: Morris, 1992.)

I tested the theory's ability to detect foraging and dispersal scales by contrasting deer mouse (*Peromyscus maniculatus*) densities along replicated live-trap transects bisecting prairie and badland habitats in southern Alberta (Morris, 1992; Figure 5.2). The isodar shown on the right of Figure 5.2 was generated along a transect from segments close to the boundary between the two habitats; the one on the left was generated from more distant segments. The slope of the isodar on the right is significantly less than that on the left, in agreement with theoretical predictions that it corresponds to the foraging scale. The isodar on the left corresponds to the dispersal scale. Regressions based on more distal segments of equal length were non-significant. This lack of correlation between distant population densities demonstrated the effective limit of density-dependent habitat selection in regulating population size. The analysis revealed a foraging scale in the order of 60 m and a dispersal scale in the order of only 140 m. Habitat selection's influence on the population regulation of deer mice occupying heterogeneous prairie landscapes was thus limited to within 70 m of the prairie–badland boundary.

Habitat selection is likely to be a potent contributor to regional deer-mouse population dynamics only if prairie and badland habitats have extensive borders resulting from complex edges or highly interspersed patches of habitat. Prairie and badland habitats are juxtaposed along sinuous, dendritic river valleys and their tributaries, suggesting that habitat selection may indeed play a major role in population regulation between the two habitats. This example demonstrates how isodar analysis can:

1. identify the spatial scales of habitat selection and population regulation, and
2. guide the choice of critical landscape features for further interpretation.

5.3.2 INFERENCES OF TEMPORAL SCALE

Landscape analyses have recently benefited from a variety of techniques that assist spatial pattern analysis (Turner *et al.*, 1991; Rossi *et al.*, 1992). But ecologists also need models and techniques that relate spatial pattern to underlying ecological processes (Fox and Morris, 1990; Kareiva, 1990; Morris, 1990; Merriam, Henein and Stuart-Smith, 1991; Fahrig, 1992). Numerous processes are doubtlessly involved in the creation of spatial pattern, and the challenge for the theorist and empiricist alike is to select those processes appropriate to the spatial and temporal scales being analyzed. Theories of density-dependent habitat selection offer substantial promise at fulfilling this need, particularly at scales corresponding to habitat disturbance and fragmentation.

Imagine a habitat that is modified by either natural or human disturbance. The objective is to predict the effect of the disturbance on local populations and communities as well as the time course of 'recovery' to ambient conditions. One way of achieving this is to consider the effect that the disturbance will have on fitness–density curves, and their resulting isodars (Figure 5.6).

Shortly after disturbance the expected fitness at any given consumer density is likely to diverge dramatically from that of an undisturbed control habitat (Figure 5.6(a)). The differences are likely to dissipate with time as the disturbance is ameliorated via ecological succession. A comparison of isodars calculated from disturbances of different ages documents the time-course of possible convergence in population dynamics (Figure 5.6(b)).

The value of this technique can be illustrated by isodar analyses on white-footed mice (*Peromyscus leucopus*) inhabiting successional and mature forest habitats in southern Ontario. Regressions of the density of

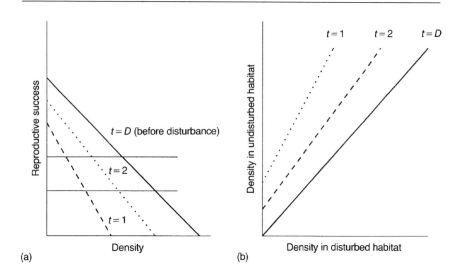

Figure 5.6 One example of how disturbance modifies habitat. (a) The disturbed habitat can be distinguished from its undisturbed neighbors by differences in the decline of fitness with population density (e.g. the line corresponding to $t=1$, the time interval since disturbance). These differences will be reduced through time ($t=2$) if the disturbed habitat becomes more similar to its neighbors. (b) The convergence of the disturbed habitat on the ambient control is reflected in a similar convergence of isodars. In both cases, the undisturbed controls are considered constant, but this assumption is not crucial to the use of isodars in the assessment of habitat convergence. (Source: Morris, 1990.)

white-footed mice occupying forest versus early succession old-field habitat revealed a consistent preference of white-footed mice for forest (isodar slope > 1.0, intercept > 0; Morris, 1988). Regressions of white-footed mouse density in 20-m tall forest versus that in mid-successional 3-m tall sumac generated an isodar with a slope very close to 1.0, and with an intercept not significantly different from zero (the densities were indistinguishable; Morris, 1988). As far as patterns in white-footed mouse population density are concerned, the sumac had converged on the forest even though the two were dramatically different in habitat structure and floristic composition (Morris, 1984). The example illustrates how isodars can be used to define habitats. Two habitats are recognized as different by a habitat-selecting species only when the isodar has an intercept different from zero, or a slope different from unity.

5.3.3 INFERENCES TO MULTISPECIES COMMUNITIES

Single-species models have obvious limitations when applied to multi-species assemblies. Isodar analysis can be easily modified to incorporate species interactions (Morris, 1989b). Figure 5.7 demonstrates the solution. Instead of simply plotting the density of one species in one habitat against its density in a second habitat, the 'multispecies isodar' incorporates the effects of potentially interacting species. These effects are scaled by multiple regression analysis by representing species interactions as additional individuals of the 'target' species. The important result is that both the single-species and multispecies isodars are predicted to be the same, as long as all relevant interactions among species are included in the multiple regression equation (Morris, 1989b).

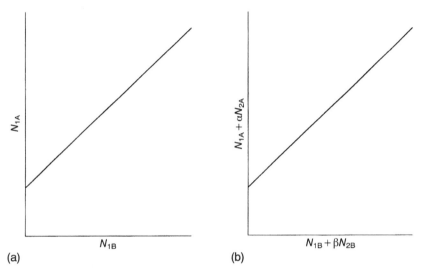

(a)   (b)

Figure 5.7 An illustration of the effect of interacting species on isodars. (a) An isodar constructed where the target species exists alone in allopatry. (b) An isodar constructed where the same species co-occupies the two habitats with a competitor in sympatry. The isodar remains unchanged as long as the appropriate species interaction is included for each habitat. (Source: Morris, 1989b.)

The prediction of equivalent isodars, whether generated from data on the target species in isolation, or from an intact set of interacting species, suggests a powerful test for species interactions. The isodar of each habitat-selecting species should be unchanged following removal of its competitors. If the isodar following species removal is different from that estimated prior to removal, we can be reasonably certain that some key or higher-order interaction was omitted in the first analysis. Perhaps the best example of a nonlinear effect is produced by distinct habitat preferences (Rosenzweig, 1979, 1981, 1989) where density-dependent habitat selection warps competitive isoclines to eliminate all evidence of competition when each species occupies only its preferred habitat. Such curved isoclines in response to habitat selection have recently been documented between pairs of competing gerbils in Israel (Abramsky, Rosenzweig and Pinshow, 1991; Abramsky, Rosenzweig and Zubach, 1992). If the theory is correct in its predictions, the single-species isodar should be reproduced with new data including the omitted interactions.

Experimental tests of the multispecies theory have not yet been published. Preliminary observational tests are encouraging but imperfect. Regressions of white-footed mouse density in either sumac or forest habitat versus that in an old field co-occupied by meadow voles (*Microtus pennsylvanicus*) detected no competitve interactions between the two species (Morris, 1989b). The mouse isodars suggested habitat partitioning related to both qualitative and quantitative differences between the two wooded habitats in comparison with the old field. Qualitative differences should lead to distinct habitat preferences (Pimm and Rosenzweig, 1981; Rosenzweig, 1985, 1987, 1989), a result in agreement with the markedly different habitats usually exploited by these two species. A reanalysis of habitat partitioning between two Arizona rodents similarly revealed density-dependent habitat selection (Morris, 1989b), but failed to confirm field manipulations demonstrating modest competitive interactions (Holbrook, 1979). Applications of isodar analysis to multispecies assemblages should therefore be interpreted with caution.

5.4 ALTERNATIVES

5.4.1 INFERENCES BASED ON HABITAT 'QUALITY'

Ecological folklore, and a good deal of theory, promulgates the view that population density is related to habitat quality. Current theories of density-dependent habitat selection argue that the density in any one habitat is also a function of the density in neighboring habitats. Modern versions of an idea dating at least to Joseph Grinnell (MacArthur, 1972) demonstrate that 'surplus' individuals produced in so-called source habitats may often spill over into unproductive sink habitat (Holt, 1985; Pulliam, 1988; Oksanen, 1990; Oksanen, Oksanen and Gyllenberg, 1992) with profound consequences on not only population size, but also on the interactions among species (Pulliam and Danielson, 1991; Danielson, 1991, 1992). This suspicion is confirmed by studies on rodents in northern latitudes where dramatic differences in population densities and population dynamics show strong correlations with habitat heterogeneity (Hansson and Henttonen, 1988). The challenge is to separate influences on population density caused by differences in habitat 'quality' (differences in the relationship between fitness and population density) from those caused by density-dependent habitat selection among different habitats.

The classic approach using multivariate statistics to infer habitat quality (for example, by regressions of population density against several independent microhabitat variables (Capen, 1981); excellent discussions of the strengths and weaknesses of this approach can be found in Verner, Morrison and Ralph (1986) and Wiens (1989)) fails to account for landscape effects that can modify population density. A simple example of such an effect can be demonstrated by biases of spatial scale. If the microhabitat variables are collected across sets of more or less homogeneous habitats and pooled for analysis, the ecologist may be misled into believing that population density is causally related to microhabitat, when in reality, animals are recognizing a much larger scale of habitat heterogeneity (Morris, 1987c, 1989c). The animals may be sampling the environment at a higher level of heterogeneity than that subsumed within the microhabitat variables (Kolasa and Rollo, 1991). Van Horne (1983) as well as Hobbs and Hanley (1990) offer several additional critiques. Isodar analysis provides little in the way of improvement because it includes no information on intrinsic habitat quality.

There are several reasons why density may not mirror habitat quality (Van Horne 1983, 1986; Maurer, 1986; Wiens, 1989; Kareiva, 1990) and thereby lead to a biased isodar analysis. One common reason may simply be that 'disconnected' populations are not at the same density relative to patch carrying capacity. Similar patches of habitat within a landscape may support quite different densities dependent upon the recent history of population growth within the patch (or nearby patches). This 'nonequilibrium' effect can be illustrated by representing the fitness–density curves for a pair of habitats by bands of parallel curves each representing a different patch (Figure 5.8(a)). The graph is drawn such that if the population could achieve equilibrium in all patches (intercepts along the abscissa), there would be a direct correspondence between population density and patch quality. Assume an ideal distribution of individuals between pairs of habitat patches. A regression of 'nonequilibrium' population density against patch quality will likely have substantial residual variation and low predictive power (Figure 5.8(b)).

An isodar plot of the same data illustrates the converse role that variation in quality can play in residual scatter about the isodar (Figure 5.8(c)). Note, however, that all error variation is eliminated if we use a hybrid technique that constructs the isodar with the residuals from the 'habitat quality' regression (Figure 5.8(d)). The slope of the original isodar, and thus our interpretation about population regulation, is different from that in the corrected residual isodar (0.72 in Figure 5.8(c); 1.0 in Figure 5.8(d)). The model presented here assumes that maximum reproductive success and equilibrium population density are perfect correlates of habitat quality (all patches have parallel fitness curves). The validity of these assumptions in any field study will depend upon

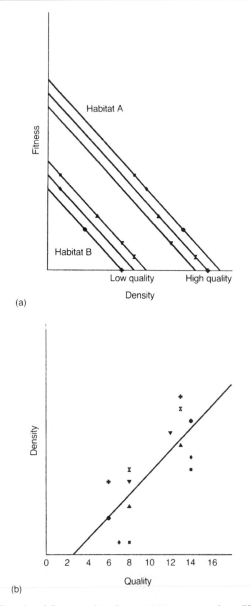

(a)

(b)

Figure 5.8 (a) Bands of fitness–density curves representing different patches of two habitats, A and B. The quality of each patch is given by its intercept with the abscissa. Paired symbols correspond to nonequilibrium ($N_i < K_i$) ideal distributions between nearby patches of the two habitats. Similar processes could act among patches within a single habitat. Interpretations about the relationship between habitat quality and population density would be complicated if the individual curves vary in slope or shape, but such differences could be detected by 'within-habitat' analyses. (b) The densities in (a) plotted against the quality of each patch.

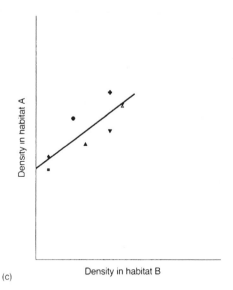

(c)

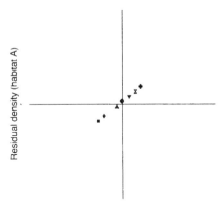

(d)

(c) An isodar of the hypothetical data presented in (a) (isodar slope = 0.72). The residual scatter is the result of differences in population density among different pairs of habitat patches isolated throughout the landscape. (d) A residual isodar created by plotting the residuals from the regression in (b) (isodar slope = 1.0; with real data we expect a significant reduction in unexplained variation about the regression rather than complete elimination of error).

consistency in the shapes and slopes of the fitness–density functions, and on the correlation between habitat features and their quality.

The two isodars give different solutions because the relative qualities of alternative habitat choices selected by individuals change among replicate estimates of population density. The mix of patches of different qualities has warped the fitness–density curves so that they appear to converge (slope < 1) rather than remain parallel. The potential for habitat quality to obscure the real relationship between the density in the two habitats (different 'slopes' in the original and residual isodars) thus depends upon the mosaic of the habitats in the landscape. It is possible that the relative differences between such pairs could remain constant. The relative abundance of species in the two habitats would be unbiased among samples and there would be no improvement in the isodar solution if one used the density/quality residuals in place of the original densities.

An effective protocol for the assessment of density-dependent habitat selection may thus first regress densities against likely correlates of habitat quality, such as those used in earlier regression and correlation studies, before subjecting the residuals to a formal isodar analysis. The potential for nonequilibrium dynamics would be implicated whenever the residuals isodar gives a better fit to the data than the isodar based on the original densities.

Differences in the slopes, intercepts and shapes of the fitness–density curves will reduce the effectiveness of the residuals analysis in assessing habitat selection. Such differences imply that the analysis is confounded by more than the two habitats of interest because a habitat can be defined by similarity in the functional relationship between fitness and density. This definition depends on the choice of scale used in the analysis, an issue of crucial interest in landscape ecology. Regional comparisons of habitat use among many landscapes may inadvertently lump habitats that individuals of a habitat-selecting species would recognize as different.

An example of the landscape effect can be found in an isodar analysis used by Knight and Morris (unpublished) to study habitat selection by red-backed voles occupying wet and dry habitats in the Hudson Bay lowland. The pattern of residuals about the regression suggested that these voles may be selecting more than just two habitats (Knight, 1993). Instead of wet and dry habitats, the voles appeared to recognize three habitats; dry ridges, wetlands without trees, and wetlands with interspersed larch and spruce. Subsequent isodar analyses confirmed the three-habitat classification.

5.4.2 HABITAT MATCHING RULES

Habitat selection theories may allow us to predict how changes in landscape composition affect population size. Pulliam and Caraco (1984) demonstrated for a special case of the ideal free distribution where each individual's fitness is proportional to its fraction of total resource, that

$$K_i/p_i = K_j/p_j, \qquad (5.1a)$$

where K is carrying capacity, and p is the number of individuals occupying patches i and j. (5.1) is the habitat matching rule that specifies how individuals should distribute themselves relative to the availability of resources. Rearrangement of (5.1a) shows that the ratio of individuals occupying different patches should be constant (Fagen, 1988)

$$p_i/p_j = K_i/K_j = \text{constant}, \qquad (5.1b)$$

or put another way, the fraction of predators in a patch should be proportional to the fraction of prey in that patch (Sutherland, 1983; Fagen, 1987; see also Kacelnik, Krebs and Bernstein, 1992; Oksanen, Oksanen and Fretwell, 1992; Kennedy and Gray, 1993). Morris (1990) used similar logic to argue for constant niche breadth for ideal habitat selectors occupying qualitatively different habitats. (The matching rule assumes that all habitats are occupied at all densities (Sutherland, 1983; Pulliam and Caraco, 1984), an assumption most likely to be met if habitats differ qualitatively.) It should be possible, therefore, to test the ideal free theory by demonstrating that changes in predator population size have no effect on the ratio of individuals occupying each habitat (Messier, Virgl and Marinelli, 1990).

In the context of landscape ecology, (5.1b) can be used to predict changes in population size with changes in habitat supply (Fagen, 1988; but see Hobbs and Hanley, 1990). As noted above, the equation applies only to an ideal free distribution when each habitat is occupied across the full range of population sizes. This assumption may often be inappropriate to landscape applications of the theory. Isodar solutions (Morris, 1994) are preferable because they can be applied to a variety of forms of habitat selection, and because they implicitly specify lower and upper limits on population size (the range of densities along the isodar).

5.5 CAVEATS AND FUTURE DIRECTIONS

Habitat selection theories that I have reviewed here specify the expected relationships between population density and reproductive success in ideal landscapes. As such, they represent appropriate null models for landscape ecology. An ecologist searching for landscape-mediated effects on population density may wish to begin the search with an isodar analysis (Morris, 1994).

My enthusiasm for application of habitat selection theory to the landscape scale is tempered by the complexity of the patterns we wish to explain. Even the most ardent advocate of isodar analysis will surely recognize its limitations at differentiating certain kinds of processes and their interactions simply by examining patterns of population density. Application of the theory also requires that reliable estimates of population density be obtained at the spatial and temporal scales appropriate to habitat selection. It could frequently be misleading, for example, to use annual or single-season estimates of density when habitat preferences vary seasonally (Van Horne, 1983). Similar biases would occur whenever density estimates are influenced by landscape processes that are not directly related to density-dependent habitat selection (e.g. passive dispersal, local extinction and recolonization).

5.5.1 SOURCE–SINK DYNAMICS

Source–sink dynamics, where average reproductive success is greater in one habitat than in another, occurs only when habitat choice follows something other than an ideal free distribution (Oksanen, Oksanen and Fretwell, 1992). Pre-emptive (Pulliam, 1988; Pulliam and Danielson, 1991) and despotic models (Fretwell and Lucas, 1970) assume a habitat selection process whereby per capita population growth rates between pairs of habitats are unlikely to be equal, and where high-quality patches may function as sources to lower-quality sinks. The stable source–sink dynamics created by these forms of habitat selection have led to some of our most dramatic insights into the role of landscape heterogeneity on species interactions and ecosystem structure (Oksanen, 1990; Pulliam and Danielson, 1991; Danielson, 1991, 1992; Dunning, Danielson and Pulliam, 1992; Oksanen, Oksanen and Gyllenberg, 1992).

The population result of source–sink dynamics, compared to an ideal free distribution, is reduced density in high-quality habitats, and inflated densities elsewhere. One way of modeling this effect for an ideal despotic distribution is to rotate the fitness–density curves of the best habitat clockwise to represent the individual's perception of reduced fitness with interference (Fretwell and Lucas, 1970; Fretwell, 1972; Morris, 1987a). This has, depending upon one's viewpoint, the desirable or undesirable effect of reducing the isodar intercept (Figure 5.9). The two habitats would appear less different quantitatively, a result of inflation in population density in the sink habitat. At any given density, aggression increases the expected reproductive success of individuals relative to the expectations from ideal free habitat use. Aggression is not without cost. The increased fitness accrued by individuals occurs at the expense of an overall reduction in population density in each habitat, and thereby in overall population size (Figure 5.9).

The desirable aspect of a reduced isodar intercept with despotic behavior is that one can, in theory, differentiate between ideal free and despotic source–sink regulation, and measure their relative magnitudes by analyzing only graphs of population density. The design of the study is crucial because other effects can also modify the intercept of the isodar (Morris, 1987a, 1988, 1990, 1992). The undesirable effect is that it may often be impossible to detect the isodar shift unless we know, or can experimentally manipulate, the fitness–density functions. This latter point would be especially crucial in those instances where a large proportion of the population is forced into sink habitat producing population densities larger than those in the source (Wiens, 1989).

Fitness curves for ideal pre-emptive distributions are easily modeled by cumulative frequency distributions of breeding-site quality (Morris, 1994). The resulting isodars have a characteristic curvilinear or nonlinear signature. We do not know how applicable these and other curvilinear models may be because data analyzed so far give a reasonable fit to the linear model.

5.5.2 QUESTIONS FOR FUTURE STUDY

Few of the assumptions and predictions of isodar analysis have been tested experimentally. Can isodars detect the qualitative and quantitative differences in habitat that the theory suggests? Are competition coefficients estimated by isodar analysis valid indicators of competitive interaction? Are the cues that individuals use to assess habitats reasonable estimates of expected fitness?

The application of habitat models to landscape predictions raises new questions. What is the correspondence, if any, between the effective spatial and temporal limits of habitat selection and the patterning of habitat patches in the landscape? What is the interaction of dispersal between habitat patches and the dynamics of metapopulations (Chapter 4)? How important is density-dependent habitat selection to population persistence in heterogeneous landscapes? How does this role vary with landscape composition and pattern?

Many other unexplored effects offer fertile ground for habitat ecologists. Among these are interactions between landscape patterns such as the interspersion and orientation of patches, the relative proportions of patches of varying quality, and the nature of patch shapes and boundaries with habitat characteristics such as the variance in habitat quality within and between patches, with variation in the form of density-dependent feedback on fitness, and with the resistance of habitats to animal movement. The question is not whether such interactions occur, but whether

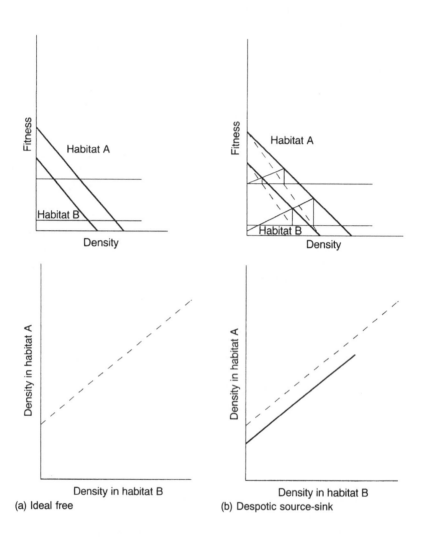

Figure 5.9 Comparison of (a) ideal free and (b) despotic source–sink regulation and their resultant isodars. Habitat B is of lower quality than habitat A. With source–sink regulation, the perceived fitness in habitat A is equal to that in habitat B (represented by dashed lines with negative slope). Unequal opportunities at reproduction inflate population densities in sink habitats (B) relative to the ideal free solution (intercepts of the two sets of solid lines). This is equivalent to reduced fitness–density functions that produce isodars with reduced intercepts (greater density in habitat B; compare the solid source–sink isodar with the dashed ideal free isodar). As shown here, the per capita decline in perceived fitness with despotic behavior is equivalent in the two habitats. If the assumption is violated the despotic isodar will have a different slope to the ideal free one.

their occurrence corroborates or invalidates the inferences we try to make from models of habitat selection.

5.6 SUMMARY

Density-dependent habitat selection is one of the key mechanisms capable of modifying the distribution and abundance of species at the landscape scale. Recent extensions of the theory based on isodars (plots of the density of individuals in pairs of habitats such that an individual's expected reproductive success is equal in both) demonstrate how we can measure habitat selection's role in spatial population regulation, and how differences between habitats can modify overall population size. One intriguing application of the theory uses patterns of density across habitat boundaries to estimate foraging and dispersal scales of habitat selection, and thereby the effective bounds of the landscape to different species. Other extensions allow us to follow the time course of community recovery following habitat disturbance, and to infer interactions between species in ecological communities. Under somewhat restricted conditions of an ideal free distribution it is possible to test for a perfect match between habitat quality and population density with only data on population density.

Observational studies generally support the theory, but definitive experiments testing assumptions at the landscape scale are lacking. The utility of habitat selection models to infer landscape processes and patterns may be compromised by nonequilibrium dynamics, by spatial differences in the quality of habitat patches, by source–sink dynamics, and by complicated relationships between fitness and population density. Many of these apparent limitations can be addressed by modified analyses that suggest productive avenues of future research at the interface between habitat selection and landscape ecology.

ACKNOWLEDGEMENTS

I thank Gray Merriam, Lennart Hansson and Lenore Fahrig for providing me this opportunity to discuss the relationships between habitat selection and landscape ecology. My views on both subjects have benefited from numerous discussions with Thomas Knight, who kindly and candidly critiqued an earlier version. An excellent review by Robert Holt helped me improve the paper substantially. I am grateful to Canada's Natural Sciences and Engineering Research Council for continuing support of my research on evolutionary and landscape ecology (Grant No. OGP0116430).

REFERENCES

Abramsky, Z., Rosenzweig, M. L. and Pinshow, B. (1991) The shape of a gerbil isocline measured using principles of optimal habitat selection. *Ecology*, **72**, 329–40.

Abramsky, Z., Rosenzweig, M. L. and Zubach, A. (1992) The shape of a gerbil isocline: an experimental field study. *Oikos*, **63**, 193–9.

Brown, J. S. and Rosenzweig, M. L. (1986) Habitat selection in slowly regenerating environments. *J. Theor. Biol.*, **123**, 151–71.

Capen, E. (ed.) (1981) The Use of Multivariate Statistics in Studies of Wildlife Habitat. *USDA Forest Service General Technical Report RM–87*.

Crowell, K. L. (1983) Islands – insight or artifact? Population dynamics and habitat utilization in insular rodents. *Oikos*, **41**, 442–54.

Danielson, B. J. (1991) Communities in a landscape: the influence of habitat heterogeneity on the interactions between species. *Am. Nat.*, **138**, 1105–20.

Danielson, B. J. (1992) Habitat selection, interspecific interactions and landscape composition. *Evol. Ecol.*, **6**, 399–411.

Dunning, J. B., Danielson, B. J. and Pulliam, H. R. (1992) Ecological processes that affect populations in complex landscapes. *Oikos*, **65**, 169–75.

Fagen, R. (1987) A generalized habitat matching rule. *Evol. Ecol.*, **1**, 5–10.

Fagen, R. (1988) Population effects of habitat change: a quantitative assessment. *J. Wildl. Manage.*, **52**, 41–6.

Fahrig, L. (1992) Relative importance of spatial and temporal scales in a patchy environment. *Theor. Pop. Biol.*, **41**, 300–14.

Fox, B. J. and Morris, D. W. (1990) Temporal changes in mammalian communities. *Oikos*, **59**, 289.

Fretwell, S. D. (1972) *Populations in a Seasonal Environment*, Princeton University Press, Princeton, NJ.

Fretwell, S. D. and Lucas, H. L. Jr (1970) On territoral behavior and other factors influencing habitat distribution in birds. I. Theoretical development. *Acta Biotheor.*, **19**, 16–36.

Hansson, L. and Henttonen, H. (1988) Rodent dynamics as community processes. *Trends Ecol. Evol.*, **3**, 195–200.

Hobbs, N. T. and Hanley, T. A. (1990) Habitat evaluation: do use/availability data reflect carrying capacity? *J. Wildl. Manage.*, **54**, 515–22.

Holbrook, S. J. (1979) Habitat utilization, competitive interactions, and coexistence of three species of cricetine rodents in east-central Arizona. *Ecology*, **60**, 758–69.

Holt, R. D. (1984) Spatial heterogeneity, indirect interactions, and the coexistence of prey species. *Am. Nat.*, **124**, 377–406.

Holt, R. D. (1985) Population dynamics in two-patch environments: some anomalous consequences of an optimal habitat distribution. *Theor. Pop. Biol.*, **28**, 181–208.

Kacelnik, A., Krebs, J. R. and Bernstein, C. (1992) The ideal free distribution and predator–prey populations. *Trends Ecol. Evol.*, **7**, 50–5.

Kareiva, P. (1990) Population dynamics in spatially complex environments: theory and data. *Phil. Trans. R. Soc. London*, **B, 330**, 175–90.

Kennedy, M. and Gray, R. D. (1993) Can ecological theory predict the distribution of foraging animals? A critical analysis of experiments on the Ideal Free Distribution. *Oikos*, **68**, 158–66.

Knight, T. W. (1993) Spatial scaling in northern landscapes: habitat selection by small mammals. MSc Thesis, Lakehead University

Kolasa, J. and Rollo, C. D. (1991) Introduction: the heterogeneity of heterogeneity: a glossary, in *Ecological Heterogeniety*, (eds J. Kolasa and S. T. A. Pickett), Springer-Verlag, New York, pp. 1–23.

Kotliar, N. B. and Wiens, J. A. (1990) Multiple scales of patchiness and patch structure: a hierarchical framework for the study of heterogeneity. *Oikos*, **59**, 253–60.

Krebs, J. R. (1971) Territory and breeding density in the great tit, *Parus major* L. *Ecology*, **52**, 2–22.

Levin, S. A. (1992) The problem of pattern and scale in ecology. *Ecology*, **73**, 1943–67.

Lomnicki, A. (1988) *Population Ecology of Individuals*, Princeton University Press, Princeton, NJ.

MacArthur, R. H. (1972) *Geographical Ecology*, Harper and Row, New York.

MacArthur, R. H. and Levins, R. (1964) Competition, habitat selection, and character displacement in a patchy environment. *Proc. Nat. Acad. Sci. USA*, **51**, 1207–10.

Maurer, B. A. (1986) Predicting habitat quality for grassland birds using density–habitat correlations. *J. Wildl. Manage.*, **50** 556–66.

Merriam, G., Henein, K. and Stuart-Smith, K. (1991) Landscape dynamics models, in *Quantitative Methods in Landscape Ecology* (eds M. G. Turner and R. H. Gardner), Springer-Verlag, New York, pp. 399–416.

Messier, F., Virgl, J. A. and Marinelli, L. (1990) Density-dependent habitat selection in muskrats: a test of the ideal free distribution model. *Oecologia*, **84**, 380–5.

Milinski, M. and Parker, G. A. (1991) Competition for resources, in *Behavioural Ecology: An Evolutionary Approach*, 3rd edn (eds J. R. Krebs and N. B. Davies), Blackwell, Oxford, pp. 137–68.

Morris, D. W. (1984) Patterns and scale of habitat use in two temperate-zone small mammal faunas. *Can. J. Zool.*, **62**, 1540–7.

Morris, D. W. (1987a) Spatial scale and the cost of density-dependent habitat selection. *Evol. Ecol.*, **1**, 379–88.

Morris, D. W. (1987b) Tests of density-dependent habitat selection in a patchy environment. *Ecol. Monogr.*, **57**, 269–81.

Morris, D. W. (1987c) Ecological scale and habitat use. *Ecology,* **68**, 362–9.

Morris, D. W. (1988) Habitat-dependent population regulation and community structure. *Evol. Ecol.,* **2**, 253–69.

Morris, D. W. (1989a) Density-dependent habitat selection: testing the theory with fitness data. *Evol. Ecol.,* **3**, 80–94.

Morris, D. W. (1989b) Habitat-dependent estimates of competitive interaction. *Oikos,* **55**, 111–20.

Morris, D. W. (1989c) The effect of spatial scale on patterns of habitat use: red-backed voles as an empirical model of local abundance for northern mammals, in *Patterns in the Structure of Mammalian Communities* (eds D. W. Morris, Z. Abramsky, B. J. Fox and M. R. Willig), Special Publications, The Museum, Texas Tech. Univ., Lubbock, pp. 23–32.

Morris, D. W. (1990) Temporal variation, habitat selection and community structure. *Oikos,* **59**, 303–12.

Morris, D. W. (1991) Fitness and patch selection by white-footed mice. *Am. Nat.,* **138**, 701–16.

Morris, D. W. (1992) Scales and costs of habitat selection in heterogeneous landscapes. *Evol. Ecol.,* **6**, 412–32.

Morris, D. W. (1994) Habitat matching: alternatives and implications to populations and communities. *Evol. Ecol,* **8**, 387–406.

Morris, D. W. and Brown, J. S. (1992) The role of habitat selection in landscape ecology. *Evol. Ecol.,* **6**, 357–9.

Oksanen, T. (1990) Exploitation ecosystems in heterogeneous habitat complexes. *Evol. Ecol.,* **4**, 220–34.

Oksanen, T., Oksanen, L. and Gyllenberg, M. (1992) Exploitation ecosystems in heterogeneous habitat complexes II: impact of small-scale heterogeneity on predator–prey dynamics. *Evol. Ecol.,* **6**, 383–98.

Oksanen, T., Oksanen, L. and Fretwell, S. D. (1992) Habitat selection and predator–prey dynamics. *Trends Ecol. Evol.,* **7**, 313.

Pimm, S. L. and Rosenzweig, M. L. (1981) Competitors and habitat use. *Oikos,* **37**, 1–6.

Pulliam, H. R. (1988) Sources, sinks, and population regulation. *Am. Nat.,* **132**, 652–61.

Pulliam, H. R. and Caraco, T. (1984) Living in groups: is there an optimal group size? in *Behavioural Ecology: An Evolutionary Approach,* 2nd edn (eds J. R. Krebs and N. B. Davies), Blackwell, Oxford, pp. 122–47.

Pulliam, H. R. and Danielson, B. J. (1991) Sources, sinks, and habitat selection: a landscape perspective on population dynamics. *Am. Nat.,* **137S**, 50–66.

Rosenzweig, M. L. (1974) On the evolution of habitat selection. Proceedings of the First International Congress of Ecology, pp. 401–4.

Rosenzweig, M. L. (1979) Optimal habitat selection in two-species competitive systems. *Fortschr. Zool.,* **25**, 283–93.

Rosenzweig, M. L. (1981) A theory of habitat selection. *Ecology,* **62**, 327–35.

Rosenzweig. M. L. (1985) Some theoretical aspects of habitat selection, in *Habitat Selection in Birds* (ed. M. L. Cody), Academic Press, Orlando, pp. 517–40.

Rosenzweig, M. L. (1987) Community organization from the point of view of habitat selectors, in *Organization of Communities* (eds J. H. R. Gee and P. S. Giller), Blackwell, Oxford, pp. 469–90.

Rosenzweig, M. L. (1989) Habitat selection, community organization and small mammal studies, in *Patterns in the Structure of Mammalian Communities* (eds D. W. Morris, Z. Abramsky, B. J. Fox and M. R. Willig), Special Publications, The Museum, Texas Tech. Univ., Lubbock, pp. 5–21.

Rosenzweig, M. L. (1991) Habitat selection and population interactions: the search for mechanism, *Am. Nat.*, **137S**, 5–28.

Rossi, R. E., Mulla, D. J., Journel, A. G. and Franz, E. H. (1992) Geostatistical tools for modelling and interpreting ecological spatial dependence. *Ecol. Monogr.*, **62**, 277–314.

Sutherland, W. J. (1983) Aggregation and the 'ideal free' distribution. *J. Anim. Ecol.*, **52**, 821–8.

Turner, M. G. and Gardner, R. H. (1991) Quantitative methods in landscape ecology: an introduction, in *Quantitative Methods in Landscape Ecology* (eds M. G. Turner and R. H. Gardner), Springer-Verlag, New York, pp. 3–14.

Turner, M. G., O'Neill, R. V., Conley, W. *et al.*, (1991) Pattern and scale: statistics for landscape ecology, in *Quantitative Methods in Landscape Ecology* (eds M. G. Turner and R. H. Gardner), Springer-Verlag, New York, pp. 17–49.

Van Horne, B. (1983) Density as a misleading indicator of habitat quality. *J. Wildl. Manage.*, **47**, 893–901.

Van Horne, B. (1986) Summary: when habitats fail as predictors – the researcher's viewpoint, in *Wildlife 2000: Modelling Habitat Relations of Terrestrial Vertebrates* (eds J. Verner, M. I. Morrison and C. J. Ralph), University of Wisconsin Press, Madison, pp. 257–8.

Verner, J., Morrison, M. L. and Ralph, C. J. (1986) *Wildlife 2000: Modelling Habitat Relationships of Terrestrial Vertebrates*, University of Wisconsin Press, Madison.

Whitham, T. G. (1978) Habitat selection by *Pemphigus* aphids in response to resource limitation and competition. *Ecology*, **59**, 1164–76.

Whitham, T. G. (1980) The theory of habitat selection: examined and extended using *Pemphigus* aphids. *Am. Nat.*, **115**, 449–66.

Wiens, J. A. (1989) *The Ecology of Bird Communities I: Foundations and Patterns*, Cambridge University Press, Cambridge.

Resource tracking in space and time

6

Michał Kozakiewicz

6.1 THE HABITAT CONCEPT – A CONCEPTUAL LINK BETWEEN ORGANISMS AND THEIR ENVIRONMENT

The concept of habitat is the conceptual link relating organisms to their external environment. Some ecologists view habitat as a principal force underlying the evolution of species (e.g. Southwood, 1977). According to Harris and Kangas (1988), there are several views of the concept of habitat and many attempts have been made to articulate a definition (e.g. Uvardy, 1959; Maguire, 1973; Whittaker, Levin and Boot, 1973, and many others).

Some authors relate habitat to a single individual and define it as a place that provides its life needs (food, cover, water, space and mates). Southwood (1976) links habitat to the range of individuals' movements:

> For any particular animal, the habitat may be defined as the area accessible to the trivial movements of the food-harvesting stages. The range of the animal's movements will therefore determine the scale of the habitat; for a *Drosophila* larva the ripe fruit of a tropical forest tree is a temporary habitat and if it survives to adulthood it will migrate to another site, but for the orang-utan the whole forest is a stable and permanent habitat.
>
> (Southwood, 1976)

Contrary to the above, some authors relate habitat to a population or species. For example, Krebs (1985) defines habitat as '. . . any part of the earth where the species can live, either temporarily or permanently'. This definition of habitat extends far beyond the requirements recognized for

Mosaic Landscapes and Ecological Processes.
Edited by Lennart Hansson, Lenore Fahrig and Gray Merriam.
Published in 1995 by Chapman & Hall, London, ISBN 0 412 45460 2

a single individual. Habitat defined in such a way must fill all requirements of the species in question and include areas sufficient at least to
maintain a viable population.

Lomnicki (1988) defines habitat as 'a part of space that is permanent
enough and large enough that the population of the species in question
may persist within its boundaries for at least one season, or for the
duration of its entire life stage' (Lomnicki, 1988).

Both spatial and temporal scales of habitat are species-specific. Consequently, landscape heterogeneity must be measured on a wide range of
spatio-temporal scales, since individuals of each species perceive it at
their own scales (e.g. Begon, Harper and Townsend, 1986). Here I will
concentrate on what Forman and Godron (1986) call the landscape scale,
i.e. on a scale measured in square kilometers. Therefore, I will restrict
the range of species discussed to those that perceive the habitat heterogeneity at that scale, such as some mammals and birds. For these species
a landscape element (as, for instance, a hedgerow, woodland or crop
field in agricultural areas) can be either a temporary or a permanent
habitat. I will concentrate on individual and population levels of biological organization.

Habitat patches may differ from each other with respect to their suitability for a given species, their durability (or persistence) and changeability over time. If all life requirements of the species are fully identified,
habitat types within a mosaic could be classified as optimal, suboptimal,
marginal or non-inhabitable.

Defining a measure of suitability of habitats is a very important point.
Usually, reproductive success of the species is considered to be a good
measure of habitat suitability: simply, 'females produce more young in
more suitable habitats than they do in less suitable habitats' (Krebs,
1985). However, the population may occupy a given habitat only temporarily and may or may not reproduce there. Thus, reproductive success
itself cannot be a sufficient measure of habitat suitability. Habitat suitability can be affected by many factors, such as food and water supply,
shelter, nesting sites and predators. It might be expected that habitat
suitability should be reflected by population density. However, many
authors have argued strongly that habitat quality cannot be assessed
through population density alone (e.g. Van Horne, 1983), particularly in
heterogeneous landscapes. In source–sink scenarios, most of the population can be found in the sink habitat, but be dependent on dispersers
from a small amount of the source habitat (Pulliam, 1988).

6.2 HABITAT HETEROGENEITY AND PERSISTENCE OF
POPULATIONS

As is pointed out by Wiens (1985; Chapter 1), natural systems are never
homogeneous – they are mosaics of patches at any scale of resolution.

Thus, most species live in heterogeneous environments composed of patches of different degrees of suitability. To live in a given environment, organisms must be able to survive, grow and reproduce successfully. Thus, only areas of sufficiently favorable conditions to make population growth possible can maintain viable populations; and the spatial distribution of suitable habitat patches can determine the distribution of species (e.g. Dunning, Danielson and Pulliam, 1992). In addition to variation in the spatial arrangement of different patches, the resource availability of each of them may have a temporal sequence driven by many factors (e.g. plant growth, seasonal climate and influence of surroundings). On the other hand, species' requirements may also have their own temporal sequences. Thus, for each species, the temporal distribution of activity in space should reflect the interactions between the temporal dynamics of the species' needs and the spatio-temporal dynamics of resources to fill these needs.

Mosaic landscapes consisting of relatively small and changeable patches offer various species a multipartite, wide array of resources. Different organisms and populations can therefore exhibit a diversity of patch use strategies. Some organisms can spend their entire lives or generations within single patches of habitat, while others may use different patches for different purposes (e.g. some patches for reproduction and others for foraging), and still others may wander over a landscape searching for patches to satisfy their particular needs.

Spatial heterogeneity of landscapes – differences in quality of habitat patches and their different positions across the landscape – may affect the distribution of species and persistence of populations. Den Boer (1968) demonstrated with a stochastic mathematical model that in heterogeneous and variable environments the chance of population survival may be increased, because the risk of extinction of the whole population can be spread throughout a heterogeneous space. Den Boer maintains that:

> ... the fluctuations of animal numbers in the population as a whole will be a resultant of the numerical fluctuations in the different places (subpopulations) ... Migration between subpopulations will generally contribute to the stabilizing tendency of spatial heterogeneity, since in this way extreme effects of some places will be levelled out more thoroughly. Hence, migration will improve the outcome of spreading of the risk in space ...

In this chapter both annual and lifetime variation in habitat use as adaptations to landscape heterogeneity will be discussed. Different strategies for survival in heterogeneous landscapes will be presented as well.

6.3 RANGING THROUGH MORE THAN ONE HABITAT TYPE

6.3.1 MULTIHABITAT INDIVIDUALS AND SPECIES

Some animals are restricted to one habitat type while others use more than one habitat type. Thomas *et al.* (1979) proposed a quantitative index, called the 'versatility index', to describe the breadth of different habitats that species use. Single-habitat species, referred to as 'alpha species' by Harris and Kangas (1988), usually occur low in the trophic pyramid (Elton, 1966). In contrast to these, species that move long distances and are not restricted to one vegetation type but depend rather on a regional landscape, are referred to by Harris and Kangas (1988) as 'gamma species'. These species usually occur high in the trophic pyramid (Elton, 1966).

Covering more than one habitat type with routine, daily movements seems to be advantageous for many species since this kind of spatial activity creates an opportunity to utilize different resources in different habitat patches. Goszczyński (1985) studied space use patterns by various forest predatory mammal and bird species in central Poland in a mosaic of woods and small fields. He distinguished two types of spatial distribution of forest predators. In the first type, predator numbers are directly dependent upon forest area. This type of spatial distribution is typical for forest specialist species, such as the pine marten (*Martes martes*). Spatial activity of this species is, in large part, restricted to the interior forest area (Figure 6.1).

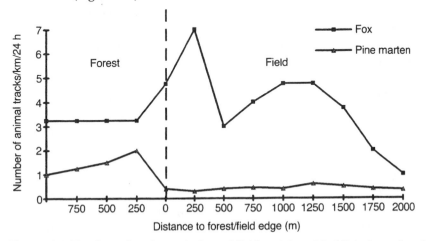

Figure 6.1 The intensity of penetration of field and forest habitats by a forest specialist species (pine marten) and a generalist species (fox). (Source: Goszczyński, 1985.)

For the second, more common type of forest predators, predator numbers depend on the length of forest/field edge. This is typical for species that use both forest and field habitats, such as foxes (*Vulpes vulpes*), badgers (*Meles meles*) and common buzzards (*Buteo buteo*). These species use fields for foraging, while having shelter and nesting places inside the forest (Figure 6.1). The proportion of field and forest species in the predators' diet shows the intensity of use of various habitats by different predators (Table 6.1).

Table 6.1 The ratios of bank voles (*Clethrionomys glareolus*) (a forest rodent species) to common vole (*Microtus arvalis*) (a field rodent species) in the diet of various predators. (Source: Goszczyński, 1985)

Species	Spring–Summer	Autumn–Winter
Buzzard (*Buteo buteo*)	0.215	0.018
Fox (*Vulpes vulpes*)	0.134	0.062
Pine marten (*Martes martes*)	1.240	1.037

Banach (1988) studied a bank vole (*Clethrionomys glareolus*) population inhabiting a mosaic of several different adjacent forest fragments: pine wood, alderwood and willow brushwood. The population dynamics of voles inhabiting areas adjoining the borders of forest fragments within a mosaic and, therefore, exploring more than one vegetation type (called 'multihabitat' voles here) were different from those inhabiting separate forest fragments. The density of multihabitat voles in a year-round cycle was almost constant, while the density of voles in the separate fragments showed clear seasonal dynamics. Multihabitat voles were sexually active adults of high body weight and high trappability. The mean size of multihabitat voles' home ranges was large (Table 6.2). Based on these characteristics, Banach (1988) suggested multihabitat voles to be social dominants in the population.

Table 6.2 Mean size of home ranges (m²) of voles inhabiting different forest habitats. (Source: Banach, 1988)

Alderwood voles	Pine forest voles	Willow brushwood voles	Multihabitat voles
2195	1706	2150	10 136*
n = 66	*n* = 40	*n* = 27	*n* = 29

* Value significantly higher than in any other habitat type (t-test $0.001 < p < 0.05$).
n, number of animals examined.

For some species the strategy of enlarging the range of individual movements and, in this way, covering several different patches of habitat with daily activity movements appears to be a good way to satisfy all life requirements in heterogeneous landscapes.

6.3.2 SELECTION OF HABITATS AND SEASONALITY IN HABITAT USE

Fretwell (1972) pointed out the possibility that certain categories of individuals are restricted to poor habitat types. Bock (1972) found that bank voles prefer *Salici–Franguletum* and *Circaeo–Alnetum* forest types, and that older, dominant adults occupy preferred habitats in the largest numbers.

Gliwicz (1989) studied dispersal dynamics in populations of bank voles inhabiting a rich alder forest and two small patches of wood (each of them less than 0.5 ha in extent). She found that young individuals born in the alder wood (optimal habitat) disperse soon after weaning into less crowded suboptimal habitats (the small patches). Young individuals (dispersers), after establishing their residence in empty suboptimal habitat patches, have a high probability of maturing and reproducing in the year of their birth (Table 6.3). Thus, the probability of maturing in the year of birth was highest in the suboptimal habitats for young dispersers. Young born later in the season did not disperse possibly because later in the season all suitable patches become 'saturated' and there was no place for dispersers to establish a new residence and start to breed. Gliwicz (1989) therefore showed that whether dispersal is advantageous or not may depend both on suitability of habitats within a mosaic and seasonal changes of population density.

Table 6.3 Probability of young males (M) and females (F) of maturing in the year of birth in optimal and suboptimal habitats. (Source: Gliwicz, 1989)

Habitat		Cohorts		
		Y_1 born in April and early May	Y_2 born in late May and June	$Y_{3, 4}$ born between July and September
Alder forest	F	(1.00)*	0.33	0.00
(optimal)	M	(1.00)	0.15	0.00
Woodlots	F	1.00	0.60	0.00
(suboptimal)	M	0.79	0.35	0.00

[a] Only 22% of Y_1 young remained in the alder forest habitat, but all of them were mature.

Both resource availability in various patches of habitat and species requirements may have their own temporal sequences. Most habitats are not permanently of the same quality for a given species; they are usually better or worse temporarily. Thus, it might be supposed that many animal species follow these temporal changes, selecting temporarily different patches of habitat. It seems, therefore, that finding the best quality patches of habitat that can satisfy a species' actual needs, might be of high importance for survival in a mosaic landscape.

Kozakiewicz and Merriam (unpublished) studied the spatial distribution and seasonal dynamics of species in spider assemblages in a mosaic of meadows, fencerows and woodlots in a farm landscape near Ottawa, Canada. Samples were taken twice a year: in late spring and in early autumn. The habitat types offered different living conditions for spiders, due to different plant growth-forms, organic litter, soil humidity, availability and diversity of shelters and probably numerous other characteristics. Species caught in more than one habitat type, at either time, constituted as much as 40% of the total species pool and they made up almost 90 % of the total number of individuals (Table 6.4).

Table 6.4 Spider species present in one and in more than one habitat type of the farmland mosaic. (Source: Kozakiewicz and Merriam, unpublished)

	Present in one habitat type	Present in two habitat types	Present in more than three habitat types	Total
Number of species		17	12	
	40		29	69
		25	17	
%	58		42	100
Number of individuals	82		580	662
%	12		88	100

Individuals of such multihabitat species probably can move, at least seasonally, between different habitat patches, selecting them according to temporal changes in their quality. It might be supposed, therefore, that the more heterogeneous the habitat mosaic, the higher the expected proportion of generalist species and the lower the proportion of habitat specialists for which preferred habitats will be more scattered.

This is supported by the findings of Kozakiewicz, Kozakiewicz and Choszczewska (1992), who studied seasonal dynamics and spatial distribution of small mammals in a lake-shore habitat surrounded by a mosaic farm landscape in north-eastern Poland. Small mammals were snap-trapped four times yearly (spring, summer, autumn and winter) in six localities situated in different parts of the lake-shore zone representing different types of vegetation cover. In order to not disturb the studied populations too much, trapping plots were relatively small (1 ha in extent each) and trapping sessions were relatively short (three days each). It was found that both density and species composition of small mammals changed clearly from season to season. The studied assemblage of small mammals showed a low number of permanent species compared to the number of temporary visitors. There were only a few species present in a given site all year round. The species dominating numerically in the total catch – striped field mouse (*Apodemus agrarius*) – was caught in all seasons at only one site, while it was only seasonally present in other trapping sites. Bank voles, the second most abundant species, were caught in all seasons at four of the six trapping sites.

6.4 KEY HABITATS

In some seasons or in particular stages of a life cycle, the species' needs can be very special, and filling these needs can be of crucial importance for population survival. Habitats in which such needs can be met may be considered 'key habitats' for population persistence. For example, many amphibian species need bodies of water or wet habitats to breed and numerous bird species have very special requirements for nesting habitats.

Marine turtles described by Harris and Kangas (1988) must find nesting beaches in which to lay eggs, despite the fact that 99.99% of their lifetime is spent in the marine environment. Thus, nesting beaches can be considered a key habitat for marine turtles since beaches are absolutely essential for species persistence. Caddis fly larvae in freshwater streams construct their cases of leaves, sticks or sand grains. Thus, particular species can live only in places where the necessary materials can be found (Cummins, 1964).

Jedraszko-Dabrowska (1991) studied the nest structure of the great reed warbler (*Acrocephalus arundinaceus*) and reed warbler (*A. scirpaceus*). Ecological requirements for nest construction by these two bird species are very specific. In habitats not containing reeds of suitable density and quality, nest construction will not support successful breeding because a high percentage of nests is destroyed by wind.

For small mammals, especially in temperate and subarctic zones, winter habitats can play a key role for population survival (e.g. Kalela *et al.*, 1971; Kaikusalo, 1972; Tast and Kaikusalo, 1976). Henderson, Merriam and Wegner (1985) recorded local winter extinctions of chipmunks (*Tamias striatus*) in almost one-third of woodlots being studied in farmland mosaic near Ottawa. Canada.

In mosaic landscapes, patches of such key habitats can be spatially scattered, with better chances for highly mobile animals to find them. Many less mobile species may become extinct in such landscapes. The processes that occur when individuals move between patches in the landscape to make use of substitutable and nonsubstitutable resources are called landscape supplementation and landscape complementation, repsectively, by Dunning, Danielson and Pulliam (1992).

6.5 STRATEGIES FOR SURVIVAL IN HETEROGENEOUS LANDSCAPES

Spatial heterogeneity can be measured in millimeters as well as in hundreds of kilometers, and temporal changeability can be measured in hours or in seasons of a year, or in geological time. The scale of spatio-temporal changeability considered for a particular environment should be appropriate for the species in question: its generation time and move-ment range of individuals.

Natural systems at any scale of resolution are mosaics of patches. These mosaics are never static; their elements are in constant temporal and spatial flux. For many species a single small and changeable patch of (temporarily suitable) habitat cannot satisfy all live requirements and cannot support the existence of a stable and viable population. At least two different strategies for survival can be distinguished: (1) the strategy of high spatial activity ('searching for various resources dissipated in space and changeable in time'); and (2) the strategy of dormancy ('being passive and waiting for better conditions').

The first strategy can be based on two different tactics concerning space use patterns: (a) active selection of the best patches of habitats relative to the species' needs; and (b) enlarging as much as possible the range of individual movements to cover with routine daily activity more than one habitat patch. The first of the two tactics requires a high fec-undity of animals and a short generation time. It also requires a high level of dispersal in a population and the ability of dispersers to travel a relatively long distance in a short time. Dispersers can, therefore, travel across the landscape searching for the best habitat patches to occupy, establish temporal residence and breed successfully there.

Many small mammal species seem to exhibit such high spatial activity. Kozakiewicz *et al.* (1993) studied movement of bank voles in homo-geneous and in heterogeneous environments by using marked bait. The mean distance between the bait station and the trapping point was almost twice as long in heterogeneous as in homogeneous habitats, i.e. 243 (SE =133) m and 135 (SE = 116) m, respectively (t = 3.66, df = 64, p < 0.05). Extremely long distances traveled by striped field mice and bank voles were also recorded in very heterogeneous suburban mosaics by Liro and Szacki (1987) and Szacki and Liro (1991). Also, Merriam (1990) and Wegner and Merriam (1990) give evidence of extremely long distances moved by white-footed mice (*Peromyscus leucopus*) in a mosaic farm landscape.

These data suggest that increased movement range may be a common behavioral response of many small mammal species to increased habitat heterogeneity. According to Kozakiewicz *et al.* (1993), in spring, bank voles disperse from patches of overwintering habitats, colonize empty patches and start to breed there. During the reproductive season animals move frequently between occupied patches; a large portion of animals use more than one habitat patch. In autumn a large number of highly mobile animals choose the best habitats for wintering, thus decreasing the probability of local winter mortality.

The second possible way to satisfy all life requirements in a heterogeneous landscape is to enlarge the range of routine daily individual movements. Foxes, which look for their prey in open areas (fields and meadows), but have their shelters and nesting sites inside forest fragments are a good example of such spatial activity (Goszczyński, 1985). The strategy of high spatial activity might be especially advantageous in heterogeneous environments characterized by non-predictable and rapid (or even catastrophic) changeability, such as human-induced landscapes.

The strategy of dormancy may be more common in less heterogeneous, more stable environments, characterized by predictable (e.g. seasonal) changeability. Species showing this strategy are more likely to be habitat specialists than generalists. According to Wegner and Henein (1991), individuals of eastern chipmunk (*Tamias striatus*) – a woodland specialist species – do not venture into agricultural fields, being strongly restricted to patchily distributed wooded areas in mosaic farm landscapes. They do not travel long distances across the landscape and are inactive throughout the winter. This strategy of 'waiting for better conditions' might be more common in reptiles and amphibians than in mammals. Trees shedding their leaves for wintering or plant seed-banks can be good examples of this strategy.

6.6 SUMMARY

For many animal species the temporal distribution of activity in space reflects the interactions between the temporal dynamics of the species' needs and spatio-temporal dynamics of resources. There are one-habitat species, restricted to a specific habitat type, and multihabitat species that move relatively long distances and depend on entire landscapes. Even for multihabitat species, there are habitat types of crucial significance for species survival, called 'key habitats' here.

Increased mobility of individuals seems to be a common strategy for survival in heterogeneous landscapes. The more heterogeneous the habitat mosaic, the higher the expected proportion of multihabitat or generalist species relative to one-habitat specialists.

ACKNOWLEDGEMENTS

I thank the editors and F. Burel, J. Dunning and E. R. Lindstrom for their constructive criticism.

REFERENCES

Banach, A. (1988) Population of the bank vole in the mosaic of forest biotopes. *Acta Theriol.*, **33**, 87–102.

Begon, M., Harper, J. L. and Townsend, C. R. (1986) *Ecology: Individual, Populations and Communities*, Sinauer, Sunderland, MA.

Bock, E. (1972) Use of forest associations by bank vole population. *Acta Theriol.*, **17**, 203–19.

Boer, P. J. den (1968) Spreading of risk and stabilization of animal numbers. *Acta Biotheor.*, **18**, 165–94.

Cummins, K. W. (1964) Factors limiting the microdistribution of larvae of the caddisflies *Pycnopsyche lepida* (Hagen) and *Pycnopsyche guttifer* (Walker) in a Michigan stream (Trichoptera: Limnephilidae). *Ecol. Monogr.*, **34**, 271–95.

Dunning, J. B., Danielson, B. J. and Pulliam, R. (1992) Ecological processes that affect populations in complex landscapes. *Oikos*, **65**, 169–75.

Elton, Ch. (1966) *The Pattern of Animal Communities*, Methuen, New York.

Forman, R. T. T. and Godron, M. (1986) *Landscape Ecology*, John Wiley, NY, Chichester, Brisbane, Toronto, Singapore.

Fretwell, S. D. (1972) *Populations in a Seasonal Environment*, Princeton University Press, Princeton, New Jersey.

Gliwicz, J. (1989) Individuals and populations of the bank vole in optimal, suboptimal and insular habitats. *J. Anim. Ecol.*, **58**, 237–47.

Goszczyński, J. (1985) *The Effect of Structural Differentiation of Ecological Landscape on the Predator–Prey Interaction*. Publications of Warsaw Agricultural University. Treatises and Monographs 46, pp. 1–80 (in Polish with English summary).

Harris, L. D. and Kangas, P. (1988) Reconsideration of the habitat concept. *Trans. 53rd N. A. Wildl. Nat. Res. Conf.*, 137–44.

Henderson, M. T., Merriam, G. and Wegner, J. (1985) Patchy environments and species survival: chipmunks in an agricultural mosaic. *Biol. Conserv.*, **31**, 95–105.

Jedraszko-Dabrowska, D. (1991) Reeds as construction supporting great reed warbler (*Acrocephalus arundinaceus* L.) and reed warbler (*A. scirpaceus* Herm.) nests. *Ekol. Pol.*, **39**, 229–42.

Kaikusalo, A. (1972) Population turnover and wintering of the bank vole, *Clethrionomys glareolus* (Schreb.) in southern and central Finland. *Ann. Zool. Fennici*, **9**, 219–24.

Kalela, O., Kilpelainen, L., Kopponen, T. and Tast, J. (1971) Seasonal differences in habitats of the Norwegian lemming, *Lemmus lemmus* L. in 1959 and 1960 at Kilpisjarvi, Finnish Lapland. *Ann. Acad. Sci. Fenn. Ser. A*, **178**, 1–22.

Kozakiewicz, A., Kozakiewicz, M. and Choszczewska, B. (1992) Small mammal communities of the nature reserve 'Luknajno Lake' coastal zone. *Parki. Nar. i Rez. Przyr.*, **11**, 121–30.

Kozakiewicz, M., Kozakiewicz, A., Lukowski, A. and Gortat, T. (1993) Use of space by bank voles (*Clethrionomys glareolus*) in a Polish farm landscape. *Landsc. Ecol.*, **8**, 19–24.

Krebs, Ch. J. (1985) *Ecology: The Experimental Analysis of Distribution and Abundance*, Harper & Row, New York.

Liro, A. and Szacki, J. (1987) Movements of field mice *Apodemus agrarius* (Pallas) in a surburban mosaic of habitats. *Oecologia*, **74**, 438–40.

Łomnicki, A. (1988) *Population Ecology of Individuals*, Princeton University Press, Princeton, New Jersey.

Maguire, B. (1973) Niche response and the analytical potentials of its relationship to the habitat. *Am. Nat.*, **107**, 213–46.

Merriam, G. (1990) Ecological processes in the time and space of farmland mosaics, in *Changing Landscape: An Ecological Perspective* (eds I. S. Zonnefeld and R. T. T. Forman), Springer Verlag, New York, Heidelberg, Berlin, pp. 121–33.

Pulliam, H. R. (1988) Sources, sinks and population regulation. *Am. Nat.*, **132**, 652–61.

Southwood, T. R. E. (1976) Bionomic strategies and population parameters, in *Theoretical Ecology. Principles and Applications* (ed. R. M. May), Blackwell, Oxford, pp. 26–48.

Southwood, T. R. E. (1977) Habitat, the templet for ecological strategies? *J. Anim. Ecol.*, **46**, 337–65.

Szacki, J. and Liro, A. (1991) Movements of small mammals in the heterogeneous landscape. *Landsc. Ecol.*, **5**, 219–24.

Tast, J. and Kaikusalo, A. (1976) Winter breeding of the root vole, *Microtus oeconomus* in 1972/1973 at Kilpisjärvi, Finnish Lapland. *Ann. Zool. Fennici*, **13**, 174–8.

Thomas, J., Miller, R., Maser, C. *et al.* (1979) Plant communities and successional stages, in *Wildlife habitats in managed forests – the Blue Mountains of Oregon and Washington*, USDA For. Serv. Agr. Handbook No. 553, pp. 22–39.

Uvardy, M. (1959) Notes on the ecological concepts of habitat, biotope and niche. *Ecology*, **40**, 725–8.

Van Horne, B. (1983) Density as a misleading indicator of habitat quality. *J. Wildl. Manage.*, **47**, 893–901.

Wegner, J. and Heinen, K. (1991) Strategies for survival: white-footed mice and eastern chipmunks in an agricultural landscape. *Proc. IALE World Congress*, Ottawa, Canada, pp. 90.

Wegner, J. and Merriam, G. (1990) Use of spatial elements in a farmland mosaic by a woodland rodent. *Biol. Conserv.*, **54**, 236–76.

Whittaker, R., Levin, S. and Root, R. (1973) Niche, habitat and ecotope. *Am. Nat.*, **107**, 321–38.

Wiens, J. (1985) Vertebrate responses to environmental patchiness, in *The Ecology of Natural Disturbance and Patch Dynamics* (eds S. T. A. Pickett and P. S. White), Academic Press, Orlando, San Diego, pp. 169–96.

Part Three

Landscape Pattern, Population Dynamics and Population Genetics

As discussed in previous chapters, landscape pattern affects the spatial distribution of populations by affecting movement between landscape elements. Landscape pattern also affects extinction and recolonization dynamics, which determine landscape-scale population dynamics. Genetic structure of populations is affected primarily by selection, genetic drift, gene flow and mating. Gene flow and mating are affected by movement through the landscape. Frequency of colonizations and extinctions determines the sizes of local founding populations and the survival time of such populations. These in turn determine the opportunity for genetic drift, and in combination with landscape composition (i.e. differences in patch quality), they also determine the opportunity for selection. Therefore, genetics of populations are inextricably linked to movement of individuals and population dynamics, which are affected by landscape pattern.

Andrew Young provides a current exploration, for plants, of the elements of landscape pattern that may influence genetic processes and genetic structure. Data for wild plant populations provide some interesting contrasts to conventional wisdom. They also offer unusual opportunities to test hypotheses regarding spatial landscape patterning and, for trees, temporal changes in landscape effects that can be recorded in the cohort structure of long-lived, non-moving organisms.

David McCauley reveals some of the major elements of the interactions between landscape spatial pattern and genetic processes in populations. Effects of landscape structure on genetic processes, through effects on landscape-scale population dynamics, have not been well-studied in the empirical literature. McCauley sets out some important elements that will improve the investigation of this long overdue synthetic view of population processes.

Landscape structure and genetic variation in plants: empirical evidence

7

Andrew Young

7.1 INTRODUCTION

Accumulating evidence from studies of a variety of morphological, allozyme and DNA markers indicates that plants generally exhibit high levels of intraspecific genetic variation (see Venable, 1984; Hamrick and Godt, 1989; Clegg, 1990 for reviews). Furthermore, this variation is often not distributed randomly among individuals but, for species with a variety of ecologies, there is evidence of significant genetic structure at various spatial and organizational scales, for example among individuals within populations (e.g. Schmitt and Gamble, 1990), among subpopulations within a region (e.g. Linhart *et al.*, 1981) and among populations across a species' range (e.g. Li and Adams, 1989). Several studies have pointed to correlations between genetic variation and environmental patterns (e.g. Snaydon and Davies, 1976; Warwick and Black, 1986; Xie and Knowles, 1992). Despite this, it is only quite recently that investigations of genetic variation and genetic processes have been more formally incorporated into the synthetic discipline of landscape ecology (e.g. Manicacci *et al.*, 1992; Foré, Hickey and Vankat, 1992). This chapter deals with the empirical evidence for the influence of landscape structure

Mosaic Landscapes and Ecological Processes.
Edited by Lennart Hansson, Lenore Fahrig and Gray Merriam.
Published in 1995 by Chapman & Hall, London. ISBN 0 412 45460 2

on genetic variation in plants and the distribution of this variation: genetic structure. Underlying genetic processes are also discussed.

For the purposes of discussion, a landscape is defined as the physical and biological space within which a species exists. Landscape structure is the type, size and spatial arrangement of environments within this space. Environments are combinations of physical and biological variables (e.g. soil type, altitude, vegetation) that confer similar probabilities of survival, reproduction and dispersal. Landscape structure is not a simple concept. Some landscapes are patchy, with sharp transitions between different environments. A good example of this is seen in the granite rock outcrop environments of south-western Australia, which form tightly defined habitat patches for the tree *Eucalyptus caesia* Benth. and contrast with intervening plateaus which support woodlands, mallee and scrub (Moran and Hopper, 1983). Other landscapes are made up of arrays of intergrading environments with few clear boundaries, for example the pastoral region in central Wales within which Bradshaw (1959) studied patterns of genetic variation in *Agrostis tenuis* Sibth. This landscape exhibited overlapping patterns of altitude, grazing regimes and vegetation. Most landscapes are probably a mixture of these two extremes. Furthermore, landscape structure is not static, but may exhibit significant temporal variation as environments are created, change and disappear under the influence of biotic and abiotic processes with varying temporal dynamics (e.g. succession, crop rotation). Finally, there is no single 'landscape scale'. The relevant scale of observation depends completely on the species under investigation (Merriam and Wegner, 1992) and there may be several levels of relevant structure within a landscape.

Two other terms that require definition are genetic variation and genetic structure. As pointed out by Ledig (1992), in the broadest sense, genetic variation encompasses the spectrum of gene diversity from differences between two individuals of the same species, to differences in species composition among communities and beyond. Here, genetic variation is used in the more limited sense of intraspecific variation and includes single and multilocus variation as well as other types of genetic variation such as differences in gene copy number. Genetic structure is the distribution of this variation at the various possible levels of organization within a species (e.g. individuals, populations). The term 'population' is used in the simple sense of a spatially contiguous group of conspecific individuals.

There are several reasons why an understanding of the effects of landscape structure on genetic variation and structure is important. Since Frankel (1970) pointed out that maintenance of genetic variation is crucial to the maintenance of a species' evolutionary potential, several workers (e.g. Wilcove, 1987; Templeton *et al.*, 1990; Saunders, Hobbs and Margules, 1991) have pointed to possible deleterious effects of current patterns

of ecosystem degradation on intraspecific genetic variation and the implications for long-term conservation. Without a thorough understanding
of relationships between landscape structure and genetic variation and,
more importantly, of the influence of landscape structure on the genetic
processes which generate observed patterns of variation, it is difficult to
predict the actual genetic and evolutionary effects of the types of degradation that are now dominant forces shaping many ecosystems throughout the world (e.g. ecosystem fragmentation). Knowledge of relationships
between landscape structure and genetic structure should also allow
more representative samples of a species' genetic variation to be included
in conservation planning. This is important in order to take account of
locally adaptive variation (Hamrick, 1983) and minimize the genetic
impact of management strategies. Indirectly, this type of information
may also provide guidelines for the maintenance of landscape structures
which provide the range of environments and spatial scales necessary
for the 'normal' operation of processes such as selection and gene flow.
Knowledge of the genetic effects of landscape structure also has considerable application in understanding processes of speciation (Levin, 1993).
Finally, information on the interaction between landscape structure and
gene flow is essential for risk assessment in field trials of transgenic
plants, where gene flow out of experimental populations and hybridiz-
ation with natural relatives is undesirable, and in the management of
cultivated populations, such as seed orchards, where pollen contami-
nation from natural populations may counter genetic gain from selective
breeding.

The objective of this chapter is to provide an overview of current
empirical knowledge regarding the effects of landscape structure on
genetic variation and structure in plants. The first section provides a brief
theoretical consideration of mechanisms whereby landscape structure can
affect genetic variation and structure. Results of empirical studies are
then presented and discussed with two main questions being addressed:

1. Does landscape structure affect genetic variation and structure in
    plants?
2. What are the genetic processes involved?

Finally some areas for further research are identified and some possible
empirical approaches are suggested.

7.2 HOW CAN LANDSCAPE STRUCTURE AFFECT GENETIC VARIATION?

The amount and distribution of genetic variation within plant species is
primarily determined by five interacting processes: mutation, random
genetic drift, selection, gene flow and reproduction, though it is now clear

that some parts of the genome (highly repeated noncoding sequences) are also affected by another class of less well understood processes which have been termed 'genomic turnover mechanisms' (Amos and Hoelzel, 1992). Any influence that landscape structure may have on genetic variation and structure must be through these processes.

Differences among environments within a landscape may affect genetic variation in several ways. Perhaps most obviously, environmental variation may exert an influence on genetic variation through differences in the intensity and direction of selection among the constituent environments of a landscape. The effects of selection on genetic variation, given environmental heterogeneity, have been the subject of extensive theoretical investigation (for a review, see Pamilo, 1988). For single locus variation, with different homozygous genotypes favored in different environments, models indicate that the maintenance of genetic polymorphisms depends on a range of factors. These include:

1. the magnitude of differences in selection coefficients among environments;
2. the relative abundances of different environments;
3. the amount of interpopulation gene flow;
4. whether heterozygotes exhibit an intermediate level of fitness with 'dominance switching', that is their fitness is always closer to that of the favored homozygote in a given environment

(Gillespie, 1978; Maynard-Smith and Hoekstra, 1980; Hoekstra, Bijlsma and Dolman, 1985; Hedrick, 1986; Pamilo, 1988). Models for quantitative variation are less well developed. Selection may act on quantitative traits in similar ways to single locus traits. However, theoretical work by Via and Lande (1985) shows that response to selection will be complicated if there are significant genotype–environment interactions for the traits being considered. In this situation selection may act independently on character states of a single trait which exhibits variable expression in different environments. Given this, the degree of genetic correlation among environmentally induced character states becomes important in determining the amount of genetic variation maintained under selection in the presence of environments which constitute different selection regimes (Via and Lande, 1987).

It is also possible that environment may directly affect mutation. It has generally been accepted that mutations occur randomly and irrespective of usefulness. However, recent data from experiments on various strains of the bacteria *Escherichia coli*, grown in stressed environments, suggest that advantageous mutations may occur more commonly than can be accounted for by a random model. This has been interpreted to indicate direct effects of environment on mutation direction and possibly rate (Cairns, Overbaugh and Miller, 1988; Hall, 1990). Several different mechanisms have been proposed to account for this phenomenon (for a review, see Hall, 1990), ranging from reverse transcription of mutant messenger RNAs (mRNAs) coding for advantageous protein sequences, to a hypermutable-state model in which stress induces a state within an individual in which mutation is more likely to occur and favorable mutations result in reversion to the normal state and subsequent growth. To date, evidence for this phenomenon is restricted to prokaryotes and yeast (Lenski and Mittler, 1993). However, a role for such directed mutation as a link between environment and genetic variation in higher eukaryotes cannot be completely discounted, although mechanisms by which it would occur are still not at all clear.

The spatial arrangement of environments within a landscape can influence genetic variation and structure by affecting the amount and composition of gene flow among populations. For wind/gravity dispersed seed and pollen, rates of gene flow may be simple functions of distance between occupied environments, possibly incorporating some directional effects such as prevailing wind direction. However, for plants with animal-mediated dispersal, interactions between dispersal agents and landscape structure may also affect gene flow, in which case the type and distribution of other environments will become more important. Theoretical and modeling treatments of genetic processes in subdivided populations (e.g. Wright, 1978; Gilpin, 1991) suggest that the amount of interpopulation gene flow exerts an important influence on genetic variation within populations through its interaction with population size in determining the balance between loss of variation through genetic drift and the generation of variation by mutation. It also affects the degree to which both of these processes and disruptive selection lead to population differentiation. The magnitude of these influences has been investigated for both infinite and finite population models by Varvio, Chakraborty and Nei (1986). If there is genetic variation among propagules in characteristics affecting dispersal ability (e.g. weight, size, shape), then landscape structure may result in gene flow becoming a selection event, such that pollen and seed that are successfully dispersed do not represent a random sample of the genetic variation of the originating populations.

The size of environments within a landscape may affect the size of populations which occupy them through imposition of upper population limits. As noted above, population size interacts with gene flow to influence genetic variation through its effect on the mutation/genetic drift balance. Depending upon species' mating systems, population size and gene flow can also influence the distribution of genetic variation among individuals within populations through their effects on levels of inbreeding.

Finally, temporal variation in landscape structure may affect the amount and distribution of genetic variation. Temporal changes in environment at a single location that result in variation in selection direction and/or intensity can have similar effects on genetic variation as differences among spatially separated environments. As discussed by Mather (1955), the actual effect will depend on the magnitude of the selection differential among time periods and the duration of different selection regimes relative to generation time. For a one-locus two-allele system exhibiting dominance, and an infinite population model with no mutation or genetic drift, Haldane and Jayakar (1963) have demonstrated that the relationship between the arithmetic and geometric mean fitness of the recessive homozygotes may also be an important factor determining the genetic effect of temporally varying selection.

Extinction and colonization dynamics generated by temporal variation in landscape structure can also influence genetic variation and differentiation among populations (Chapter 8). Such dynamics might result from periodic and drastic modifications of environments, such as those imposed by agricultural practices, for example rotational cropping. Theoretical treatments by Slatkin (1977) suggest that while founder effects at colonization are important in generating genetic differentiation among populations, colonizations may also effectively result in interpopulation gene flow, resulting in reduced among-population differentiation. Extensions of Slatkin's models by Wade and McCauley (1988) (Chapter 8) indicate that the overall effect of extinction and colonization will depend critically on two factors:

1. the origin of colonists: if colonists are drawn from only a single population, founder effects are severe and differentiation is always increased; however, when colonists are drawn from the pool of all populations within a landscape, the effect of colonization and extinction depends on 2 below.
2. number of colonists compared to the amount of gene flow: if gene flow is low, colonization and extinction may actually result in reduced differentiation among populations.

Theoretically then, there are a wide variety of processes that may provide links between landscape structure and genetic variation and structure for plants. Which, if any of these, are actually important processes for a particular species is dependent on the structure of the landscape under consideration and the ecological traits of that species which, as shown by Hamrick and Godt (1989), in themselves exert considerable influence on the amount and distribution of intraspecific genetic variation.

7.3 EMPIRICAL EVIDENCE FOR THE EFFECTS OF LANDSCAPE STRUCTURE ON GENETIC VARIATION

There is nothing particularly novel about the processes considered above. All of the ideas discussed have permeated the literature, several to such an extent that they have largely become accepted as real processes rather than as theoretical possibilities. In this section the results of a variety of empirical studies are presented and two main questions are addressed:

1. Does landscape structure affect genetic variation and structure in plants?
2. What are the genetic processes involved?

7.3.1 ENVIRONMENTAL VARIATION

Evidence for the influence of environmental variation on genetic variation comes from a wide variety of studies. By far the most common process considered is selection. At the simplest level, genetic structure observed at various scales, both within and among natural populations of plants, has been ascribed a posteriori to differences in selection regimes among different environments on the basis of general correlation between environmental and genetic patterns, e.g. Cuguen *et al.* (1985). Though obviously such *post hoc* pattern analyses do not provide explicit tests of the role of selection, if environments are well defined, correlations are high and realistic mechanisms by which selection could operate on observed variation exist, such arguments are not intuitively unreasonable. For example, Ledig and Korbobo (1983) have suggested that differences in photosynthetic and respiratory rates and leaf weight/area ratios among *Acer saccharum* Marsh. (sugar maple) seedlings, grown from seed collected along an altitudinal gradient, may represent the result of selec-

Table 7.1 rRNA gene copy number for eight populations of *Pinus rigida*. (Source: Govindaraju and Cullis, 1992)

Population	Mean copy number	Standard deviation
White's Bog	4438	1368
Lebanon Lake	5416	1231
Penn Forest	5239	1528
Bass River	4598	1399
West Plains[a]	1754	1427
East Plains[a]	1434	528
Wayne Forest[a]	1743	703
Atsion[a]	1459	1054

[a] Presumed to occupy stressed environment.

tion for increased photosynthetic efficiency at high altitudes which provide shorter growing seasons. More recently, a survey of ribosomal RNA (rRNA) gene copy number among 77 individuals from eight populations of *Pinus rigida* Mill. (pitch pine), in the environmentally heterogeneous landscape of the Pine Barrens of New Jersey (USA), revealed a fourfold variation in mean gene copy numbers among populations (Govindaraju and Cullis, 1992). Significantly lower gene copy numbers were observed in populations thought to be stressed owing to local fire regimes than in other populations (Table 7.1). Here, the central role of rRNA in protein synthesis was suggested as providing a mechanism for selection through effects on growth and development.

Often, however, the possible effects of selection are confounded with effects of other genetic processes, such as gene flow, which may share similar spatial dynamics. Warwick and Black (1986) surveyed genetic variation among 39 populations of *Abutilon theophrasti* Medic (velvetleaf) distributed from Ohio (USA) to Ontario (Canada). Of the 51 morphological and life history characters measured on seed grown under controlled conditions, 33 showed significant differences among populations. Variation in several characters (e.g. seed dormancy, leaf blade length, inflorescence dry weight) showed correlations with climate as estimated by degree-days, suggesting the possibility of selection along a climatic gradient. However, Warwick and Black (1986) also point out that the climatic gradient over which observations were made coincides with a spatial gradient in time since colonization, and so effects of genetic drift and gene flow in generating the observed genetic structure cannot be discounted. Similar results have been obtained by Comps *et al.* (1990) in a large-scale allozyme study of 140 natural populations of *Fagus sylvatica* L. (beech) in Europe. The clearest results were that allele frequencies and amount of variation at two peroxidase loci varied with climatic region suggesting a role for selection. However, spatial variation in population age, owing to patterns of post-glacial migration, also provided some explanation of genetic structure.

Several attempts have been made to partition out the effects of selection among different environments from those of other processes, using the same types of genetic and environmental data as discussed above. Sokal and Wartenberg (1983) and Epperson (1990) have suggested that as selection only operates on part of the genome, while processes such as gene flow and genetic drift affect the whole of the genome equally, simultaneous examination of genetic patterns for several loci or characters may provide some evidence regarding the relative roles of these processes in generating observed genetic variation and structure. Several recent studies have adopted this approach. In an investigation of genetic structure in *Impatiens capensis* Meerb. (jewelweed), Argyres and Schmitt (1991) found significant spatial variation for height-related quantitative characters among individuals within a population, but not for any of the other measured characters. Based on this they suggest that the observed spatial cline in these characters is likely to be the result of selection, possibly owing to microgeographic variation in soil moisture environment. Similarly, Pigliucci, Benedettelli and Villani (1990) investigated allozyme variation at 15 loci in 18 populations of *Castanea sativa* Mill. (chestnut) across its geographic range in Italy. Their observations of significant spatial genetic structure for only three of the 15 loci examined led them to suggest that selection along an undefined environmental gradient was the most likely underlying process, though effects of gene flow were not completely discounted.

These types of analyses provide quite good evidence for the influence of environmental variation on genetic variation through selection. However, initial assumptions about the testing structure of such analyses must be considered carefully, as they have a major impact on the subsequent interpretation of results. If the comparisons of genetic variation among sampling units (e.g. individuals, populations), provided by each genetic marker, are considered to be tests of structure only for that marker, then differences among results may indeed be evidence for selection. If, on the other hand, comparisons are viewed as multiple tests of the same hypothesis, that overall there is genetic pattern among sampling units, then some variation among results for different markers is to be expected simply by chance, even in the absence of selection.

Bradshaw (1972) and Ennos (1983) have suggested that effects of selection might also be separated from the influences of other processes by examining genetic variation among cohorts within populations. Young cohorts will presumably reflect the action of mutation, gene flow and any sampling effects due to genetic drift, while later cohorts will incorporate these and subsequent effects of selection. Using this method, Young (1993) compared allozyme variation in two populations of *A. saccharum* occupying different environments (continuous forest and forest fragment). Observed differences in among cohort patterns of abundance of heterozygous genotypes at an alcohol dehydrogenase locus provided some evidence of selection for different genotypes in the two environments (Table 7.2).

Table 7.2 Coefficients of heterozygote deficiency (inbreeding coefficients) at an alcohol dehydrogenase locus among four cohorts in two populations of *Acer saccharum* occupying different environments. (Source: Young, 1993)

Cohort	Continuous forest population	Forest fragment population
Seedlings	0.375	−0.002
Juveniles	0.025	0.174
Poles	0.169	0.279
Reproductive	0.290	0.475

When environments within a landscape are defined, or particular environmental parameters are identified as being likely to exert a selective influence, a priori, examination of relationships between genetic variation and environment provide more powerful tests for possible effects of selection, allowing null hypotheses of no correlation to be constructed. The most common parameters used to define environments in investigations of selection in plants have been soil chemistry variables. This is probably owing to the ease with which they can be theoretically linked to important physiological processes such as nutrient uptake, photosynthesis and respiration.

In an investigation of the effect of soil environment on genetic variation in *Pinus banksia* Lamb. (jack pine), Xie and Knowles (1992) examined spatial relationships between alleles at 11 allozyme loci and 16 soil nutrients. They observed significant relationships between single alleles at each of two loci (aldolase and acid phosphatase) and the concentration of the metal ions manganese, magnesium, lead, aluminium, iron and zinc. Such direct correlation is suggestive of a role for environmental variation in determining genetic structure through selection; more so when, as was the case in this situation, the physiological functions of the enzymes involved are quite well understood and, for acid phosphatase, are associated with plant root cell walls and the permeability of cell membranes. At a larger spatial scale, Furnier and Adams (1986) compared allozyme variation between seven populations of *Pinus jeffreyi* Grev. & Balf. (jeffrey pine) restricted to infertile ultramafic soils in the northern part of the species' range in Oregon and seven other populations occupying a broader range of more fertile soils in California and Nevada (USA). Significant allele frequency differences were found between the two population groups at 11 of 18 polymorphic loci and, on average, heterozygosity was approximately 40% lower for populations on infertile soils. Furnier and Adams (1986) suggest that these differences most likely reflect the effect of strong directional selection operating within populations occupying infertile soils.

These sorts of data do represent evidence for the importance of selection and environmental variation in determining genetic variation and structure within plants. However, care must be taken to account for the effects of the genetic and ecological traits of the species under consideration when interpreting results. For example, Nevo, Beiles and Krugman (1988) reported evidence for effects of selection between two soil environments on allozyme variation in the highly selfing allotetraploid *Triticum diccoides* (Koern.) Koern. ex Schweinf (wild wheat). Though the observed genetic differentiation may well represent a real effect of selection between soil types, multiple seed samples from single individuals were used to provide estimates of genetic variation. In a plant exhibiting high natural levels of selfing, such a sampling strategy may artificially reduce observed within-population variation, possibly generating spurious significances for observed among-population, and therefore among-environment, differences.

Possibly the most robust evidence for effects of selection comes from situations in which there is apparent genetic differentiation between individuals in different environments even when populations are large and the potential for among-population gene flow is high. Snaydon and Davies (1976) compared quantitative genetic variation between two adjacent populations of *Anthoxanthum odoratum* L. occupying soils subject to different nutrient regimes, using both tillers and seed grown in common environments. *Anthoxanthum odoratum* is short-lived and wind pollinated, suggesting that the potential for interpopulation gene flow is high. There were clear differences between populations for several of the measured characters (e.g. panicle height, dry weight), though not all responded to the same degree, suggesting differences in selection pressures for various characters. In subsequent experiments (Davies and Snaydon, 1976), in which plants from several different soil nutrient environments were reciprocally transplanted, selection coefficients against transplanted individuals, based on survival after 18 months, averaged 0.36. Similar results were obtained by Warwick and Briggs (1978a) who observed clear and heritable differences in growth form (prostrate versus erect) of *Poa annua* L. individuals between adjacent populations subject to different mowing regimes, with prostrate individuals being more common in the mowed populations and erect plants more common in unmowed populations. Following Bradshaw (1972), within-population comparisons of the numbers of seedling and adult individuals exhibiting different growth forms in the two environments provided more evidence for the action of selection in eliminating the erect growth form from the mowed environment (Warwick and Briggs, 1978b) (Table 7.3). Based on results of an artificial clipping experiment, selection coefficients against the erect growth form in the mowed environment ranged from 0.61 to 0.68 depending on the biomass component considered (Warwick and Briggs, 1978b).

Table 7.3 Numbers (and percentage) of *Poa annua* seedlings and adults exhibiting prostrate and erect growth forms in environments subject to different mowing regimes. (Source: Warwick and Briggs, 1978b)

Growth form	Mowed environment		Unmowed environment	
	Seed	Adult	Seed	Adult
Prostrate	630 (84)	47 (94)	89 (47)	17 (34)
Erect	116 (16)	3 (6)	99 (53)	33 (66)
Total	746	50	188	50

The results discussed above suggest that selection is an important mechanism directly linking environmental and genetic variation within landscapes. However, environmental differences within a landscape may also influence genetic variation less directly. Certainly differences among environments which impact reproductive phenology have considerable potential to affect patterns of gene flow among populations. Schuster, Alles and Mitton (1989) investigated the timing of pollen release and megastrobilus closure among eight populations of *Pinus flexilis* James (limber pine) along an altitudinal gradient of 1700 m in the Rocky Mountains of Colorado (USA). Duration of pollen release and megastrobilus receptivity were roughly synchronous within populations, but there was clinal variation among populations. Generally populations separated by more than 400 m of elevation did not have overlapping pollination periods, suggesting that direct interpopulation gene flow over such distances is impeded. Furthermore, as open microstrobili may retain viable pollen after megastrobili within a population have closed, there may also be a directional effect of altitude (environment) on gene flow among populations, with movement of genes upslope being less restricted than downslope. Generally these observations were concordant with genetic data which indicated differentiation between the highest and lowest populations for eight of ten allozyme loci. Billington, Mortimer and McNeilly (1988) have also noted restriction of gene flow owing to differences in flowering time between adjacent populations of *Holcus lanatus* L. in fields subject to different grazing and cutting regimes in Wales. Differences in quantitative genetic variation were also observed between the two populations.

In another study illustrating possible indirect effects of different environments on genetic variation, Boyle, Liengsiri and Piewluang (1990) compared the amount and spatial distribution of allozyme variation among individuals between two *Picea mariana* (Mill.) B.S.P. (black spruce) populations occupying contrasting upland and lowland environments in Ontario (Canada), which differed in fire-related disturbance regimes. Allele frequencies for the two populations were similar. However, while the undisturbed lowland population exhibited spatial genetic structure, with significant clumping of individuals with similar genotypes, the fire disturbed upland population showed none. Boyle, Liengsiri and Piewluang (1990) suggest that this is the result of effects of disturbance regime on patterns of gene flow associated with regeneration, with spatial structure in the undisturbed population resulting from regeneration through short-distance dispersal of seed from mature trees, while in the population that had regenerated after fire there may have been more potential for long-distance seed dispersal.

Overall these results suggest that differences among environments within a landscape do influence the amount and distribution of genetic variation observed within plants. Although it would appear that selection is probably an important process underlying these effects, it should be noted that the most direct evidence for selection comes from studies of populations in modified environments, such as those defined by artificial management regimes (e.g. Snaydon and Davies, 1976; Warwick and Briggs, 1978a, b). The evidence for the action of selection in generating genetic structure in more natural landscapes is primarily indirect. Certainly, the impacts of environmental variation on other processes that may influence genetic variation, such as gene flow, through effects on reproductive phenology and dispersal, should not be disregarded.

7.3.2 ENVIRONMENT SIZE

As discussed previously, size of environments may influence genetic variation by affecting population size, which interacts with gene flow in determining the balance between genetic drift and mutation, as well as possibly affecting levels of inbreeding. Empirical data dealing with relationships between population size and genetic variation, in situations where population size may be directly limited by environment, are few. Several examples come from studies of plants which are habitat specialists, or from landscapes where environments suitable for maintaining plant life in general are restricted.

Moran and Hopper (1983) examined allozyme variation in 13 different-sized populations of *Eucalyptus caesia*. This species is entirely restricted to granite rock outcrop environments in south-western Australia, which are separated by plateaus supporting other vegetation types. The number of polymorphic loci per population was found to be positively correlated with population size, possibly reflecting increased effects of genetic drift in small populations, which have probably been isolated for many thousands of years. Similarly, Billington (1991) reported positive correlations between allozyme variation and estimated population size for 17 populations of *Halocarpus bidwilli* (Kirk) Quinn, a subalpine shrub restricted to bogs and stony alluvial river terraces south of latitude 39° in New Zealand. McClenaghan and Beauchamp (1986) also detected a significant positive relationship between population size and allozyme variation for 16 populations of *Washingtonia filifera* (Linden) H. Wendel (California fan palm) restricted to isolated moist environments in canyons and around springs and seeps in the Colorado Desert of southern California (USA). Population size was positively correlated with polymorphism, but not with heterozygosity, suggesting that small populations may have low genetic variation owing to greater effects of random genetic drift, but that population size has little apparent effect on inbreeding.

Perhaps the best evidence for an effect of environment size on genetic variation in plants comes from a recent study of allozyme variation in populations of *Eucalyptus albens* Benth. (white box) in south-eastern Australia (Prober and Brown, 1994). Populations of this species have been reduced in size over the past few hundred years as their woodland environments have been destroyed and fragmented by pastoral development. Given this, current population sizes are primarily attributable to the size of their remnant environments. The number of alleles, polymorphism and heterozygosity at 18 allozyme loci were all positively correlated with the number of reproductively mature individuals within populations which ranged from 14 to > 10 000 (Figure 7.1). As estimated levels of inbreeding in *E. albens* populations were not correlated with population size, bottleneck effects and increased genetic drift probably best account for the lower genetic variation in the smaller populations.

Though, when compared with studies of the effects of environmental variation, the evidence is limited, these data do lend support to the idea that, at least for long-lived plant species with mixed mating systems, environment size may impact on genetic variation through effects on population size. Based on the results from two of the above studies (McClenaghan and Beauchamp, 1986; Prober and Brown, 1994), random genetic drift seems more likely than inbreeding to be the process underlying this effect.

7.3.3 SPATIAL ARRANGEMENT OF ENVIRONMENTS

The simplest way for the spatial arrangement of environments within a landscape to exert influence on genetic variation and structure is through its effect on patterns of gene flow. Given the difficulty involved in observing gene flow directly, current empirical evidence for such an effect is indirect.

Several studies of genetic differentiation among populations have revealed correlations between geographic distances separating populations and measures of genetic similarity (negative) or genetic distance (positive) between them (e.g. Yeh and O'Malley, 1980; Young, Warwick and Merriam, 1993). Based on such correlations it has been suggested that the spatial arrangement of populations does affect gene flow and subsequently genetic structure, through isolation by distance. However, if not all possible environments are full, which is probably the general case, such observations are not strong evidence for genetic effects of the spatial arrangement of environments. Furthermore, when considering clinal environmental variation, distances among populations may be correlated with the similarity of the environments they occupy, in which case genetic similarities among populations may be due to selection. However, at least two studies have described relationships between geographic and genetic distances between populations of plants that are strict habitat specialists. These probably reflect effects of environmental

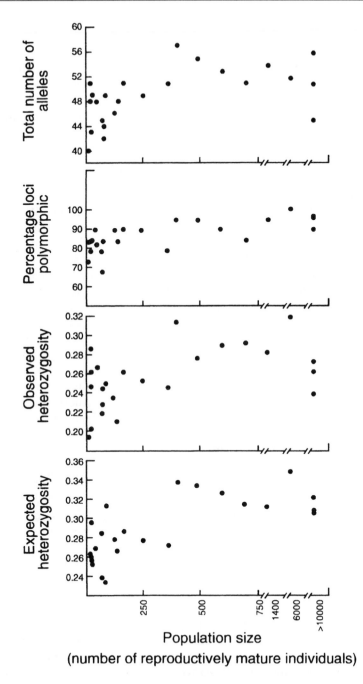

Figure 7.1 Relationships between allozyme variation and reproductive population size for *Eucalyptus albens*. (Source: Prober and Brown, 1994.)

arrangement on genetic structure. Godt and Hamrick (1993) found a very strong positive relationship between genetic distance and linear geographic distance among 13 populations of *Tradescantia hirsuticaulis* Small (hairy-stemmed spiderwort), a plant restricted to spatially disjunct habitats of rock outcrops and dry woodlands in the south-eastern United States. Similarly, Moran and Hopper (1983) observed a positive relationship between genetic and geographic distances separating populations of *Eucalyptus caesia*, which are restricted to spatially discrete granite outcrop environments in south-western Australia.

Two recent studies have directly addressed the genetic effects of spatial arrangement of environments through investigations of the effects of changes in environmental arrangement brought about by forest fragmentation (Foré, Hickey and Vankat, 1992; Young, Merriam and Warwick, 1993). In both of these studies the species considered was *A. saccharum*. *Acer saccharum* is a common forest tree throughout the north-eastern United States and south-eastern Canada which has been subject to severe fragmentation of its forest environment over the past 100–200 years. Foré, Hickey and Vankat (1992) examined variation at eight allozyme loci in both canopy and seedling individuals of 15 populations occurring in forest patches in Ohio (USA). These patches were created at least 50 years previously by forest fragmentation. Comparisons between these cohorts were used to compare pre-fragmentation (canopy) and post-fragmentation (seedling) spatial patterns of genetic variation. Results revealed significant effects of the changed spatial arrangement of the forest environment, which has gone from largely continuous to spatially disjunct. Post-fragmentation cohorts exhibited less interpopulation genetic differentiation than pre-fragmentation cohorts. Foré, Hickey and Vankat (1992) suggest that this may be the result of increased interpopulation gene flow since forest fragmentation for this wind dispersed tree. However, these results should be treated cautiously as observed differences among cohorts may also reflect effects of selection, and sample sizes for canopy individuals were often small (6 populations $n < 10$).

In a similar study, Young, Merriam and Warwick (1993) compared the amount and distribution of allozyme variation (17 loci) in eight populations of *A. saccharum* in continuous forest in Québec (Canada) with that in eight populations occupying nearby forest patches in Ontario created 100–200 years previously by forest fragmentation. Young, Merriam and Warwick (1993) observed slightly higher genetic variation within forest patch individuals and populations despite patches as a whole having six less alleles than the continuous forest populations (Table 7.4). Based on this and the observation that, in forest patch populations, a greater number of alleles were spread over more populations than in continuous forest populations (Figure 7.2), Young, Merriam and Warwick (1993) also suggested that the changed spatial arrangement of

A. saccharum populations, resulting from fragmentation of the forest environment, has increased interpopulation gene flow. A major assumption in this study is that the current amounts and patterns of genetic

Table 7.4 Allozyme variation in continuous forest and forest patch populations of *Acer saccharum* (*n* = 8). (Source: Young, Merriam and Warwick, 1993)

	Forest patch populations	Continuous forest populations
Total alleles	48	54
Mean polymorphism	0.588 (0.019)[a]	0.537 (0.017)
Mean alleles per locus	2.10 (0.04)	2.03 (0.06)
Mean heterozygosity	0.121 (0.006)	0.109 (0.004)

[a] Standard error in parentheses.

variation in continuous forest populations are representative of the pre-fragmentation variation of forest patch populations.

Taken overall, these two main lines of evidence, though generally less direct than for the effects of the two other components of landscape structure discussed, do suggest that the spatial arrangement of environments within landscapes can exert an influence on the amount and distribution of genetic variation within plants. Presumably this influence

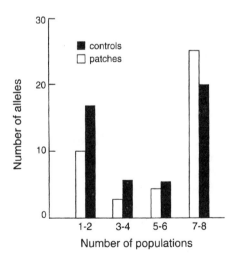

Figure 7.2 Distribution of alleles among eight continuous forest populations (controls) and eight forest patch populations (patches) of *Acer saccharum*. (Source: Young, Merriam and Warwick, 1993.)

is through effects on gene flow. However, evidence from direct measurements of gene flow is notably lacking.

The increased availability of detailed environmental information through geographic information systems (GIS), combined with developments in statistical techniques for analysis of spatial patterns (e.g. Epperson, 1992), provide the potential for indirectly assessing the relative effects of selection and gene flow through partitioning observed genetic variation into that explained by environmental variation (selection) and spatial autocorrelation among sampling points (gene flow). However, such an approach will also require the careful partitioning out of effects of environmental spatial correlations and the investigation of indirect environmental effects (e.g. effects on gene flow), if initial assumptions about the processes being observed are not to severely limit interpretation of the results.

More generally, it is important that investigations of genetic variation and/or processes be carried out for single species at several spatial or organizational scales within a single landscape. Very few studies (e.g. Cuguen *et al.*, 1985; Young, Warwick and Merriam, 1993) have done this but those which have have provided interesting, if qualitative, insights into the relative importance of different genetic processes at different scales.

Finally, considering landscapes as physical entities with real scales, within which species have to be managed, and on which management decisions are often made as a unit, rather than as the scaleless and theoretical constructs previously defined, it is important to try and understand any generalities of the effects of landscape structure of genetic variation. To this end, comparative studies of genetic variation, structure and processes, on species with varying ecologies, within the same landscape, should prove instructive.

7.4 DISCUSSION

Based on the results of the empirical studies discussed above, the answer to the question, 'Does landscape structure affect genetic variation and structure in plants?' is probably yes. Certainly there is evidence from a variety of different types of studies to indicate effects of the three components of landscape structure discussed: type, size and spatial arrangement of environments, on genetic variation and structure for plant species with varying ecologies, existing in a wide array of different landscapes. There are, at present, no empirical data that specifically address the genetic effects of the temporal dynamics of landscape structure, though some comparative studies of genetic variation in populations resulting from colonization of newly formed environments relative to populations in established environments have been undertaken recently (e.g. Antrobus and Lack, 1993).

Answering the question, 'What are the genetic processes involved?' is more difficult. There is both direct and indirect empirical evidence corroborating the effect of selection, and indirect evidence for effects of random genetic drift and gene flow, in providing links between landscape structure and genetic variation and structure for plants. However, categorically establishing the effects of particular genetic processes in natural populations is problematic and assessing the relative magnitudes of the roles of different processes has so far proved to be difficult.

Determining the relative roles of different processes in generating genetic patterns within landscapes is a major challenge in population/ecological genetics. Identification of general relationships between genetic variation and environmental pattern (if such exist) may provide broad landscape management principles for short-term genetic conservation. However, true predictive ability regarding the genetic implications of landscape structure can only come from a thorough understanding of the functioning of genetic processes within landscapes. An essential step towards this understanding is direct empirical investigation of the spatial and temporal dynamics of genetic processes in relation to landscape structure.

Developments in several fields are providing tools which may help in this rather daunting task. The recent development of high resolution DNA markers, such as randomly amplified polymorphic DNA (RAPDS), minisatellites and microsatellites (Rogstad, Patton and Schaal, 1988; Welsh and McClelland, 1990; Queller, Strassmann and Hughes, 1993), and their application to paternity analysis in plants (e.g. Nybom, 1990), should provide the potential for direct observation of gene flow events among individuals and populations within landscapes. Unfortunately, even with such genetic markers, the sampling required for detecting mating events in natural populations, over any but the smallest scales, may limit direct practical application. However, small-scale intensive studies using these methods can still play a valuable role in verifying gene flow estimates obtained using less direct methodologies, for example Slatkin's rare alleles method (Slatkin, 1985) and even studies of pollinator movements. This will allow such methods to be more rigorously 'calibrated' and provide some tests of their underlying assumptions.

Perhaps of more direct application, it has recently been suggested (Petit, Kremer and Wagner, 1992) that comparison of genetic patterns for DNA markers with uniparental inheritance (e.g. chloroplast DNA) with those for markers that undergo biparental inheritance, may allow partitioning of effects of pollen versus seed mediated gene flow. This would allow assessment of differential impacts of landscape structure on these two modes of genetic interaction among populations. Similarly, comparisons of genetic patterns for neutral markers with patterns for markers thought to be associated with fitness, should provide some insight into the overall importance of selection as a process generating genetic structure. The greatest problem here is deciding just which classes of markers are affected by selection and which are not. Certainly the results of studies discussed in previous sections indicate that this is not easy to do a priori and there is still a good deal of debate regarding this issue, particularly with regard to the neutrality of allozymes (Mitton, 1989).

7.5 SUMMARY

Many plant species exhibit high levels of intraspecific genetic variation and, for species with a variety of ecologies, there is evidence of genetic structure at various organizational and/or spatial scales (e.g. individuals, populations). Results of empirical studies suggest that three components of landscape structure – environment type, environment size and spatial arrangement of environments – may all affect the amount and distribution of this genetic variation. There is evidence for effects of selection, random genetic drift and gene flow in providing links between landscape structure and genetic variation and structure; however, the relative magnitudes of the roles of these processes are poorly understood. Increased empirical knowledge of the spatial and temporal dynamics of genetic processes, in relation to components of landscape structure, is required to achieve predictive ability regarding the genetic implications of landscape structure for plants.

REFERENCES

Amos, B. and Hoelzel, A. R. (1992) Applications of molecular genetic techniques to the conservation of small populations. *Biol. Conserv.*, **61**, 133–44.

Antrobus, S. and Lack, A. J. (1993) Genetics of colonising and established populations of *Primula veris. Heredity*, **71**, 252–8.

Argyres, A. Z. and Schmitt, J. (1991) Microgeographic genetic structure of morphological and life history traits in a natural population of *Impatiens capensis. Evolution*, **45**, 178–89.

Billington, H. L. (1991) Effect of population size on genetic variation in a dioecious conifer. *Biol. Conserv.*, **5**, 115–19.

Billington, H. L., Mortimer, A. M. and McNeilly, T. (1988) Divergence and genetic structure in adjacent grass populations. 1. Quantitative genetics. *Evolution*, **42**, 1267–77.

Boyle, T., Liengsiri, C. and Piewluang, C. (1990) Genetic structure of black spruce on two contrasting sites. *Heredity*, **65**, 393–9.

Bradshaw, A. D. (1959) Population differentiation in *Agrostis tenuis* Sibth. 1. Morphological differentiation. *New Phytol.*, **58**, 208–27.

Bradshaw, A. D. (1972) Some of the evolutionary consequences of being a plant. *Evol. Biol.*, **5**, 25–47.

Cairns, J., Overbaugh, J. and Miller, S. (1988) The origin of mutants. *Nature*, **335**, 142–5.

Clegg, M. T. (1990) Molecular diversity in plant populations, in *Plant Population Genetics, Breeding, and Genetic Resources* (eds A. H. D. Brown, M. T. Clegg, A. L. Kahler and B. S. Weir), Sinauer Press, Sunderland, pp. 98–115.

Comps, B., Thiebaut, B., Paule, L., Merzeau, D. and Letouzey, J. (1990) Allozymic variability in beechwoods (*Fagus sylvatica* L.) over central Europe: Spatial differentiation among and within populations. *Heredity*, **65**, 407–17.

Cuguen, J., Thiebaut, B., Ntsiba, F. and Barriere, G. (1985) Enzymatic variability of beechstands (*Fagus sylvatica* L.) on three scales in Europe: Evolutionary significance, in *Genetic Differentiation and Dispersal in Plants* (eds P. Jacquard, G. Heim and J. Antonovics), Springer-Verlag, Berlin, pp. 17–39.

Davies, M. S. and Snaydon, R. W. (1976) Rapid population differentiation in a mosaic environment. 3. Measures of selection pressures. *Heredity*, **36**, 59–66.

Ennos, R. A. (1983) Maintenance of genetic variation in plant populations. *Evol. Biol.*, **16**, 129–55.

Epperson, B. K. (1990) Spatial autocorrelation of genotypes under directional selection. *Genetics*, **124**, 757–71.

Epperson, B. K. (1992) Spatial structure of genetic variation within populations of forest trees, in *Population Genetics of Forest Trees*, (eds W. T. Adams, S. H. Strauss, D. L. Copes and A. R. Griffin), Kluwer Academic, pp. 257–78.

Foré, S. A., Hickey, R. J. and Vankat, J. L. (1992) Genetic structure after forest fragmentation: a landscape ecology perspective on *Acer saccharum*. *Can. J. Botany*, **70**, 1659–68.

Frankel, O. H. (1970) Variation, the essence of life. *Proceedings of the Linnean Society of New South Wales*, **95**, 158–69.

Furnier, G. R. and Adams, W. T. (1986) Geographic patterns of allozyme variation in jeffrey pine. *Am. J. Botany*, **73**, 1009–15.

Gillespie, J. H. (1978) A general model to account for enzyme variation in natural populations. 5. The SAS-CFF model. *Theor. Pop. Biol.*, **14**, 1–45.

Gilpin, M. (1991) The genetic effective size of a metapopulation. *Biol. J. Linn. Soc.*, **42**, 165–75.

Godt, M. J. W. and Hamrick, J. L. (1993) Genetic diversity and population structure in *Tradescantia hirsuticaulis* (Commelinaceae). *Am. J. Botany,* **80**, 959–66.

Govindaraju, D. R. and Cullis, C. A. (1992) Ribosomal DNA variation among populations of a *Pinus rigida* Mill. (pitch pine) ecosystem. 1. Distribution of copy numbers. *Heredity,* **69**, 133–40.

Haldane, J. B. S. and Jayakar, S. D. (1963) Polymorphism due to selection of varying direction. *J. Genetics,* **58**, 237–42.

Hall, B. G. (1990) Spontaneous point mutations that occur more often when advantageous than when neutral. *Genetics,* **126**, 5–16.

Hamrick, J. L. (1983) The distribution of genetic variation within and among natural plant populations, in *Genetics and Conservation* (eds C. M. Schonewald-Cox, S. M. Chambers, B. MacBryde and W. L. Thomas), Benjamin/Cummings, London, pp. 335–44.

Hamrick, J. L. and Godt, M. J. W. (1989) Allozyme diversity in plant species, in *Plant Population Genetics, Breeding and Genetic Resources* (eds A. H. D. Brown, M. T. Clegg, A. L. Kahler and B. S. Weir), Sinauer Press, Sunderland, MA, pp. 43–63.

Hedrick, P. W. (1986) Genetic polymorphism in heterogeneous environments: A decade later. *Ann. Rev. Ecol. Syst.,* **17**, 535–66.

Hoekstra, R. F., Bijlsma, R. and Dolman, A. J. (1985) Polymorphism from environmental heterogeneity: Models are only robust if the heterozygote is close in fitness to the favoured homozygote in each environment. *Genet. Res. Cambridge,* **45**, 299–314.

Ledig, F. T. (1992) Human impacts on genetic diversity in forest ecosystems. *Oikos,* **63**, 87–108.

Ledig, F. T. and Korbobo, D. R. (1983) Adaptation of sugar maple populations along altitudinal gradients: Photosynthesis, respiration and specific leaf weight. *Am. J. Botany,* **70**, 256–65.

Lenski, R. E. and Mittler, J. E. (1993) The directed mutation controversy and Neo-Darwinism. *Science,* **259**, 188–94.

Levin, D. A. (1993) Local speciation in plants: The rule not the exception. *Syst. Botany,* **18**, 197–208.

Li, P. and Adams, W. T. (1989) Range-wide patterns of allozyme variation in Douglas-fir (*Psuedotsuga menziesii*). *Can. J. For. Res.,* **19**, 149–61.

Linhart, Y. B., Mitton, J. B., Sturgeon, K. B. and Davis, M. L. (1981) Genetic variation in space and time in a population of ponderosa pine. *Heredity,* **46**, 407–26.

Manicacci, D., Olivieri, I., Perrot, V. *et al.* (1992) Landscape ecology: Population genetics at the metapopulation level. *Landsc. Ecology,* **6**, 147–59.

Mather, K. (1955) Polymorphism as an outcome of disruptive selection. *Evolution,* **9**, 52–61.

Maynard-Smith, J. and Hoekstra, R. (1980) Polymorphism in a varied environment: How robust are the models? *Gen. Res. Cambridge*, **35**, 45–57.

McClenaghan, L. R. and Beauchamp, A. C. (1986) Low genetic differentiation among isolated populations of the California fan palm (*Washingtonia filifera*). *Evolution*, **40**, 315–22.

Merriam, H. G. and Wegner, J. (1992) Local extinctions, habitat fragmentation, and ecotones, in *Landscape Boundaries: Consequences for Biotic Diversity and Ecological Flows* (eds A. J. Hansen and F. di Castri), Springer-Verlag, New York, pp. 150–69.

Mitton, J. B. (1989) Physiological and demographic variation associated with allozyme variation, in *Isozymes in Plant Biology* (eds D. E. Soltis and P. S. Soltis), Dioscorides, Portland, pp. 127–45.

Moran, G. F. and Hopper, S. D. (1983) Genetic diversity and the insular population structure of the rare granite rock species, *Eucalyptus caesia* Benth. *Aust. J. Botany*, **31**, 161–172.

Nevo, E., Beiles, A. and Krugman, T. (1988) Natural selection of allozyme polymorphisms: a microgeographical differentiation by edaphic, topological, and temporal factors in wild emmer wheat (*Triticum dicoccoides*). *Theor. Appl. Genetics*, **76**, 737–52.

Nybom, H. (1990) Genetic variation in ornamental apple trees and their seedlings (*Malus*, Rosaceae) revealed by DNA fingerprinting with M13 repeat probe. *Hereditas*, **113**, 17–28.

Pamilo, P. (1988) Genetic variation in heterogeneous environments. *Ann. Zool. Fennici*, **25**, 99–106.

Petit, R. J., Kremer, A. and Wagner, D. B. (1992) The measurement of both nucelar and cytoplasmic genetic differentiation allows the estimation of the relative seed and pollen flow, in *Abstracts of the International Symposium on Population Genetics and Gene Conservation of Forest Trees*, INRA, Bordeaux-Cestas.

Pigliucci, M., Benedettelli, S. and Villani, F. (1990) Spatial patterns of genetic variability in Italian chestnut (*Castanea sativa*). *Can. J. Botany*, **68**, 1962–7.

Prober, S. M. and Brown, A. H. D. (1994) Conservation of the grassy white box woodlands: Population genetics and fragmentation of *Eucalyptus albens* Benth. *Conserv. Biol.* (in press).

Queller, D. C., Strassmann, J. E. and Hughes, C. R. (1993) Microsatellites and kinship. *Trends Ecol. Evol.*, **8**, 285–8.

Rogstad, S. H., Patton, J. C. and Schaal, B. A. (1988) M13 repeat probe detects DNA minisatellite-like sequences in gymnosperms and angiosperms. *Proceedings of the National Academy of Sciences USA*, **85**, 9176–8.

Saunders, D. A., Hobbs, R. J. and Margules, C. R. (1991) The biological consequences of ecosystem fragmentation: A review. *Conserv. Biol.*, 18–32.

Schmitt, J. and Gamble, S. E. (1990) The effect of distance from the parental site on offspring performance and inbreeding depression in *Impatiens capensis*: A test of the local adaptation hypothesis. *Evolution*, **44**, 2022–30.

Schuster, W. S., Alles, D. L. and Mitton, J. B. (1989) Gene flow in limber pine: Evidence from pollination phenology and genetic differentiation along an elevational transect. *Am. J. Botany*, **76**, 1395–403.

Slatkin, M. (1977) Gene flow and genetic drift in a species subject to frequent local extinctions. *Theor. Pop. Biol.*, **12**, 253–62.

Slatkin, M. (1985) Rare alleles as indicators of gene flow. *Evolution*, **39**, 53–65.

Snaydon, R. W. and Davies, M. S. (1976) Rapid population differentiation in a mosaic environment. 4. Populations of *Anthoxanthum odoratum* at sharp boundaries. *Heredity*, **37**, 9–25.

Sokal, R. R. and Wartenberg, D. E. (1983) A test of spatial autocorrelation analysis using an isolation-by-distance model. *Genetics*, **105**, 219–37.

Templeton, A. R., Shaw, K., Routman, E. and Davis, S. K. (1990) The genetic consequences of habitat fragmentation. *Ann. Missouri Botanic Gardens*, **77**, 13–27.

Varvio, S., Chakraborty, R. and Nei, M. (1986) Genetic variation in subdivided populations and conservation genetics. *Heredity*, **57**, 189–98.

Venable, D. L. (1984) Using intraspecific variation to study the ecological significance and evolution of plant life histories, in *Perspectives on Plant Population Ecology* (eds R. Dirzo and J. Sarukhan), Sinauer Press, Sunderland, pp. 166–87.

Via, S. and Lande, R. (1985) Genotype-environment interaction and the evolution of phenotypic plasticity. *Evolution*, **39**, 505–22.

Via, S. and Lande, R. (1987) Evolution of genetic variability in a spatially heterogenous environment: Effects of genotype-environment interaction. *Gen. Res. Cambridge*, **49**, 147–56.

Wade, M. J. and McCauley, D. E. (1988) Extinction and recolonisation: Their effects on the genetic differentiation of local populations. *Evolution*, **42**, 995–1005.

Warwick, S. I. and Black, L. D. (1986) Genecological variation in recently established populations of *Abutilon theophrasti* (velvetleaf). *Can. J. Botany*, **64**, 1632–43.

Warwick, S. I. and Briggs, D. (1978a) The genecology of lawn weeds. 1. Population differentiation in *Poa annua* L. in a mosaic environment of bowling green lawns and flower beds. *New Phytol.*, **81**, 711–23.

Warwick, S. I. and Briggs, D. (1978b) The genecology of lawn weeds. 2. Evidence for disruptive selection in *Poa annua* L. in a mosaic environment of bowling green lawns and flower beds. *New Phytol.*, **81**, 725–37.

Welsh, J. and McClelland, M. (1990) Fingerprinting genomes using PCR with arbitrary primers. *Nucleic Acids Res.*, **18**, 7213–18.

Wilcove, D. S. (1987) From fragmentation to extinction. *Nat. Areas J.*, **7**, 23–9.

Wright, S. (1978) *Evolution and the Genetics of Populations, Vol. 4. Variability Within and Among Natural Populations*, University of Chicago Press, Chicago.

Xie, C. Y. and Knowles, P. (1992) Associations between allozyme phenotypes and soil nutrients in a natural population of jack pine (*Pinus banksiana*). *Biochem. Syst. Ecol.*, **20**, 179–85.

Yeh, F. C. H. and O'Malley, D. 1980. Enzyme variations in natural populations of Douglas-fir, *Psuedotsuga menziesii* (Mirb.) Franco, from British Columbia. 1. Genetic variation patterns in coastal populations. *Silvae Genetica*, **29**, 83–92.

Young, A. G. (1993) The Effects of Forest Fragmentation on the Population Genetics of *Acer saccharum* Marsh. (sugar maple). PhD Thesis, Carleton University, Ottawa.

Young, A. G., Merriam, H. G. and Warwick, S. I. (1993) The effects of forest fragmentation on genetic variation in *Acer saccharum* Marsh. (sugar maple) populations. *Heredity*, **71**, 277–89.

Young, A. G., Warwick, S. I. and Merriam, H. G. (1993) Genetic variation and structure at three spatial scales for *Acer saccharum* (sugar maple) in Canada and the implications for conservation. *Can. J. For. Res.*, **23**, 2568–2578.

Effects of population dynamics on genetics in mosaic landscapes

8

David E. McCauley

8.1 INTRODUCTION

Most terrestrial landscapes consist of a mosaic of habitat patches of varying size and composition. This is perhaps most evident when the natural biome has been fragmented by human impact, as in the mixture of woodlots, cultivated fields and pastures so familiar to residents of eastern North America and northern Europe. One consequence of this patchwork of habitats is to subdivide many species into numerous localized collections of individuals. When an organism's dispersal potential is limited relative to the dispersion pattern of the patches it can occupy, the landscape could be said to impose a population structure on the species. Here, a local population is being defined loosely as a collection of conspecifics more likely to interact with each other than with individuals from some other locality. The range of possible population structures can be described as a continuum, with highly interdependent populations connected by high rates of migration lying at one extreme and highly isolated, independent populations lying at the other. The way in which a particular landscape imposes a population structure depends on the movement ecology of the species in question, as well as on the spatial arrangement of patch types (Chapter 7). For example, the same landscape could impose very different population structures on snails and sparrows.

The population structure is thus a reflection of the spatial scale over

Mosaic Landscapes and Ecological Processes.
Edited by Lennart Hansson, Lenore Fahrig and Gray Merriam.
Published in 1995 by Chapman & Hall, London. ISBN 0 412 45460 2

which populations are organized, as well as the relative influences of local and regional processes on their properties. To the degree that regional processes are important, little can be learned from considering individual populations in isolation. For example, consider the size of a population. It is obvious that in a highly self-contained population, numbers are determined by the relationship between local birth and death rates. That is, population size is determined entirely by local processes. The population will only persist if the rate of mortality is matched or exceeded by the birth rate. In contrast, in populations subject to high rates of immigration, numbers might be under regional control, according to the rate at which individuals move in from outside the patch. The numerical properties of one patch are influenced by birth, death and dispersal events in other patches. In fact, under 'source–sink' dynamics the local population could remain numerically stable even when the mortality rate exceeds the birth rate by a considerable margin (Pulliam, 1988).

As another example, the field of population genetics has long been concerned with the effect of population structure on the genetic properties of populations (Wright, 1931; Crow and Kimura, 1970; Wade and McCauley, 1984; Slatkin, 1985; Hartl and Clark, 1989; Barton, 1992). When a species is subdivided into some number of populations, each could be characterized statistically by allele frequencies at some number of loci. The set of populations could also be characterized, most simply by averaging allele frequencies across populations, but more completely by also considering population to population allele frequency variation. The local allele frequencies are determined by the tension between within-population processes, such as selection and genetic drift, which are driven by birth and death events, within the patch, and gene flow, a regional phenomenon driven by the migration of individuals among patches. It is well known that gene flow can limit the degree to which populations can diverge genetically from one another, i.e. that regional processes can overwhelm local control.

One could also define a continuum of population structures based on the stability of the local population subunits. When a species is subdivided into some number of local populations of finite size, according to the distribution of favorable habitat across the landscape, the demographic stability and persistence of individual populations can be limited. For example, in small populations demographic stochasticity could result in local extinctions that are uncorrelated from patch to patch. Localized catastrophic events could also cause patch-specific extinctions, a process known as environmental stochasticity. Both of these events become more likely when patches are small or are occupied by a small number of individuals. When local populations suffer a high probability of extinction the persistence of the entire system is also in question, should the rate of loss of populations exceed the rate at which new populations become established for a sustained period of time.

Figure 8.1 describes four arbitrary points along this continuum of local and regional stability. In the first (Figure 8.1(a)), the individual populations are stable and capable of indefinite persistence. Their demographic and genetic properties are largely determined by internal processes, but can be modified by the importation of migrants from other populations. I will call this the 'equilibrium-island' model. In the second (Figure 8.1(b)), some patches support stable populations capable of long-term persistance while others can only support ephemeral populations under constant threat of extinction. This is similar to the 'mainland–island' model discussed by Harrison (1991). The system persists owing to the aforementioned stability of populations occupying some of the patches (the continents) and because a fraction of the remaining patches (the islands) are occupied at any one time owing to recent colonization. In the third (Figure 8.1(c)), all local populations are unstable and capable of only limited persistence. If vacant patches are colonized at a rate approximately equal to the rate of local extinction, however, the system will persist much longer than its individual elements. This scenario has been called a 'metapopulation' (Levins, 1969, 1970). Finally (Figure 8.1(d)), the local extinction rate could exceed the colonization rate to the point that the entire system collapses. In this case the landscape cannot support the species in question even if the total area of favorable habitat is relatively large.

This chapter will focus on the conditions outlined in Figure 8.1(c), that is on the properties of metapopulations, particularly the genetic consequences of frequent extinction and colonization. It will be primarily a discussion of attempts to model the genetic structure of ephemeral populations, and the relationship of these models to more ecologically oriented metapopulation models. It is hoped that this modeling approach will provide a conceptual framework for the empirical study of natural populations. The metapopulation construct should be of particular interest to landscape ecologists owing to the dual local and regional control of population processes; local control owing to birth, death and extinction events and regional control owing to immigration into patches that are either empty (colonization) or occupied (migration). To the degree that the field of landscape ecology addresses the genetic or evolutionary component of biotic processes, the genetic properties of metapopulations must be considered. It has been argued recently that the general problem of pattern and scale is central to many questions in ecology (Levin, 1992). It is hoped that this chapter will contribute to an understanding of how landscape level processes, particularly the frequent turnover of populations, might contribute to the pattern and scale that organize genetic variation.

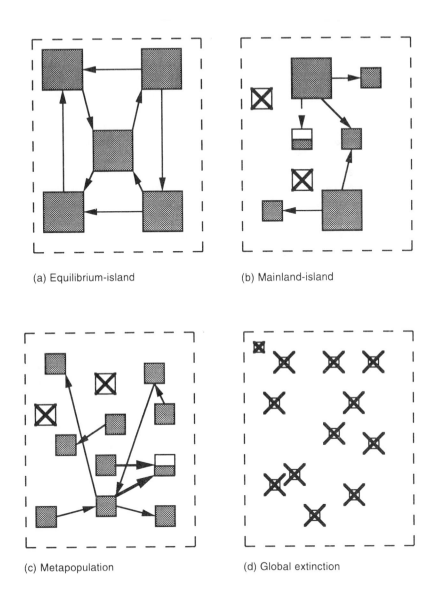

(a) Equilibrium-island

(b) Mainland-island

(c) Metapopulation

(d) Global extinction

Figure 8.1 Four stages along a continuum of population stability. (a) All populations persist indefinitely, though they may exchange migrants. (b) Some populations persist and serve as a source for others which turn over at a high rate. (c) All populations ultimately go extinct but are replaced at a rate that permits species persistence. (d) The extinction rate exceeds the colonization rate and all populations are ultimately lost.

8.2 METAPOPULATION DYNAMICS: AN OVERVIEW

Before considering population genetic processes, it may be worthwhile to consider metapopulations from a purely demographic perspective. The metapopulation concept recognizes that species can be subdivided into many populations that are ephemeral, but that also can give rise to new populations via dispersing propagules. Under this scenario one might investigate the conditions under which the species can persist despite the constant loss of its constituent populations; that is, the conditions under which the colonization process can equal or outstrip the extinction process. The history and current status of ecologically oriented metapopulation models have been reviewed recently (Hanski, 1991; Hanski and Gilpin, 1991) and will be outlined here primarily with regard to their relationship to the genetic models.

Much current thinking about the dynamics of metapopulations can be traced to a model developed by Richard Levins about 25 years ago (Levins, 1969, 1970). Assume a large number of equivalent habitat patches, of which a fraction p are occupied by the populations of the species in question, leaving $1-p$ unoccupied. Assume further that populations go extinct randomly with a probability e. Obviously, without replacement of lost populations the entire metapopulation would eventually go extinct. This need not be the case, however, if unoccupied patches can be colonized or recolonized via the dispersal of individuals from the remaining extant populations. If the rate at which new populations are founded, m, is assumed to be proportional to p, then an equilibrium value of p is found to be $1-e/m$. If $m>e$ then the system can persist indefinitely even if individual populations suffer a high probability of extinction.

If the number of patches is finite, however, there is some probability that, by chance, all occupied patches will go extinct simultaneously owing to either demographic or environmental stochasticity, even if extinction events are independent of one another (Nisbet and Gurney, 1982; Hanski, 1989). Generally, that probability increases as the number of patches decreases. Under the demographic stochasticity scenario, the expected persistence of the metapopulation would be less than expected were all local populations to be pooled into one large unit. Under the environmental stochasticity model the subdivided metapopulation would persist longer than a single large population (Quinn and Hastings, 1987). This need not be true if the assumption of independence is relaxed and local extinctions are correlated (Harrison and Quinn, 1989; Gilpin, 1990).

The most simple models ignore within-population dynamics and the spatial relationship among populations. Several recent models have focused on the relationship between the short-term dynamics of individuals within local populations, dispersal, and the long-term dynamics of the metapopulation, or are spatially explicit. Examples include Hastings (1991), Gyllenberg and Hanski (1992), Mangel and Tier (1993), and Perry and Gonzalez-Andujar (1993). One approach that appears promising for empirical biologists (Hanski, 1994) uses patch characteristics and incidence functions to predict the dynamics of metapopulations.

8.3 GENETIC CONSEQUENCES OF POPULATION TURNOVER

Classically, genetic variation is defined as the amount of allelic diversity present at some number of gene loci, and the relative frequency of those alleles. It is an important characterization of populations because their short-term evolutionary potential is determined by levels of genetic variation (Fisher, 1958; Falconer, 1989) and because erosion of genetic variation is thought to threaten population viability (Gilpin and Soule, 1986). When considered across numerous populations, total variation can be partitioned into a within-population component reflecting the probability that individuals from the same population will carry the same allele and an among-population component reflecting the probability that individuals from different populations will carry the same allele. In many mathematical models the genetic consequences of population structure are evaluated in terms of the contrasting effects of random genetic drift, the random genetic divergence of populations associated with finite local population size, and gene flow, the genetic homogenization of populations associated with migration. Assuming a large number of populations of constant size and migration rate, these forces can become balanced, resulting in an equilibrium distribution of genetic variation within and among populations (Slatkin, 1985). This distribution is often quantified by Wright's F statistics or by the probability that alleles are identical by descent. Wright's F_{st} (Wright, 1977, 1978) is a measure of the among-population component of variation and can range from 0 when all populations are genetically equivalent (no allele frequency differences) to 1 when populations display fixed allelic differences. The conditions necessary for genetic drift and gene flow to come into balance (numerical constancy and long-term population persistence) are most closely met by the population structure outlined in Figure 8.1(a). Certainly these conditions will be approximated in many species, but what of the cases in which local populations are much less stable?

When populations are subject to a high rate of turnover (i.e. when extinction and colonization events are common; Figure 8.1(c)) most populations are but a few generations removed from their respective founding events. Newly established populations are almost certainly going to have demographic properties very different from those of mature or well-established populations. Most obviously, a previously empty patch is likely to be colonized by far fewer individuals than it can ultimately support. At first glance, a colonization event might then appear to be equivalent to a so-called population bottleneck and provide an opportunity for a sampling episode akin to genetic drift. However, the colonists are not derived *in situ*, but rather have migrated to the empty patch from some other locality or localities. In that sense colonization also represents an opportunity for the movement of genes in a process similar to gene flow. This dual nature of colonization raises two questions about systems that display a continual replacement of established populations with those that are newly founded. First, how might the turnover of populations be incorporated into models of genetic structure? Second, do these models suggest that metapopulations have genetic properties different from systems in which the component local populations are capable of longer-term persistence? In approaching this question consider that such models would have to evaluate the genetic structure that can be derived from colonization events and also how that structure might be modified in the interval between establishment and extinction by the more traditional forces of genetic drift and gene flow.

There have been several attempts to incorporate the turnover of populations into models of genetic structure (Wright, 1940; Slatkin, 1977; Maryuma and Kimura, 1980; Ewens *et al.*, 1987; Wade and McCauley, 1988; Gilpin, 1991; Whitlock and McCauley, 1990; Lande, 1992). They have addressed two related issues: the effect of extinction/colonization on (a) the pattern and scale by which genetic variation is distributed within and among populations and (b) the magnitude of genetic diversity that can be maintained by the metapopulation as a whole. The following discussion will focus on these two issues by addressing some of the conclusions, rather than the mathematical details, of the models. Some of the parameters incorporated into the models, and their influence on genetic structure, are listed in Table 8.1.

The first issue has been addressed most cogently in a model developed by Slatkin (1977). The properties of this model have been discussed in detail elsewhere (Wade and McCauley, 1988; McCauley, 1991, 1993) and will be summarized here. One version of the model assumes a large number of local populations of size N. These populations exchange 'island-type' migrants at a rate m per generation. If these populations were to persist indefinitely the forces of drift and gene flow would come to an equilibrium with regard to the distribution of genetic variation at a rate determined by N, m and the initial conditions (Varvio, Chakraborty and Nei, 1986; Whitlock, 1992a). If F_{st} represents the proportion of genetic variation that is distributed among populations, then at equilibrium F_{st} equals approximately $1/(4Nm+1)$. However, Slatkin (1977) further assumes that a proportion, e, of the populations go extinct at random

Table 8.1 Some parameters used in the models of extinction/colonization mentioned in the text, their biological interpretation, and some comments on their influence on the behavior of the models

Parameter	Interpretation	Comments
F_{st}	The among-population component of genetic variation	See below
k	Size of colonizing group	F_{st} increases with decreasing k
N	Size of established population	Determines strength of genetic drift
m	Rate of migration among populations	Determines rate of gene flow
Nm	Number of individuals that move between populations (see above)	Determines F_{st} in the absence of population turnover
ø	Probability that two founders come from the same source	F_{st} increases with ø
e	Probability of population extinction	F_{st} increases with e
n	Number of local populations	Genetic variation lost from metapopulation with small n

each generation and are replaced by an equal number of new populations, each founded by k individuals. The populations then grow immediately to size N.

This continual turnover of populations can be thought of as a chronic disturbance capable of keeping the metapopulation from obtaining the theoretical equilibrium between drift and gene flow. The founding events generate some amount of genetic variation among new populations owing to a sampling process associated with the finite size of k. As a sampling phenomenon, the magnitude of this variation is obviously inversely proportional to k. It is also determined, however, by the way in which dispersing individuals are distributed onto empty patches. Slatkin (1977) and Wade (1978) contrast two extreme modes of colony formation. In one, the 'migrant-pool' mode, each gene copy entering a given empty patch is drawn at random from all possible extant populations. Thus, a colonizing group represents a mix of genotypes drawn from the metapopulation at large. Variation among new populations derives from the binomial sampling of the genetic variation contained within the entire metapopulation and is proportional to $1/2k$. Genetic variation among new populations can be greater or less than that found among previously established populations, depending on the relationship between k, N and m (Wade and McCauley, 1988).

At the other extreme, the 'propagule-pool' mode of colony formation, each set of $2k$ gene copies involved in founding a given population is drawn from just one of the possible source populations, though different founding events can represent draws from different sources. Thus, the sampling process always amplifies any standing variation already found among established populations, provided $k<N$. Figure 8.2(a)-(c) illustrates the two modes of colonization along with a case intermediate to the two extremes that is addressed by Whitlock and McCauley (1990). In that paper a term, ø, is introduced to the models as the probability that two gene copies present in a founding group are drawn from the same source population. This represents a generalized extension of the Slatkin model: ø of zero equals the migrant pool, ø of one the propagule pool, and $1>ø>0$ cases intermediate to the two extremes. Whitlock and McCauley were able to show that genetic variation among new populations will exceed that among long-established populations provided $k<[2Nm/(1-ø)]+1/2$.

The net effect of extinction/colonization on the genetic structure of metapopulations depends not only on the genetic differentiation of new populations as set up by founding events but also on what happens to that differentiation in the generations subsequent to colonization. A group of newly established populations can be thought of as a cohort whose membership declines with time according to the extinction rate. After founding, the genetic characteristics of that cohort are influenced by the standard processes of genetic drift (a function of N) and gene flow (a function of m). This means that an age cohort of populations can converge genetically after founding, or diverge further. If all populations were to persist indefinitely they would reach the approximate steady state $F_{st}=1/(4Nm+1)$. However, because they are subject to some probability of extinction this will not happen unless e is very small. In the Slatkin model (and the Wade–Whitlock–McCauley extension) the metapopulation has a population level age distribution that is determined by the extinction rate. The metapopulation-wide F_{st} is the weighted average of the cohort-specific F_{st} values. With high e (say greater than 0.10), the metapopulation consists primarily of newly or recently established populations and its genetic characteristics are largely derived from the founding process. With low e (say less than 0.01) the consequences of founding events are largely ameliorated by the subsequent effects of drift and migration.

The most significant conclusion that can be drawn from the models is that in metapopulations the sampling associated with colonization can play a major role in determining genetic structure. The nature of that sampling depends not just on the size of the colonizing group but also where the colonists come from. The net influence of colonizing events on genetic structure is proportional to the extinction rate, since that determines the population level age structure. The longer populations persist, the greater the opportunity for the genetic structure set up by

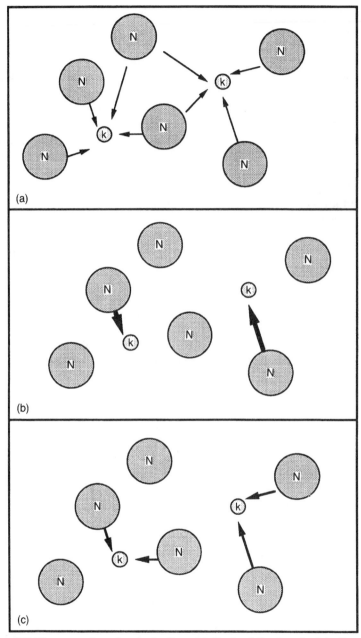

Figure 8.2 Three modes of colonization. (a) Each of the *k* colonists (*k*=4) that found populations are drawn individually from one of the source populations (ø=0, this is the migrant-pool mode). (b) All four individuals are drawn from a common source (ø=1, this is the propagule-pool mode). (c) Two individuals are drawn from each of two sources and combine to found populations (ø=0.5).

colonization to be modified by the combined effects of drift and gene flow; processes determined by Nm.

The preceding paragraphs focus on the way in which genetic variation is partitioned within and among a large number of populations. Genetic drift and/or colonization events convert some amount of the within-population component of variation to among-population variation. A related question that is also raised by population geneticists asks how much total genetic variation can be maintained by a system at large, regardless of the manner in which it is partitioned among its component units. In structured populations this becomes important when the number of populations, n, is limited. If n and N are both relatively small numbers then the total number of individuals in the system, nN, can be small enough to allow for a second type of genetic drift, drift at the level of the whole system. That is, one must be concerned with the absolute loss of genetic variation from the system, as well as from its component populations. This is of particular concern in metapopulations since the pooled effective population size is not nN, but rather some smaller number owing to the continual population reductions that are associated with local founding events.

Several models have addressed this issue. Maruyama and Kimura (1980) developed a model with propagule type colonization that shows that population turnover can greatly reduce the overall effective population size and limit the amount of genetic (allelic) diversity that can be maintained by a metapopulation, especially when k is small. Ewens et al. (1987) reiterate this point, also using a propagule-pool approach in modelling colonization. They emphasize that the loss of genetic variation is a two stage process – loss from within populations and loss from all populations. They further point out that eventually every population will trace its ancestry to just one of the original populations through a kind of coalescence process. Gilpin (1991) uses the computer simulation approach to reach qualitatively the same conclusion. Finally, McCauley (1991) used a version of the Slatkin (1977) model to show that this metapopulation-wide drift may not be as severe when colonization is of the migrant-pool type. In any event metapopulations with a small number of subunits and a high turnover rate cannot support extensive genetic variation.

When considering metapopulations with a limited number of patches it is important to consider an apparent paradox. That is, frequent extinction and colonization can increase the among-population component of genetic variation, as measured by F_{st}, yet limit the absolute magnitude of that differentiation. This is because F_{st} is a measure of the *relative* distribution of genetic variation within and among populations, not a measure of absolute differentiation. When drift acts at the level of the entire metapopulation to reduce genetic variation it also limits the absolute magnitude of interpopulational differentiation that is possible. If even a limited amount of variation is distributed primarily among populations, however, then F_{st} will still be high. In the extreme, should all genetic variation be lost from a metapopulation, no differentiation is possible and F_{st} becomes undefined.

8.3.1 BROADENING THE ECOLOGICAL CONTEXT

Theory suggests that the spatio-temporal processes associated with the continued turnover of populations can have an important impact on the genetic structure of species whose populations are subdivided across mosaic landscapes (Chapter 7). Theory is limited, of course, by the many assumptions required to achieve generality or mathematical tractability. One implicit assumption of the genetic models is that the metapopulation will persist for a period sufficient for the expression of genetic effects of population turnover. In contrast, the more ecologically oriented models recognize that the conditions that allow metapopulations to persist for long periods can be limited. Indeed, identifying the conditions that allow for persistence has been the major motivation for the development of these models. This section will ask how the ecological requirement that the condition of persistence be met might influence our view of genetic structure.

Conservation biologists have been concerned with the erosion of genetic variation in populations small enough to be subject to severe drift and inbreeding (Frankel and Soule, 1981; Schonewald-Cox et al., 1983; Gilpin and Soule, 1986). In theory, the loss of genetic variation could be a real threat to small populations. Lande (1988) questioned the significance of these genetic issues for the viability of real populations, however, because populations small enough for genetic drift to be a threat to viability are also threatened with extinction owing to demographic factors, often on a time-scale much shorter than that over which the negative effects of genetic drift might be felt. Extending this logic by analogy to metapopulations, one must ask whether the characteristics of extinction and colonization suggested by the models to have the greatest influence on genetic structure would also threaten the persistence of the metapopulation itself. This might be answered by combining insights gained from the genetic and the ecological models (Chapter 12).

One assumption of most of the genetic metapopulation models is that newly extinct populations are replaced one for one each generation by the colonization of empty patches. That is, the metapopulation is guaranteed to persist no matter what the extinction rate. In contrast, the Levins metapopulation model allows extinction and colonization to occur with different probabilities. This is important because in the Levins model the colonization rate has to equal or exceed the extinction rate in order for the metapopulation to persist. The requirement that colonization exceeds extinction suggests that patches must show a fairly high degree of connectivity or that dispersal is common – conditions that generally limit genetic differentiation.

In the genetic models the effects of extinction/colonization are generally maximized when k is small and e is large. It may be difficult to obtain these conditions and still allow for long-term persistence. Colonization events must be frequent (high dispersal to offset high e) yet involve small numbers of individuals (small k), an apparent contradiction. In Gilpin's (1991) simulations, extinction and colonization rates were decoupled. In that case the effect of population turnover on genetic structure was limited by the instability of the metapopulation over much of the parameter space. This problem becomes worse when the number of patches is small, the case that allows for the loss of genetic variation from the metapopulation at large. The ecological approach recognizes the possibility that when the number of patches is limited, all patches could go extinct simultaneously, even when extinction events are uncorrelated (Hanski, 1989).

Another assumption of the genetic models is that all populations have equal and independent probabilities of extinction. While this was also an assumption of the original Levins model, a second generation of ecological models has recognized that this need not be the case. For example, it is also possible that extinction events could be correlated across populations. This possibility has been investigated recently because correlated extinctions could greatly reduce metapopulation stability (Hanski, 1989). This would be relevant to the genetic models because again a condition that would appear to promote the greatest genetic effect (high local extinction rate) might in fact simply cause the entire metapopulation to collapse. In a related issue, Harrison (1991) points out that in nature, populations typically vary tremendously in their extinction probabilities; for example when patch size or quality varies. It may be that a common scenario in nature is one in which a fraction of patches support populations capable of long-term persistence while others wink in and out at a higher rate. This is much closer to the 'mainland–island' model than a true metapopulation, at least as that term is used here. In this case the genetic properties of the stable populations would be controlled by within-patch demographic properties while ephemeral patches would be under a more regional control.

A third issue that merits discussion is the assumption of both the genetic and Levins models that populations grow to their maximum size immediately after founding ($k{\rightarrow}N$ in one generation). The ecological consequences of relaxing this assumption have been addressed recently (Ebenhard, 1991). If N is much larger than k then reaching carry capacity in one generation is unlikely in all but the most fecund organisms.

Otherwise, newly established populations must necessarily pass through some number of generations in which their numbers fall between k and N. Relatively small population size in the generations immediately following founding might have two implications for the genetics of metapopulations. First is the role demographic stochasticity might play in the early history of populations. When populations are very small there is some probability that they will go extinct each generation by chance, for example when all offspring happen to be of the same sex. Generally, populations have to be quite small for this to be of great concern. This might appear to limit the significance of demographic stochasticity as an agent of extinction. This may well be the case once populations become established. The possibility of extinction in the early history of populations could be a very real possibility, however, if the founding group is small in number and if population growth is limited for a number of generations after colonization (Richter-Dyn and Goel, 1972; Ebenhard, 1991). In particular, with slow population growth, colonizing groups below some critical size would have little chance of establishing successful populations. The significance of this for genetic structure is that, while the theoretical minimum k in sexual species is two, it may be that only colonizing events involving a much larger number of individuals actually contribute to the persistence of the metapopulation. Recall that it is when k is small that founding events are most likely to generate among-population genetic structure.

One might also consider that the influence of gene flow could vary with population size. At first glance it would seem that slow population growth in the generations immediately following colonization would provide an opportunity for genetic drift and promote among-population differentiation. This would certainly be true in the absense of migration. When immigration is considered, however, the effect of small population size on genetic differentiation is more problematic. This is because population size and immigration rate could covary. A positive covariation might arise when there is some type of conspecific attraction (Smith and Peacock, 1990). Most simply, larger populations might be easier for dispersing individuals to find. In that case, recently established populations might be effectively more isolated, as well as smaller, than mature populations. This would allow for exceptional opportunities for genetic drift, at least for a few generations. A negative covariation is also possible, however. Consider the case in which a constant number of immigrants move into a patch per generation. That constant number would represent a larger proportion of a smaller population than a large one. In that case high rates of gene flow early in the history of populations would limit their divergence.

Finally, dispersal out of occupied patches can result in two types of events: movement into unoccupied patches (colonization) and movement into another occupied patch (migration). In the Levins model, dispersing individuals are divided among the two types of patches in proportion to their frequency. In terms of the genetic models this is equivalent to letting $k=Nm$. (Recall that Wade and McCauley (1988) and Whitlock and McCauley (1990) pointed out the importance of the relative sizes k and Nm for genetic structure.) Ray, Gilpin and Smith (1991) develop a model that allows dispersers to move preferentially into occupied patches as a result of conspecific attraction. This is one mechanism by which k can be considerably smaller than Nm, allowing for genetic structure in a system that otherwise displays considerable gene flow. Ray, Gilpin and Smith (1991) point out, however, that conspecific attraction tends to reduce metapopulation persistence.

8.3.2 CODA

One motivation for developing metapopulation models is to provide a conceptual framework for empirical studies. Certainly studies of the distribution of genetic variation within and among natural populations have been the mainstay of empirical population genetics for the last 25 years. Included in this literature are various attempts to link the descriptive genetics of natural populations to ecological processes such as the interpopulational movement that is the prerequisite for gene flow (Nevo, 1978; McCauley and Eanes, 1987; Hamrick and Godt, 1990; Barton, 1992). In general, there is a negative correspondence between the perceived vagility of the species and the amount of population differentiation, as measured by the distribution of genetic variants such as allozymes. Theoretical population genetics has provided much of the framework for this attempt (Slatkin, 1985). The models described here suggest that in some species the metapopulation dynamics of extinction and colonization must also be investigated as causative agents when attempting to link genetics and ecology (Chapter 12). Despite the appeal of the metapopulation construct, relatively little is known about the turnover of natural populations, though empirical studies of this process are increasing in number (e.g. Harrison, Murphy and Ehrlich, 1988; Schoener and Spiller, 1987, 1992; McCauley, 1989; Peltonen and Hanski, 1991; Sjogren, 1991; Kindvall and Ahlen, 1992; Hanski, Kuussaari and Nieminen, 1994). The addition of a genetic component to the study of natural metapopulations is rare, but Whitlock (1992b) and Dybdahl (1994) present studies that link the metapopulation dynamics of two invertebrate species to their genetic structure. One very interesting current line of inquiry addresses the coevolution of plants and their pathogens in a metapopulation context (Burdon and Jarosz, 1991; Antonovics et al., 1994).

It is not yet clear how often the demographic properties of a species place it in a parameter space in which population turnover is an important determinant of genetic structure. Evaluating this question for a particular species is largely an empirical issue. Most obviously one must ask how often populations turn over. The models suggest that it is equally important to investigate the conditions that permit the establishment of new populations. How many individuals are required to start a population and how does the texture of the landscape influence where they come from? These questions should be approached by a combination of experimental and observational field studies. The incidence function approach of Hanski (1994) may prove useful in this regard, as it has been used successfully in the analysis of a butterfly metapopulation (Hanski, Kuussaari and Nieminen, 1994). One broader goal of empirical studies might be to identify those life history or taxonomic affiliations common to species that fall in this category.

Undoubtedly empirical studies will show the real world to be far more complex than even the most elaborate model. In real landscapes patches must vary in their size and degree of isolation from one another. Patches themselves come and go, not just the species that occupy them. They are likely to be organized temporally and spatially into some hierarchical arrangement much more complex than the simple metapopulation scenario described here – 'meta-metapopulations' if you will. Patches must also vary in time and space in those characteristics that define the 'quality' of the patch to the organism in question (Chapter 5). In order to persist, a species must cope with this variation. Incorporation of a genetic component into more spatially explicit ecological models may be helpful.

The question of how species adapt to heterogeneous environments has long been a central topic in evolutionary biology (Levene, 1953; Levins, 1968). Adaptation requires selection. It is not yet clear how natural selection will act in metapopulations, given the complex spatio-temporal variation in the environmental parameters that determine fitness. The models discussed in this chapter that address the behavior of selectively neutral genetic variation simply provide a point of reference. The real issue is how the demographic dynamics imposed by mosaic landscapes influence adaptation. Perhaps insights from the field of landscape ecology will help to focus this question.

8.4 SUMMARY

When a species is limited to a subset of patch types in a mosaic landscape, the patch structure may subdivide it into a number of local populations interconnected by some amount of migration. If so, both the demographic and genetic characteristics of those populations are under dual local (due to birth and death events within the patch) and regional (due to immigration from other patches) control. If local populations suffer a high rate of extinction, but are replaced by the simultaneous colonization of empty patches, the system is known as a metapopulation. Attempts to model the genetic characteristics of metapopulations have focused on whether the sampling processes associated with the founding of populations differ from the generation to generation processes of genetic drift and gene flow that are associated with more stable populations. Two major conclusions are evident. First, with frequent turnover of populations the distribution of genetic variation within and among patches is determined largely by the colonization process and less so by events after colonization. Equally important are the size of colonizing groups and whether the individuals comprising a colonizing group are drawn together from a single source patch or separately from several sources. Second, when the number of occupied patches is relatively small, the frequent turnover of populations enhances the rate at which genetic variation is lost from the metapopulation as a whole.

The genetic models assume that extinction and colonization are balanced such that the metapopulation will persist indefinitely. Ecological models of metapopulations focus on the conditions that will permit that persistence. This chapter questions whether the conditions under which the genetic models predict that extinction and colonization would have their greatest effects are those that would also permit persistence.

ACKNOWLEDGEMENTS

I would like to thank Michael Wade for his comments and I. Hanski and M. Dybdahl for access to unpublished manuscripts. This work was supported by NSF award DEB–9221175.

REFERENCES

Antonovics, J., Thrall, P., Jarosz, A. and Stratton, D. (1994) Ecological genetics of metapopulations: The *Silene–Ustilago* plant–pathogen system, in *Ecological Genetics* (ed. L. Real), Princeton University Press, Princeton, NJ, pp. 146–170.

Barton, N. H. (1992) The genetic consequences of dispersal, in *Animal Dispersal Small Mammals as a Model* (eds N. C. Stenseth and W. L. Lidicker Jr), Chapman & Hall, New York, pp. 37–59.

Burdon, J. J. and Jarosz, A. M. (1992) Temporal variation in the racial structure of flax rust (*Melampsora lini*) populations growing on natural stands of wild flax (*Linum marginale*): local versus metapopulation dynamics. *Plant Path.*, **41**, 165–79.

Crow, J. F. and Kimura, M. (1970) *An Introduction to Population Genetic Theory*, Harper and Row, New York.

Dybdahl, M. F. (1994) Extinction, recolonization, and the genetic structure of tidepool copepod populations. *Evol. Ecol.*, **8**, 113–124.

Ebenhard, T. (1991) Colonization in metapopulations: A review of theory and observations. *Biol. J. Linn. Soc.*, **42**, 105–21.

Ewens, W. J., Brockwell, P. J., Gani, J. M. and Resnick, S. I. (1987) Minimum viable population size in the presence of catastrophes, in *Viable Populations for Conservation* (ed. M. E. Soule), Cambridge University Press, Cambridge, pp. 59–68.

Falconer, D. (1989) *Introduction to Quantitative Genetics*, 3rd edn, Longman, New York.

Fisher, R. A. (1958) *The Genetical Theory of Natural Selection*, Dover, New York.

Frankel, O. H. and Soule, M. E. (1981) *Conservation and Evolution*, Cambridge University Press, Cambridge.

Gilpin, M. E. (1990) Extinction of finite metapopulations in correlated environments, in *Living in a Patchy Environment* (eds B. Shorrocks and I. Swingland), Oxford University Press, Oxford, pp. 177–86.

Gilpin, M. E. (1991) The genetic effective size of a metapopulation. *Biol. J. Linn. Soc.*, **42**, 165–75.

Gilpin, M. E. and Soule, M. E. (1986) Minimum viable populations: Processes of species extinction, in *Conservation Biology the Science of Scarcity and Diversity* (ed. M. E. Soule), Sinauer Sunderland, MA, pp. 19–34.

Gyllenberg, M. and Hanski, I. (1992) Single-species metapopulation dynamics: A structured model. *Theor. Pop. Biol.*, **42**, 35–61.

Hamrick, J. L. and Godt, M. J. (1990) Allozyme diversity in plant species, in *Plant Population Genetics, Breeding, and Genetic Resources* (eds A. H. D. Brown, M. T. Clegg, A. L. Kahler and B. S. Weir), Sinauer, Sunderland, MA, pp. 43–63.

Hanski, I. (1989) Metapopulation dynamics: does it help to have more of the same? *Trends Ecol. Evol.*, **4**, 113–14.

Hanski, I. (1991) Single-species metapopulation dynamics: concepts, models and observations. *Biol. J. Linn. Soc.*, **42**, 17–38.

Hanski, I. (1994) A practical model of metapopulation dynamics. *J. Anim. Ecol.*, **63**, 151–62.

Hanski, I. and Gilpin, M. (1991) Metapopulation dynamics: brief history and conceptual domain. *Biol. J. Linn. Soc.*, **42**, 3–16.

Hanski, I., Kuussaari, M. and Nieminen, M. (1994) Metapopulation structure and migration in the butterfly *Melitaea cinxia*. *Ecology*, **75**, 747–61.

Harrison, S. (1991) Local extinction in a metapopulation context: an empirical evaluation. *Biol. J. Linn. Soc.*, **42**, 73–88.

Harrison, S. and Quinn, J. F. (1989) Correlated environments and the persistence of metapopulations. *Oikos*, **56**, 293–8.

Harrison, S., Murphy, D. D. and Ehrlich, P. R. (1988) Distribution of the bay checkerspot butterfly, *Euphydryas editha bayensis*: Evidence for a metapopulation model. *Am. Nat.*, **132**, 360–82.

Hartl, D. L. and Clark, A. G. (1989) *Principles of Population Genetics*, Sinauer, Sunderland, MA.

Hastings, A. (1991) Structured models of metapopulation dynamics. *Biol. J. Linn. Soc.*, **42**, 57–71.

Kindvall, O. and Ahlen, I. (1992) Geometrical factors and metapopulation dynamics of the bush cricket, *Metrioptera bicolor* Philippi (Orthoptera: Tettigoniidae). *Conserv. Biol.*, **6**, 520–9.

Lande, R. (1988) Genetics and demography in biological conservation. *Science*, **241**, 1455–60.

Lande, R. (1992) Neutral theory of quantitative genetic variation in an island model with local extinction and recolonization. *Evolution*, **46**, 381–9.

Levene, H. (1953) Genetic equilibrium when more than one ecological niche is available. *Am. Nat.*, **87**, 331–3.

Levin, S. A. (1992) The problem of pattern and scale in ecology. *Ecology*, **73**, 1943–67.

Levins, R. (1968) *Evolution in Changing Environments*, Princeton University Press, Princeton, NJ.

Levins, R. (1969) Some demographic and genetic consequences of environmental heterogeneity for biological control. *Bull. Ent. Soc. Am.*, **15**, 237–40.

Levins, R. (1970) Extinction, in *Some Mathematical Problems in Biology* (ed. M. Gerstenhaber), American Mathematical Society, Providence, RI, pp. 77–107.

Mangel, M. and Tier, C. (1993) Dynamics of metapopulations with demographic stochasticity and environmental catastrophes. *Theor. Pop. Biol.*, **44**, 1–31.

Maruyama, T. and Kimura, M. (1980) Genetic variability and the effective size when local extinction and recolonization are frequent. *Proc. Nat. Acad. Sci. USA*, **77**, 6710–14.

McCauley, D. E. (1989) Extinction, colonization, and population structure: A study of a milkweed beetle. *Am. Nat.*, **134**, 365–76.

McCauley, D. E. (1991) Genetic consequences of local population extinction and recolonization. *Trends Ecol. Evol.*, **6**, 5–8.

McCauley, D. E. (1993) Genetic consequences of extinction and colonization in fragmented habitats, in *Biotic Interactions and Global Change* (eds P. M. Kareiva, J. G. Kingsolver and R. B. Huey), Sinauer Associates, Sunderland, MA, pp. 217–33.

McCauley, D. E. and Eanes, W. F. (1987) Hierarchical population structure analysis of the milkweed beetle, *Tetraopes tetraophthalmus. Heredity*, **58**, 193–201.

Nevo, E. (1978) Genetic variation in natural populations: pattern and theory. *Theor. Pop. Biol.*, **13**, 121–77.

Nisbet, R. M. and Gurney, W. S. C. (1982) *Modelling Fluctuating Populations*, John Wiley and Sons.

Quinn, J. F. and Hastings, A. (1987) Extinction in subdivided habitats. *Conserv. Biol.*, **1**, 198–208.

Peltonen, A. and Hanski, I. (1991) Patterns of island occupancy explained by colonization and extinction rates in shrews. *Ecology*, **72**, 1698–708.

Perry, J. N. and Gonzalez-Andujar, J. L. (1993) Dispersal in a metapopulation neighborhood model of an annual plant with a seedbank. *J. Ecol.*, **81**, 453–63.

Pulliam, H. R. (1988) Sources, sinks, and population regulation. *Am. Nat.*, **132**, 652–61.

Ray, C., Gilpin, M. E. and Smith, A. T. (1991) The effect of conspecific attraction on metapopulation dynamics. *Biol. J. Linn. Soc.*, **42**, 123–34.

Richter-Dyn, N. and Goel, N. S. (1972) On the extinction of a colonizing species. *Theor. Pop. Biol.*, **3**, 406–33.

Schoener, T. W. and Spiller, D. A. (1987) High population persistence in a system with high turnover. *Nature*, **330**, 474–7.

Schoener, T. W. and Spiller, D. A. (1992) Is temporal variability in population size related to extinction rate? An empirical answer for orb spiders. *Am. Nat.*, **139**, 1176–207.

Schonewald-Cox, C. M., Chambers, S. M., MacBryde, F. and Thomas, L. (eds) (1983) *Genetics and Conservation: A Reference for Managing Wild Animal and Plant Populations*, Benjamin Cummings, Menlo Park, CA.

Sjogren, P. (1991) Extinction and isolation gradients in metapopulations: the case of the pool frog (*Rana lessonae*). *Biol. J. Linn. Soc.*, **42**, 135–47.

Slatkin, M. (1977) Gene flow and genetic drift in a species subject to frequent local extinctions. *Theor. Pop. Biol.*, **12**, 253–62.

Slatkin, M. (1985) Gene flow in natural populations. *Ann. Rev. Ecol. Syst.*, **16**, 393–430.

Smith, A. T. and Peacock, M. M. (1990) Conspecific attraction and the determination of metapopulation colonization rates. *Conserv. Biol.*, **4**, 320–2.

Varvio, S.-L., Chakraborty, R. and Nei, M. (1986) Genetic variation in subdivided populations and conservation genetics. *Heredity*, **57**, 189–98.

Wade, M. J. (1978) A critical review of the models of group selection. *Q. Rev. Biol.*, **53**, 101–14.

Wade, M. J. and McCauley, D. E. (1984) Group selection: The interaction of local deme size and migration in the differentiation of small populations. *Evolution*, **38**, 1047–58.

Wade, M. J. and McCauley, D. E. (1988) Extinction and recolonization: Their effects on the genetic differentiation of local populations. *Evolution*, **42**, 995–1005.

Whitlock, M. C. (1992a) Temporal fluctuations in demographic parameters and in genetic variance among populations. *Evolution*, **46**, 608–15.

Whitlock, M. C. (1992b) Non-equilibrium population structure in forked fungus beetles: Extinction, colonization, and the genetic variance among populations. *Am. Nat.*, **139**, 952–70.

Whitlock, M. C. and McCauley, D. E. (1990) Some population genetic consequences of colony formation and extinction: genetic correlations within founding groups. *Evolution*, **44** 1717–24.

Wright, S. (1931) Evolution in mendelian populations. *Genetics*, **16**, 97–159.

Wright, S. (1940) Breeding structure of populations in relation to speciation. *Am. Nat.*, **74**, 232–48.

Wright, S. (1977) *Evolution and the Genetics of Populations. Vol. 3, Experimental Results and Evolutionary Deductions*, University of Chicago Press, Chicago.

Wright, S. (1978) *Evolution and the Genetics of Populations. Vol. 4, Variability Within and Among Natural Populations*, University of Chicago Press, Chicago.

Part Four

Effects of Landscape Pattern on Species Interactions

As discussed in the previous sections, effects of landscape pattern on individuals, population dynamics and population genetics are complex. Spatial landscape pattern affects movement of organisms, which in turn affects spatial patterns of populations and genetics. Spatio-temporal landscape pattern affects resource tracking and population dynamics, which in turn affect genetic processes.

Similar species will use similar resources and potentially occur in similar habitat patches. Competition will ensue under certain conditions. Resource-tracking species are also food for predators, or may have mutualistic relationships with other organisms. Thus, the landscape pattern and composition will affect species interactions and whole biotic communities. The chapters in this section discuss effects of landscape pattern on species interactions, including competition, predation and mutualism. Ilkka Hanski reviews the variety of possible effects, suggested over the past three decades, of spatial and spatio-temporal patterns of resource distribution on the outcome of competitive interactions. Predator–prey interactions are affected by the landscape composition, productivity of resources patches, and the distances among resources patches in the landscape. The chapter by Henrik Andrén demonstrates the importance of landscape composition in determining local predation rates. Finally, Judith Bronstein presents a novel view of plant–pollinator systems as landscapes in which the spatio-temporal dynamics of plant phenology determine pollinator survival.

Effects of landscape pattern on competitive interactions

9

Ilkka Hanski

9.1 INTRODUCTION

Interspecific competition for shared and limiting resources is widely thought to be one of the forces, perhaps even the main force, that has shaped biodiversity in the past and continues to shape it in the present. In the course of evolution, new taxa have replaced old ones (Simpson, 1949; Stanley, 1979; for a particular taxon, see e.g. Cambefort, 1991a), and although we remain largely ignorant about the actual mechanisms, competition is most likely involved. Natural invasions and human introductions of alien species have often led to a significant reduction in abundances (Elton, 1958) or complete elimination of native species (Greenway, 1967; Ebenhard, 1988), although how frequently aliens have competitively eliminated natives remains a matter of debate (e.g. Simberloff, 1981, versus Herbold and Moyle, 1986). One likely reason why aliens do not routinely cause a rapid extinction of native species is the arena of their competition, a diverse landscape mosaic, which in various ways helps species to coexist – the subject matter of this chapter.

The term 'landscape mosaic' implies an entity, the landscape, which consists of entities at a lower hierarchical level, often called habitat patches (or habitat fragments). Real landscapes typically consist of habitat patches which vary in many ways, including their size and

Mosaic Landscapes and Ecological Processes.
Edited by Lennart Hansson, Lenore Fahrig and Gray Merriam.
Published in 1995 by Chapman & Hall, London. ISBN 0 412 45460 2

geometry, quality, isolation and, especially nowadays, rates of change in these parameters. The term 'landscape mosaic' implies the presence of at least two habitat types. For the purpose of conceptual clarity, I have structured my discussion of landscape-level competition under three different scenarios (Table 9.1). In the first scenario, the landscape consists of many, but identical, habitat patches. This special case is challenging conceptually and theoretically. The key question here is, given that two or more competing species cannot coexist locally, in a single habitat patch, under which conditions can they coexist in a landscape with many 'replicate' patches of the same habitat type? The fact that real landscapes typically consist of many qualitatively different habitat types does not mean that this question would be uninteresting, as the differences in habitat type may not always be significant for coexistence: species may specialize on the same habitat type. Table 9.1 lists three different mechanisms, discussed in section 9.2, which may facilitate coexistence in landscapes with identical habitat patches.

Table 9.1 Mechanisms of coexistence of competitors at the landscape level when local coexistence within a single habitat patch is not possible

Landscape consists of many habitat patches of. . .	Mechanism of coexistence
1. The same type	a. Fugitive coexistence
	b. Aggregated spatial distributions
	c. Alternative local equilibria
2. Quantitatively different types	a. Trade-offs related to differential sensitivity to resource levels
	b. Differences in dispersal rate
3. Qualitatively different types	a. Habitat selection

Note: a lower-level mechanism, e.g. 1a, may facilitate coexistence in a more complex environment, e.g. 3, but not vice versa.

The second scenario introduces quantitative differences among the patches. The patches are still all of the same type, having the same relative abundances of different resources, but some patches are more productive (have more of the same) than others. This is another important special case, discussed in section 9.3.

In the third scenario, described in section 9.4, there are qualitative differences among the habitat patches, which now have different mixtures of resources. This must be the most frequent case in reality, but it is perhaps the least intriguing scenario for this chapter, because the mechanism of coexistence typically involves classical niche differences, as does coexistence within a single habitat patch; only the spatial scale is different. This is not to say that landscape structure would make no difference – it does. For instance, a species which is an intrinsically superior competitor in a rare habitat type may become excluded if that habitat is merged in a landscape mosaic (section 9.4). Finally, though I emphasize throughout this chapter mechanisms of coexistence other than classical resource partitioning, I hasten to add that some ecological differences among the competing species are necessary in all potential mechanisms of coexistence, although these differences do not need to involve resource use (Chesson, 1991).

Before turning to the three scenarios, let us be reminded of a classical result about interspecific competition. The coexistence of two competing species is possible, under any mechanism and scenario of competition, if an individual's competitive action has a more harmful per capita effect on conspecifics than on heterospecifics (Gause, 1934; if the competitive effects are nonlinear, the competition coefficients should be defined accordingly, as discussed by Holt, 1985). To see why this is the case, consider two competing species of which one is common and one happens to be, temporarily, very rare. If they are to coexist, the rare species must be able to increase in relative abundance; otherwise it becomes extinct. Now, because of the rarity of the rare species, individuals of both species mostly experience competition with individuals of the common species. Because these actions harm more of the common than rare species (our assumption above), the rare species experiences less competition because of its rarity and will increase in numbers, until the advantage due to rarity disappears. I call the need for relatively more intraspecific than interspecific competition the general principle of coexistence of competitors, and will refer to it while discussing particular mechanisms below.

9.2 LANDSCAPES WITH IDENTICAL HABITAT PATCHES

9.2.1 IDENTICAL SPECIES

A key result of the classical competition theory (Volterra, 1926), often called Gause's principle, or the principle of competitive exclusion (Hardin, 1960; Hutchinson, 1978), states that two identical species cannot coexist. Identical species cannot coexist because an individual's competitive action harms equally strongly both conspecifics and heterospecifics, in violation of the general principle of coexistence. The relative abundances of two identical competitors are expected to follow a random walk until one of them happens to become extinct. This is theory. Testing Gause's principle empirically is hard, partly because no two species are likely to be exactly identical (and we will never know whether they are), and partly because we generally have no way of telling which differences among species are consequential in competition. Luckily, Gause's principle, by its nature, is not in need of empirical testing.

Surprisingly, and in an apparent violation of Gause's principle, some metapopulation models of competition make the claim that identical species may in fact coexist in landscapes consisting of identical habitat patches (Levins and Culver, 1971; Slatkin, 1974; Christiansen and Fenchel, 1977; Hanski, 1983; Shorrocks, 1990). Does landscape heterogeneity really promote coexistence so much that a fundamental result of the competition theory is overturned?

The metapopulation models with which interspecific competition has been analyzed are 'patch' models, in which local dynamics are ignored, for the sake of mathematical tractability, and only changes in the fractions of habitat patches occupied by the two species are modeled. In the case of two competing species, there are four kinds of patches, namely those occupied by species 1 only, patches occupied by species 2 only, patches jointly occupied by species 1 and 2, and empty patches. The models assume that the fractions of occupied patches are given by a balance between stochastic extinctions and recolonizations, as assumed in Levins's (1969) metapopulation model for a single species. Interspecific competition is assumed to decrease the rate of colonization, or to increase the rate of extinction, or both, and thereby competition may affect this balance.

At a two-species stochastic steady state, a certain fraction of habitat patches is empty, because of the assumed stochastic extinctions. The presence of empty patches means that individuals of both species are spatially aggregated: high density in the occupied patches, zero density in the empty patches. Spatially aggregated distributions may allow coexistence (section 9.2.3), but only if the spatial distributions of the competitors are not completely correlated. Another way of expressing the same is that coexistence in patch models is based on both species enjoying a temporary refuge in the patches in which the competitor happens to be absent. Chesson (1991) has argued that two identical species should generally become identically distributed (have completely correlated spatial distributions), which would remove the 'refuges'. There must, therefore, be some mechanism in the model which prevents the two species from occupying the same set of patches. This mechanism is clearly local competition, which reduces the fraction of jointly occupied patches, because of increased extinction rate in the presence of the competitor, and generates an excess of single-species patches. But does this not violate the general principle of coexistence?

In the patch models, competition may reduce the probability of successful colonization (Slatkin, 1974; Hanski, 1983), but once a species succeeds in colonizing a patch already occupied by its competitor, it will instantly grow to the same equilibrium size as the resident species (after which they both increase each others' probabilities of local extinction). Here is the crux of the matter. This simplifying assumption gives an unfair advantage to the species which happens to be regionally rarer. If the two species are really identical, it is difficult to see why the species to arrive first at a patch would not have a higher chance of winning the patch than the later arriving species, as the first species will enter into competition with a higher local abundance. Because the regionally commoner species is more likely to be the first one in a particular patch, it should also win more patches in local competition. I conclude that ignoring local dynamics, and assuming that a local population jumps instantly to the local equilibrium following colonization, is a sensible first approximation in a single-species model, but this assumption may be badly misleading in the case of two species. In summary, two identical species cannot coexist locally nor in a landscape of many habitat patches, in spite of stochastic extinctions and recolonizations.

9.2.2 FUGITIVE SPECIES

Let us now turn to cases where there are competitive asymmetries between two (or more) species, preventing local coexistence, and where these asymmetries in competitive ability are negatively correlated with another asymmetry in colonization ability. In other words, inferior competitors are superior colonizers and vice versa. Assuming further that there are local extinctions, due to stochastic processes or successional changes in habitat quality, it may be shown theoretically that two such species may coexist, in a mosaic landscape, the inferior species finding a temporary refuge in the habitat patches which the superior competitor, but inferior colonizer, has not yet reached. This landscape mechanism of coexistence is the classical scenario of fugitive coexistence, first described by Hutchinson (1951) and Skellam (1951), and later analyzed, using patch models, by Levins and Culver (1971), Horn and MacArthur (1972), Slatkin (1974), Hanski and Ranta (1983), Hanski (1983) and Nee and May (1991). (For a model including local population sizes, see Hanski and Zhang, 1993.)

A possible example of fugitive coexistence is provided by three species of *Daphnia* water fleas living in small rock pools on islands off the coasts of Finland and Sweden, studied in detail by Ranta (1979), Hanski and Ranta (1983), Pajunen (1986) and Bengtsson (1986, 1988, 1989, 1991). Bengtsson's (1988) studies clearly demonstrate that the species compete severely locally, in a single rock pool. There is plenty of evidence to demonstrate local extinctions (Hanski and Ranta, 1983; Pajunen, 1986; Bengtsson, 1989), which occur at a rate of around 10% per year per population in rock pools occupied by a single species (Table 9.2). Pajunen (1986) and Bengtsson (1988) found no evidence for lowered colonization rate in occupied versus empty pools, but the presence of another species in the same rock pool increases the risk of local extinction (Table 9.2).

The question then is: What allows two or even three species of *Daphnia* to coexist in the same network of rock pools?

Table 9.2 Extinction rate in three species of *Daphnia* water fleas in rock pools on islands off the coasts of Finland and Sweden in the Baltic. The figures are probabilities of extinction per population per year (±SD, with the number of possible extinction events given in brackets). (Sources: Bengtsson, 1988, 1989; Pajunen, 1986)

Locality	Single-species pools	Two-species pools
Flatholmen	0.13 ± 0.037 (82)	0.15 ± 0.046 (58)
Mönster	0.12 ± 0.038 (74)	0.42 ± 0.140 (12)
Ängskär	0.10 ± 0.025 (143)	0.17 ± 0.051 (54)
Tvärminne	0.11 ± 0.028 (123)	0.16 ± 0.052 (50)

Hanski and Ranta (1983) suggested that landscape-level coexistence in these species is based on colonization–extinction dynamics. We also hypothesized size-dependent, inverse hierarchies in the competitive and colonization abilities of the three species, the largest species being the weakest competitor but the best colonizer. However, subsequent observations and experiments, summarized by Bengtsson (1991), do not give strong support to this hypothesis: there appear to exist no consistent differences in the colonization abilities of the three species, whereas their relative competitive strengths depend on the prevailing environmental conditions, especially temperature and the level of food availability. In a laboratory experiment, the smaller species *D. pulex* and *D. longispina* were superior competitors to the large *D. magna* at high temperatures and low food availabilities (Bengtsson, 1991), which may in fact be typical conditions in small rock pools. The smallest species, *D. longispina*, typically occurs only on islands with many rock pools, which is consistent with low colonization ability, but there are other possible explanations.

If not classical fugitive coexistence, what other mechanism might explain the coexistence of the three *Daphnia* species in spite of frequent extinctions? Some islands may have large pools with more persistent populations, which may function as sources of colonists (Pajunen, 1986), but many other islands lack such 'mainland' pools. Another possibility is that there are, in fact, some niche differences among the species, which would give competitive advantage to particular species in particular pools. None the less, the fact remains that on many islands, long-term persistence of the species depends on recurrent colonizations, and that interspecific competition increases the rate of local extinctions. Increased extinction rate due to competition would necessitate more frequent recolonizations for coexistence, and would, hence, restrict the presence of competing species to networks of habitat patches where colonization is easier. In the case of *Daphnia*, colonization rate per empty pool may be expected to depend on the number and density of rock pools on an island. Hence, I predict that species number per network of habitat patches increases with the number of rock pools. This is observed in a comparison of islands with different numbers of rock pools (Figure 9.1(a)). Furthermore, competition should decrease the fraction of pools occupied by each species at equilibrium, which is also observed (Figure 9.1(b)), and competition can explain the observed negative correlations

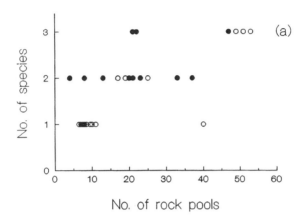

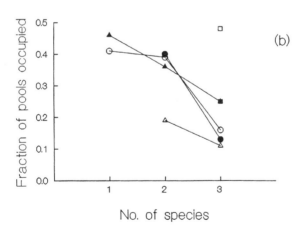

Figure 9.1 (a) The relationship between the number of *Daphnia* species on an island and the number of suitable rock pools. Solid symbols are from islands off the coast of Finland (Source: Hanski and Ranta, 1983); open symbols are from islands off the coast of Sweden (Source: Bengtsson, 1991). (b) The fraction of suitable rock pools occupied against the number of *Daphnia* species on islands in two archipelagoes. Symbols: circles, *D. magna*; triangles, *D. pulex*; squares, *D. longispina* (data from the same sources as above).

in the fractions of occupied pools in two species on a set of islands (Hanski and Ranta, 1983; Bengtsson, 1991). In conclusion, although the well-studied *Daphnia* system may not fit the classical scenario of fugitive coexistence, it seems still probable that the long-term coexistence of these species in systems of rock pools hinges on the landscape-level dynamics – stochastic extinctions and recolonizations.

Two species of corixids, *Arctocorisa carinata* and *Callicorixa producta*, live in the same rock pools as are also used by *Daphnia*. A long series of investigations by Pajunen has revealed that resource competition is often severe and leads to high mortality (Pajunen, 1979). A replacement series experiment revealed a complete overlap of resource use between the two species (Pajunen, 1982). However, strong interference competition (cannibalism) by the larger species (*A. carinata*) gives it a decisive edge in interspecific competition, summarized by the competition coefficients of 2.6 (effect of one *A. carinata* on one *C. producta*) and 0.4 (effect of one *C. producta* on one *A. carinata*; Pajunen, 1979). Locally, the larger *A. carinata* would soon replace the smaller *C. producta*. But the smaller species has about twice the migration rate of the superior competitor. Hence, the inferior species is typically more frequent in newly-filled rock pools (Pajunen, 1979). This example provides very strong evidence for the fugitive mechanism: long-term coexistence of these two species is almost certainly based on inverse asymmetries in their competitive and migration abilities.

Two studies on fungi and mosses have supported the fugitive coexistence mechanism. Armstrong (1976) studied competition between *Aspergillus* and *Penicillium* growing on agar-filled Petri dishes in the laboratory. *Penicillium* matured quickly and produced colonists in abundance, while the competitively superior *Aspergillus* matured slowly and produced colonists at a lower rate. Armstrong (1976) was able to show that the two species coexisted in an environment in which a fraction of the Petri dishes were replaced at intervals. Marino (1988) studied competition and coexistence in four species of mosses in the family Splachnaceae, which grow on dung pats. Individual dung pats in the field are typically dominated by a single species because of competition. Marino (1988) demonstrated that of two species of *Splachnum*, *S. luteum*, the superior competitor locally, produced fewer spores to be carried by flies to fresh dung pats than its competitor, *S. ampullaceum*. One could continue with more examples of this kind, which suggest, but do not critically demonstrate, that the fugitive mechanism is decisive for coexistence.

A community-wide extension of the fugitive mechanism is the so-called 'intermediate disturbance hypothesis' (Grubb, 1977; Connell, 1978; Lubchenco, 1978; Tilman, 1982), which is generally applied to organisms competing for space. If the disturbance rate is low, the few competitively dominant taxa predominate; if the disturbance rate is very high, only species with high growth rate and exceptional colonization ability survive; while at intermediate rates there are more opportunities for species to exist/coexist. Lubchenko's (1978) study of algal species richness in tidal pools with different numbers of herbivores is an oft-cited example. However, it is unclear to what extent her results are due to local as opposed to landscape-level processes.

9.2.3 AGGREGATED SPATIAL DISTRIBUTIONS

A general mechanism of landscape-level coexistence of competitors is based on their spatial distributions. Hanski (1981) and Atkinson and Shorrocks (1981) originally demonstrated that two competitors which could not coexist in a single habitat patch would often coexist in a network of habitat patches in which the frequencies of the two species varied from patch to patch, independently of resource availability. The explanation of coexistence in this case is very straightforward: because of the more or less independently aggregated spatial distributions and localized competitive interactions, the per capita number of conspecific interactions exceeds the per capita number of heterospecific interactions, and coexistence is facilitated by the general principle of coexistence of competitors. (For formal analyses and reviews of the theoretical literature, see Ives and May, 1985; Hanski, 1987a; Ives, 1988, 1991; Shorrocks, 1990.)

The aggregation model describes a general mechanism of coexistence of competitors in mosaic landscapes. Most of the theoretical work has been motivated by, and the experiments have been conducted with, systems at small spatial scales, such as *Drosophila* breeding in mushrooms (Shorrocks, 1990) and blowflies breeding in animal carcasses (Hanski, 1987b), but the general principle applies to any spatial scale. However, it is, perhaps, unlikely that there is much spatial variation in density, independent of variation in the availability of essential resources, at large spatial scales, where chance effects, so important at small scales, have relatively little effect on expected population sizes. The exception is situations in which some habitat patches are entirely empty because of local extinctions.

Spatial aggregation will facilitate coexistence only if the competing species have not completely correlated spatial distributions. Hence any factors decreasing spatial correlation in the densities of the competing species make their coexistence easier. Because identical species are expected to have completely correlated spatial distributions (Chesson, 1991), this mechanism does not predict their coexistence (see section on identical species above). However, a very important point is that interspecific differences in any biological character, not only in characters affecting resource use, can significantly facilitate coexistence, as long as the character somehow affects the spatial distributions of the species. For instance, Cambefort (1991b) describes patterns of spatial distribution of dung beetles colonizing cattle pats in West African savanna. Cambefort (1991b) found that interspecific spatial correlations between pairs of species decreased with an increasing size difference among the species; was smaller in species pairs with different diel activities than in species pairs with the same diel activity; and interspecific correlation was smaller in species belonging to different tribes than in species belonging to the same tribe. Body size, diel activity and tribe (related to morphology, behavior and many other factors) affect the movement behavior of beetles, and any differences in the movement behavior are likely to influence the numbers of individuals colonizing and exploiting particular resource patches (Hanski, 1987a).

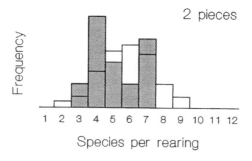

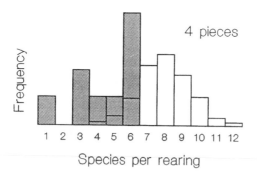

Figure 9.2 Frequency histograms of the number of blowfly species emerging from sets of rearings in which maggots were reared in individual carcasses (control, open histogram) or in groups of two or four carcasses (stippled histogram), allowing interactions among all the maggots in the combined carcasses and thus removing patchiness (Source: Hanski, 1987a.)

Experiments on blowflies (Hanski, 1987a; Ives, 1988, 1991) have provided convincing evidence for the significance of the aggregation mechanism in promoting coexistence of competitors in patchy environments. Figure 9.2 describes the result of an experiment in which the degree of aggregation of competing individuals was manipulated. Adult blowflies were allowed to oviposit on pieces of resource (liver) placed on a regular grid in a homogeneous field. Following a three-day oviposition period, the 50-g pieces of liver were placed in rearing pots either singly or in groups of two or four pieces. In the latter rearing pots, developing maggots could easily crawl from one piece of resource to another – resource patchiness had been removed! Figure 9.2 shows that removing patchiness, and thereby making the environment in which the individuals competed more homogeneous, significantly decreased the number of species which eventually emerged from the rearings, as the inferior competitors lost the advantage of having resource patches with low density of the superior competitor (Hanski, 1987a). The experiment only lasted for one generation, but it is clear that if the same process were continued for several generations, species number would only further decline in an environment with reduced patchiness. This experiment provides perhaps the most direct evidence ever published on the relationship between landscape patchiness and species richness in a guild of competitors, albeit on a small spatial scale.

For other empirical examples, mostly based on observational studies, see Hanski (1987a), Ives (1988, 1991), Ståhls, Ribeiro and Hanski (1989) and Shorrocks (1990).

9.2.4 ALTERNATIVE LOCAL EQUILIBRIA

The most intriguing of the four possible outcomes of competition in the classical Lotka–Volterra competition model is the one with an unstable internal (two-species) equilibrium, with two alternative, stable equilibria with only one species present. This is the case where two competing species cannot coexist locally, but where the outcome of local competition depends on the initial numbers of individuals. The 'initial numbers' argument translates to a local priority effect: whichever species occurs first in a habitat patch is likely to outcompete the other species.

If two or more such competitors occur in a landscape with many habitat patches, and if different species have become established in different patches, for whatever reasons, species richness at the landscape level is enhanced by the mechanism of alternative local equilibria (Levin, 1974). An exceptionally convincing example of alternative local equilibria is provided by mutual competition/predation by rock lobsters and whelks on two small islands off the coast of South Africa, separated by only 4 km. The benthic communities on the Malgas island are dominated by seaweeds and rock lobsters, on the Marcus islands by whelks and mussels. Barkai and McQuaid (1988) report observations and experiments strongly suggesting that rock lobsters at high density prevent colonization by whelks and vice versa, giving rise to two completely different communities on otherwise comparable islands.

Landscape-level coexistence due to alternative stable equilibria at the local level rests on the assumptions that there is not enough migration among the patches to drive the system away from the local equilibria (Levin, 1974), and that the single-species local equilibria are really equilibria; that is, that there is no significant population turnover due to, for example, stochastic extinctions. The situation becomes radically different with population turnover, in which case the local 'equilibria' are no long-term equilibria and the 'alternative local equilibria' mechanism turns to the model of lottery competition.

Sale (1977, 1979, 1982) suggested that a mosaic landscape would facilitate coexistence of competitors if there was a strong priority effect, the first species to arrive at a patch being able to exclude all later immigrants, for instance because of strong interspecific territoriality as in his examples on coral reef fishes. However, it is now known that the lottery mechanism itself is not sufficient to allow coexistence (Chesson and Warren, 1981; Hanski, 1987a; Chesson, 1991), but it leads to a random walk process in which one of the species is sooner or later lost – later if the species are similar and the number of habitat patches is large, as suggested by Hubbell (Hubbell, 1979; Hubbell and Foster, 1986) for tropical forest tree species competing for microsites for germination and growth.

9.3 LANDSCAPES WITH QUANTITATIVELY DIFFERENT HABITAT PATCHES

I will now turn to the more usual situations, landscape mosaics consisting of different sorts of habitat patches (cf. also Chapter 6). In this section, I focus on quantitative differences, such that resource availability is higher in some patches ('good' patches) than in some others ('poor' patches). Landscapes with qualitatively different patches are covered in the next section.

The following conceptual model of coexistence with two critical elements has been developed independently by several ecologists (Kotler and Brown, 1988; Hanski, 1989, 1992; Erlinge and Sandell, 1988). First, because foraging rate is ultimately limited by food availability, there exists a critical minimum level of food availability for reproduction, and another, still lower level, for survival. Because per capita resource requirement increases with body size, these critical minimum levels must also increase with body size. Therefore, comparing otherwise similar species but of different body size, in some habitat patches food availability may be high enough for the small but not for the large species. The second key element is the species' success in interference competition, which generally increases with body size. Taking into account these two consequences of body size in exploitative and interference competition leads to the prediction that, in a mosaic landscape, the larger and competitively superior species predominates in the more productive habitat patches, whereas the smaller species is restricted, by interference competition, to the less productive patches. Figure 9.3 illustrates, with a simple model, how species composition is expected to vary along an environmental gradient from the least to the most productive habitat patches.

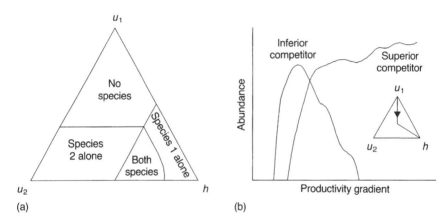

(a)   (b)

Figure 9.3 A model of habitat selection. u_1, u_2, and h are the fractions of habitat patches which are not suitable for any species (u_1), which are suitable for species 2 (an inferior competitor) but unsuitable for species 1 (a superior competitor; u_2), and are suitable for both species (h). The model is constructed for semelparous species and it is described in Hanski and Kaikusalo (1989). (a) An example of how environments with different values of u_1, u_2 and h are expected to have no species (least productive environments, u_1 large), species 2. alone, species 1 alone (most productive environments, h large), or both species 1 and 2. (b) An example of varying relative abundances of species 1 and 2 along an environmental gradient of increasing productivity, as indicated in the insert.

The above-described mechanism of landscape-level coexistence, or some variant of it, has been invoked frequently in studies of vertebrate competitors (Frye, 1983; Bowers, Thompson and Brown, 1987; Erlinge and Sandell, 1988; Hanski, 1989, 1992). Figure 9.4 shows a distributional pattern that is consistent with this hypothesis: in a guild of competing shrew species, the larger species dominate in the more productive habitat types, whereas the smaller species are numerically more dominant in the less productive habitat types. Unfortunately, observational results of this type do not provide critical evidence for coexistence based on a body

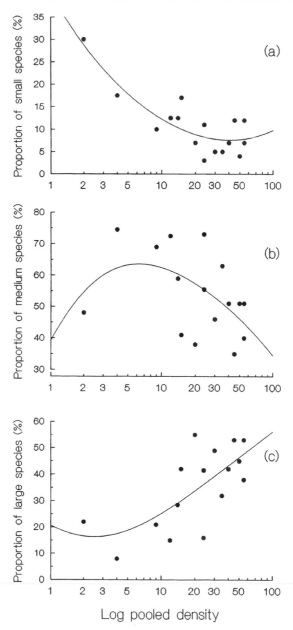

Figure 9.4 Relative abundances of (a) small (adult weight <6 g, four species), (b) medium (7–9 g, three species), and (c) large species (>10 g, four species) of *Sorex* shrews against the logarithm of their pooled density, a measure of habitat productivity for shrews. The data are average values for 16 habitat types (on the horizontal axis), calculated from results for 93 localities in northern Europe and Siberia. (Source: Hanski, 1992.)

size-mediated trade-off between exploitative and interference competitive abilities. A critical test would be an experimental comparison of competition in two landscape structures with and without the poor-quality patches, but nobody appears to have yet attempted such a test. In any case, this mechanism suggests a simple explanation for the often-observed humped relationship between species richness and productivity, though other explanations are also possible (Tilman, 1982; Rosenzweig, 1992). For a small mammal example of a humped species richness–productivity relationship, see Abramsky and Rosenzweig (1984).

The level of food availability in habitat patches is set by the processes of food production and food consumption. Therefore, in principle, foraging animals themselves may generate variability in food availability to an extent that would facilitate coexistence, as suggested by Kotler and Brown (1988). Kotler and Brown (1988) assume that there is a trade-off between travelling cost (or speed) and foraging efficiency in habitat patches. Generally, one could expect that larger species are able to move faster but are less efficient foragers than small species. The species with low travelling cost may be expected to visit many patches and preferentially exploit the ones with currently highest food availabilities. Such behavior should decrease spatial variation in food availability and leave a relatively high mean value. Another species with high cost of movement does not visit as many habitat patches, but exploits the visited patches to a low level of food availability; hence it generates a less even distribution of food availabilities. Both species thus tend to change the distribution of food availabilities in the habitat patches in the direction that is relatively more beneficial to the competitor, and by the general principle, coexistence becomes more likely. Kotler and Brown (1988) describe some putative empirical examples from desert rodents.

Quantitative differences among habitat patches may interact with interspecific differences in dispersal rate to facilitate coexistence even in equally strong competitors. Generally, the species with a higher dispersal rate should become relatively more common in small (less productive) patches, while the low-dispersal species becomes relatively more common in large patches. Both species thereby compete more with conspecifics than heterospecifics, and they may coexist in a mosaic landscape (McPeek and Holt, 1992).

9.4 LANDSCAPES WITH QUALITATIVELY DIFFERENT HABITAT PATCHES

Most real landscapes are complex mosaics of qualitatively different kinds of habitat patches and, undoubtedly, the most significant landscape-level factor promoting coexistence of competitors is habitat selection (Chapter 5). Furthermore, interspecific competition is likely to select for and maintain differences in habitat selection (for reviews, see Colwell and Fuentes, 1975; Connell, 1983; Schoener, 1983; Connor and Bowers, 1987), though a critical demonstration of this is very difficult (Underwood, 1986).

Schoener (1974) compared, in a pioneering paper on resource partitioning, the frequencies of habitat, food and temporal factors in apparently 'separating' potentially competing species. Habitat was the most significant factor most frequently (55% of the cases, compared with 40% and 5% for food and time respectively), and some habitat separation was detected in 90% of the studies reviewed. Such prevalence of macrohabitat separation in a guild of potentially competing species is not unexpected. Since two other pioneering studies by MacArthur and Pianka (1966) and Levins (1968), ecologists have been aware of the importance of 'environmental grain' in promoting/inhibiting specialization. An individual may easily spend all or most of its life within one or a few large macrohabitat patches, which facilitates specialization, whereas different food items are typically encountered in a more fine-grained manner, making specialization less likely or less profitable. Abramsky and Rosenzweig have explored, in a series of theoretical and empirical papers, the consequences of density-dependent habitat selection on competition in a heterogeneous environment (Rosenzweig 1981; Abramsky, Rosenzweig and Brand, 1985; Rosenzweig and Abramsky, 1985, 1986; Abramsky *et al.*, 1990; Abramsky, Rosenzweig and Pinshow, 1991; Abramsky, Rosenzweig and Zubach, 1992). Their work is particularly important in exploring the consequences of individual foraging behavior to interspecific competition in a heterogeneous environment.

Generally, whenever species compete in a heterogeneous environment, the outcome of competition at the landscape level may depend not only on the performance of the species in individual habitat patches but also on the mixture of different kinds of patches (Danielson, 1991). Consider a species which is specialized to a rare or low-productivity habitat, in which it is competitively superior to another species. Thus, in the absence of outside influence, the superior species would competitively exclude the latter species. In a heterogeneous environment, the population of the competitively inferior species may however be augmented by migration from another, nearby habitat, up to the point where the intrinsically superior competitor in the focal habitat is excluded (Christiansen and Fenchel, 1977). This possibility serves to highlight the point that although habitat heterogeneity generally facilitates coexistence, coupling of populations via migration in a landscape mosaic may also have the opposite effect.

9.5 DISCUSSION

One of the difficulties in applying the various ideas and models described in the previous sections to species competing in real landscapes is that two or more mechanisms may be involved simultaneously. Take for instance the rock pool *Daphnia*, one of the putative examples of fugitive coexistence (section 9.2.2). Not surprisingly, the different species are somewhat different in their ecologies (Ranta, 1979; Bengtsson, 1988), and as there are also differences in the attributes of different habitat patches (rock pools), it is quite possible that coexistence is, to some extent at least, facilitated by habitat separation: different species do best in different kinds of rock pools. Life would be much simpler for ecologists if we could isolate a single critical factor, or mechanism, but often this is simply not possible.

One of the reasons for the current interest in metapopulation dynamics (Gilpin and Hanski, 1991; Chapter 12) is the notion that a species may survive regionally in spite of stochastic local extinctions. What is the role of stochastic extinctions in the coexistence of competitors in landscape mosaics? Interestingly, no simple answer can be given. In the standard metapopulation scenarios, population turnover, stochastic extinctions and recolonizations are essential for coexistence at the landscape level (the fugitive mechanism). But in the case of alternative local equilibria, stochastic extinctions will actually lead to the opposite result, coexistence becoming more difficult with increasing rate of extinctions and colonizations. The reason for this difference is that while in the case of fugitive coexistence one species is assumed to be competitively superior locally, regardless of which species happens to arrive first, in the case of alternative local equilibria both species may win locally, and obtain a safe site from competition, which advantage may, however, be lost due to extinctions.

My final remark is about the general significance of spatial variation in population densities. Intraspecific aggregation generally makes coexistence easier (section 9.2). Aggregation is typically greater in small species than in large species; for instance, aggregation is widespread in insects but not always in vertebrates, in which territoriality often reduces aggregation. On this basis one may predict that coexistence of competitors is generally easier in insects than in vertebrates. To return to the question of aliens outcompeting natives, I would expect that this is more frequent among vertebrate competitors than among insect competitors.

9.6 SUMMARY

This chapter reviews several landscape-level mechanisms which may play a role in interspecific competition, using the following principle of coexistence as a guide: coexistence of two competing species is possible, under any mechanism and scenario of competition, if an individual's competitive action has a more harmful per capita effect on conspecifics than on heterospecifics. In landscapes consisting of similar habitat patches, coexistence is facilitated by a high migration/colonization rate in inferior competitors (the fugitive coexistence mechanism); by intraspecific spatial aggregation of the superior competitors (which increases the relative strength of intraspecific competition in relation to interspecific competition); and by alternative local equilibria. Contrary to what is commonly assumed, all these mechanisms involve some ecological differences among the competitors. In landscapes with more and less productive habitat patches of qualitatively the same type, coexistence may be based on a negative correlation between species' performance in exploitation and interference competition, for instance due to body size differences. Finally, landscapes consisting of different kinds of habitat patches typically facilitate coexistence due to differences in species' habitat selection. The various scenarios of landscape-level coexistence of competitors have been illustrated with selected empirical examples.

ACKNOWLEDGEMENTS

I thank Brent Danielson and Bob Holt for comments on the manuscript.

REFERENCES

Abramsky, Z. and Rosenzweig, M. L. (1984) Tilman's predicted productivity–diversity relationship shown by desert rodents. *Nature,* **309**, 150–1.

Abramsky, Z., Rosenzweig, M. L. and Brand, S. (1985) Habitat selection in Israeli desert rodents: comparison of a traditional and a new method of analysis. *Oikos,* **45**, 79–88.

Abramsky, Z., Rosenzweig, M. L. and Pinshow, B. (1991) The shape of a gerbil isocline measured using principles of optimal habitat selection. *Ecology,* **72**, 329–40.

Abramsky, Z., Rosenzweig, M. L., Pinshow, B. *et al.* (1990) Habitat selection: an experimental field test with two gerbil species. *Ecology,* **71**, 2358–69.

Abramsky, Z., Rosenzweig, M. L. and Zubach, A. (1992) The shape of a gerbil isocline: an experimental field study. *Oikos,* **63**, 193–9.

Armstrong, R. A. (1976) Fugitive species: experiments with fungi and some theoretical considerations. *Ecology*, **57**, 953–63.

Atkinson, W. D. and Şhorrocks, B. (1981) Competition on a divided and ephemeral resource: a simulation model. *J. Anim. Ecol.*, **54**, 507–18.

Barkai, A. and McQuaid, C. (1988) Predator–prey role reversal in a marine benthic ecosystem. *Science*, **242**, 62–4.

Bengtsson, J. (1986) Life histories and interspecific competition between three *Daphnia* species in rockpools. *J. Anim. Ecol.*, **55**, 641–55.

Bengtsson, J. (1988) Life Histories, Interspecific Competition and Regional Distribution of three Rockpool Daphnia Species. PhD thesis, Uppsala Univ., Sweden.

Bengtsson, J. (1989) Interspecific competition increases local extinction rate in a metapopulation system. *Nature*, **340**, 713–15.

Bengtsson, J. (1991) Interspecific competition in metapopulations, in *Metapopulation Dynamics* (eds M. Gilpin and I. Hanski), Academic Press, London, pp. 219–37.

Bowers, M. A., Thompson, D. B. and Brown, J. H. (1987) Foraging and microhabitat use in desert rodents: the role of a dominant competitor. *Oecologia*, **7**, 77–82.

Cambefort, Y. (1991a) Biogeography and evolution, in *Dung Beetle Ecology* (eds I. Hanski and Y. Cambefort), Princeton University Press, Princeton, NJ, pp. 51–68.

Cambefort, Y. (1991b) Dung beetles in tropical savannas, *Dung Beetle Ecology* (eds I. Hanski and Y. Cambefort), Princeton University Press, Princeton, NJ, pp. 156–78.

Chesson, P. (1991) A need for niches? *Trends Ecol. Evol.*, **6**, 26–8.

Chesson, P. L. and Warren, R. R. (1981) Environmental variability promotes coexistence in lottery competitive systems. *Am. Nat.*, **117**, 923–43.

Christiansen, F. B. and Fenchel, T. M. (1977) *Theories of Populations and Biological Communities*, Springer-Verlag, Berlin.

Colwell, R. K. and Fuentes, E. R. (1975) Experimental studies of the niche. *Ann. Rev. Ecol. Syst.*, **6**, 281–310.

Connell, J. H. (1978) Diversity in tropical forests and coral reefs. *Science*, **199**, 1302–10.

Connell, J. H. (1983) On the prevalence and relative importance of interspecific competition: evidence from field experiments. *Am. Nat.*, **122**, 661–96.

Connor, E. F. and Bowers, M. A. (1987) The spatial consequences of interspecific competition. *Ann. Zool. Fennici*, **24**, 213–26.

Danielson, B. J. (1991) Communities in a landscape: the influence of habitat heterogeneity on the interactions between species. *Am. Nat.*, **138**, 1105–20.

Ebenhard, T. (1988) Introduced birds and mammals and their ecological effects. *Swed. Wildl. Res.*, **13**, 1–107.

Elton, C. S. (1958) *The Ecology of Invasions by Animals and Plants*, Chapman and Hall, London.

Erlinge, S. and Sandell, M. (1988) Co-existence of stoat (*Mustela erminea*) and weasel (*Mustela nivalis*): social dominance, scent communication and reciprocal distribution. *Oikos*, 53, 242–6.

Frye, R. J. (1983) Experimental field evidence of interspecific aggression between two species of kangaroo rat (*Dipodomys*). *Oecologia*, 59, 74–8.

Gause, G. F. (1934) *The Struggle for Existence*, Hafner, New York.

Gilpin, M. and Hanski, I. (eds) (1991) *Metapopulation Dynamics*, Academic Press, London.

Greenway, J. C., Jr (1967) *Extinct and Vanishing Birds of the World*, Dover Books, New York.

Grubb, P. (1977) The maintenance of species richness in plant communities: the importance of the regeneration niche. *Biol. Rev.*, 52, 107–45.

Hanski, I. (1981) Coexistence of competitors in patchy environments with and without predation. *Oikos*, 37, 306–12.

Hanski, I. (1983) Coexistence of competitors in patchy environments. *Ecology*, 64, 493–500.

Hanski, I. (1987a) Colonization of ephemeral habitats, in *Colonization, Succession and Stability* (eds A. J. Gray, M. J. Crawley and P. J. Edwards), Blackwell, Oxford, pp. 155–86.

Hanski, I. (1987b) Carrion fly community dynamics: patchiness, seasonality and coexistence. *Ecol. Ent.*, 12, 257–66.

Hanski, I. (1989) Population biology of Eurasian shrews: Towards a synthesis. *Ann. Zool. Fennici*, 26, 469–79.

Hanski, I., (1992) Insectivorous mammals, in *Natural Enemies* (ed. M. J. Crawley), Blackwell, Oxford, pp. 163–87.

Hanski, I. and Kaikusalo, A. (1989) Distribution and habitat selection of shrews in Finland. *Ann. Zool. Fennici*, 26, 339–48.

Hanski, I. and Ranta, E. (1983) Coexistence in a patchy environment: three species of *Daphnia* in rock pools. *J. Anim. Ecol.*, 52, 263–79.

Hanski, I. and Zhang, D.-Y. (1993) Migration, metapopulation dynamics and fugitive coexistence. *J. Theor. Biol.*, 163, 491–504.

Hardin, G. (1960) The competitive exclusion principle. *Science*, 131, 1292–7.

Herbold, B. and Moyle, P. B. (1986) Introduced species and vacant niches. *Am. Nat.*, 128, 751–60.

Holt, R. D. (1985) Density-independent mortality, non-linear competitive interactions, and species coexistence. *J. Theor. Biol.*, 116, 479–93.

Horn, H. S. and MacArthur, R. H. (1972) Competition among fugitive species in a harlequin environment. *Ecology*, 53, 749–52.

Hubbell, S. P. (1979) Tree dispersion, abundance, and diversity in a tropical dry forest. *Science*, 203, 1299–309.

Hubbell, S. P. and Foster, R. B. (1986) Biology, chance, and history and the structure of tropical rain forest tree communities, in *Community Ecology* (eds T. J. Case and J. Diamond), Harper & Row, New York, pp. 314–29.

Hutchinson, G. E. (1951) Copepodology for the ornithologist. *Ecology*, **32**, 571–7.

Hutchinson, G. E. (1978) *An Introduction to Population Ecology*, Yale University Press, New Haven.

Ives, A. R. (1988) Aggregation and the coexistence of competitors. *Ann. Zool. Fennici*, **25**, 75–88.

Ives, A. R. (1991) Aggregation and coexistence in a carrion fly community. *Ecol. Monogr.*, **61**, 75–94.

Ives, A. R. and May, R. M. (1985) Competition within and between species in a patchy environment: relations between microscopic and macroscopic models. *J. Theor. Biol.*, **115**, 65–92.

Kotler, B. J. and Brown, J. S. (1988) Environmental heterogeneity and the coexistence of desert rodents. *Ann. Rev. Ecol. Syst.*, **19**, 281–308.

Levin, S. A. (1974) Dispersion and population interactions. *Am. Nat.*, **108**, 207–28.

Levins, R. (1968) *Evolution in Changing Environments*, Princeton University Press, Princeton, NJ.

Levins, R. (1969) Some demographic consequences of environmental heterogeneity for biological control. *Bull. Ent. Soc. Am.*, **15**, 237–40.

Levins, R. and Culver, D. (1971) Regional coexistence of species and competition between rare species. *Proc. Nat. Acad. Sci. USA*, **68**, 1246–8.

Lubchenko, J. (1978) Plant species diversity in a marine intertidal community: importance of herbivore food preference and algal competitive abilities. *Am. Nat.*, **112**, 23–39.

MacArthur, R. H. and Pianka, E. (1966) On the optimal use of a patchy environment. *Am. Nat.*, **100**, 603–9.

Marino, P. C. (1988) Coexistence on divided habitats: mosses in the family Splachnaceae. *Ann. Zool. Fennici*, **25**, 89–98.

McPeek, M. A. and Holt, R. D. (1992) The evolution of dispersal in spatially and temporally varying environments. *Am. Nat.*, **140**, 1010–27.

Nee, S. and May, R. M. (1992) Dynamics of metapopulations: habitat destruction and competitive coexistence. *J. Anim. Ecol.*, **61**, 37–40.

Pajunen, V. I. (1979) Competition between rock pool corixids. *Ann. Zool. Fennici*, **16**, 138–43.

Pajunen, V. I. (1982) Replacement analysis of non-equilibrium competition between rock pool corixids (Hemiptera, Corixidae). *Oecologia*, **52**, 153–5.

Pajunen, V. I. (1986) Distributional dynamics of *Daphnia* species in a rockpool environment. *Ann. Zool. Fennici*, **23**, 131–40.

Ranta, E. (1979) Niche of *Daphnia* species in rockpools. *Ann. Zool. Fennici*, **19**, 337–47.

Rosenzweig, M. L. (1981) A theory of habitat selection. *Ecology*, **62**, 327–35.

Rosenzweig, M. L. (1992) Species diversity gradients: we know more and less than we thought. *J. Mamm.*, **73**, 715–30.

Rosenzweig, M. L. and Abramsky, Z. (1985) Detecting density-dependent habitat selection. *Am. Nat.*, **126**, 400–17.

Rosenzweig, M. L. and Abramsky, Z. (1986) Centrifugal community organization. *Oikos*, **45**, 79–88.

Sale, P. F. (1977) Maintenance of high diversity in coral reef fish communities. *Am. Nat.*, **111**, 337–59.

Sale, P. F. (1979) Recruitment, loss, and coexistence in a guild of territorial coral reef fishes. *Oecologia*, **42**, 159–77.

Sale, P. F. (1982) Stock–recruitment relationships and regional coexistence in a lottery competitive system: a simulation study. *Am. Nat.*, **120**, 139–59.

Schoener, T. W. (1974) Resource partitioning in ecological communities. *Science*, **185**, 27–39.

Schoener, T. W. (1983) Field experiments on interspecific competition. *Am. Nat.*, **12**, 240–85.

Shorrocks, B. (1990) Coexistence in a patchy environment, in *Living in a Patchy Environment* (eds B. Shorrocks and I. R. Swingland), Oxford University Press, Oxford, pp. 91–106.

Simberloff, D. (1981) Community effects of introduced species, in *Biotic Crises in Ecological and Evolutionary Time* (ed. H. Nitecki), Academic Press, New York, pp. 53–81.

Simpson, G. G. (1949) *The Meaning of Evolution*, Yale University Press, New Haven.

Skellam, J. G. (1951) Random dispersal in theoretical populations. *Biometrika*, **38**, 196–218.

Slatkin, M. (1974) Competition and regional coexistence. *Ecology*, **55**, 126–34.

Stanley, S. M. (1979) *Macroevolution*, Freeman, San Francisco.

Ståhls, G., Ribeiro, E. and Hanski, I. (1989) Fungivorous *Pegomya*: spatial and temporal variation in a guild of competitors. *Ann. Zool. Fennici*, **26**, 103–12.

Tilman, D. (1982) *Resource Competition and Community Structure*, Princeton University Press, Princeton, NJ.

Underwood, T. (1986) The analysis of competition by field experiments, in *Community Ecology* (eds J. Kikkawa and D. J. Anderson), Blackwell, Oxford.

Volterra, V. (1926) Variazioni e fluttuazioni del numero d'individui in specie animali conviventi. *Mem. R. Acad. Naz. dei Lincei*, **2**, 31–113.

Effects of landscape composition on predation rates at habitat edges

10

Henrik Andrén

10.1 INTRODUCTION

All landscapes are mosaics of habitat patches of different types (Forman and Godron, 1981). Therefore, there will always be edges between habitat patches in a landscape. These habitat edges may be associated with a higher diversity of plants and animals, traditionally called an **edge effect** (Odum, 1971). However, the current use of the edge-effect concept summarizes a diverse group of phenomena that occur in habitat edges (Harris, 1988; Yahner, 1988; Angelstam, 1992). For example, interaction between different organisms living in different habitat patches in the landscape mosaic may sometimes cause effects related to habitat edges (Janzen, 1986; Wilcove, McLellan and Dobson, 1986). Predator–prey interaction might be related to habitat edges. However, the relationship between predation rate and habitat edge depends on the specific habitat use and density of the potential nest predators, as well as the configuration of habitat patches in the landscape. Predators might increase the predation rate in habitat edges for several different reasons. First, the habitat edges might have a higher density of prey and the predators might view the edges as good foraging patches (Gates and Gysel, 1978). Secondly, the predators might use the habitat edges as traveling lines (Bider, 1968; Chapter 4) and therefore spend more time in the habitat

Mosaic Landscapes and Ecological Processes.
Edited by Lennart Hansson, Lenore Fahrig and Gray Merriam.
Published in 1995 by Chapman & Hall, London. ISBN 0 412 45460 2.

edge than in other parts of the habitat patch. The predation might be incidental, i.e. the predators might find the prey by chance while looking for other food items (Angelstam, 1986; Vickery, Hunter and Wells, 1992). Finally, predators living in one type of habitat might penetrate into the neighboring less-preferred habitats. The activity in the less-preferred habitat will be higher close to the habitat edge (Wilcove, 1985a; Angelstam, 1986). This last explanation assumes that the predators use several habitats in the landscape, i.e. they are habitat generalists and they experience the landscape as being heterogeneously undivided (Addicott et al., 1987). Furthermore, they should easily move across habitat edges (soft edges; Stamps, Buechner and Krishnam, 1987). On the other hand, in the two first explanations the predators do not necessarily have to use the habitat on both sides of the edge. Thus, the predators can be restricted to only one type of habitat, i.e. they perceive the landscape as divided (Addicott et al., 1987) and the habitat edges can be a barrier (hard edge; Stamps, Buechner and Krishnam, 1987). Such species can be described as habitat specialists. But they can still be attracted by the habitat edge within the habitat patch they prefer. If the predation rate in habitat edges is related to a high density of prey in edges attracting predators or related to predators using habitat edges as traveling lines, then the edge-related increase in predation rate might occur on both sides of the edges, but it will be confined to the very edge. On the other hand, if predation rate in habitat edges is related to predators entering from the surrounding, the effect can penetrate considerable distances into the habitat patch. Therefore, habitat generalists, that penetrate into habitat patches, have the largest potential to cause an increasing predation pressure in habitat patches and in habitat edges as a habitat becomes fragmented and surrounded by another habitat (Andrén, 1992). Hence, the type of predator (habitat specialist or habitat generalist) involved will have a very large influence on the outcome of an edge-related predator–prey interaction. Therefore, to understand the effect of habitat fragmentation, as well as the effect of different landscape mosaics, on predator–prey interaction, it is important to study the interaction among organisms living in different habitat patches in the landscape (Forman, 1981; Wiens, Crawford and Gosz, 1985; Addicott et al., 1987; Chapter 6).

In this chapter I will review studies performed to test whether predation rates on birds' nests are higher close to habitat edges than in the interior parts of habitat patches. Some studies have tested the effect of patch size on predation rate. However, the results have often been discussed in terms of predation in relation to edge, because the proportion of edge in a patch increases as the patch becomes smaller (Levenson, 1981).

10.2 WHEN IS PREDATION RELATED TO HABITAT EDGE AND PATCH SIZE?

The effect of habitat edge and patch size on nest predation, both on real birds' nests and on dummy nests, has been tested several times. I have found 40 such studies; 18 from North America, 19 from Europe and three from other parts of the world, namely Costa Rica, Belize and Panama. Some studies have found an effect of distance to edge and/or patch size, whereas others have not (Table 10.1, Figures 10.1 and 10.2).

Table 10.1 Studies that have tested whether predation rates on bird nests are higher in small than in large habitat patches and/or higher close to the habitat edge than inside the patch

Description of the landscape	Nest type	Higher predation in small than in large habitat patches (range)	Higher predation close to habitat edge than inside the patch (range)	Source, study site
Predation rate measured in forest habitats neighboring farmland, meadows, abounded fields, etc.				
Deciduous forest patches surrounded by farmland. Farmland dominated the landscape	Real open nests above ground	–	Yes 0–123 m	Gates and Gysel (1978) Michigan, USA
Deciduous forest –field edges along transmission-lines. Forest dominated the landscape	Real open nests above ground	–	Yes 0–80 m	Chasko and Gates (1982) Maryland, USA
Deciduous forest –open habitat edge	Real open nests above ground	–	Yes 0–>300 m	Brittingham and Temple (1983) Wisconsin, USA
Deciduous forest patches surrounded by farmland. Farmland dominated the landscape	Dummy nests both on ground and above ground	Yes 3.8–209 000 ha	–	Wilcove (1985) Maryland and Tennessee, USA

Table 10.1 Continued

Description of the landscape	Nest type	Higher predation in small than in large habitat patches (range)	Higher predation close to habitat edge than inside the patch (range)	Source, study site
Coniferous forest surrounding farmland (87% forest)	Dummy nests on ground	–	No 0–>1500 m	Angelstam (1986) South-central Sweden
Deciduous forest patches surrounded by farmland. Farmland dominated the landscape	Dummy nests both on ground and above ground	–	Yes 0–1000 m	Wilcove, McLellan and Dobson (1986) Maryland and Tennessee, USA
Coniferous forest patches surrounded by farmland (58% farmland)	Dummy nests on ground	–	Yes 0–500 m	Andrén and Angelstam (1988)[a] South-central Sweden
Coniferous forest –open habitat edge (e.g. lake, field, bog; 65% forest)	Treecreeper in nest-boxes	–	Yes 0–>200 m	Kuitunen and Helle (1988) Southern Finland
Deciduous forest patches surrounded by farmland (98% farmland)	Real blackbird nests and old ones with dummy eggs	Yes 0.05–3.61 ha	–	Møller (1988) Denmark
Mixed forest– open habitat edge (e.g. field, water, road; 66–85% forest)	Open dummy nests both on ground and above ground	Yes 20–1040 ha	–	Small and Hunter (1988) Maine, USA
Deciduous forest –farmland edge	Real open nests above ground	–	Yes 0–>200 m	Temple and Cary (1988) Wisconsin, USA

Table 10.1 Continued

Description of the landscape	Nest type	Higher predation in small than in large habitat patches (range)	Higher predation close to habitat edge than inside the patch (range)	Source, study site
Deciduous forest patches surrounded by farmland (98% farmland)	Dummy nests both on ground and above ground	–	Yes 0–100 m	Møller (1989)[a] Denmark
Deciduous forest –farmland edge	Scarlet rosefinch nests	–	Yes 0–100 m	Björklund (1990) Central Sweden
Deciduous forest patches surrounded by pasture	Open dummy nests above ground	Yes 1.1–190 ha	No 5–>500 m	Gibbs (1991)[a] Costa Rica
Deciduous forest patches surrounded by farmland (98% farmland)	Real open nests above ground	Yes 0.05–3.61 ha	–	Møller (1991) Denmark
Deciduous forest –farmland edge	Dummy eggs in natural cavities	–	Yes 0–>20 m	Sandström (1991) South-central Sweden
Deciduous forest –farmland edge. Forest dominated the landscape	Dummy nests both on ground and above ground	–	No 0–100 m	Santos and Telleria (1992) Central Spain
Deciduous forest patches surrounded by farmland. Forest dominated the landscape	Dummy nests on ground	Yes 0.3–350 ha	–	Telleria and Santos (1992) Central Spain
Rainforest–field edge along a gravel road. Forest dominated the landscape	Dummy nests on ground	–	Yes 30–500 m	Burkey (1993) North-western Belize

Table 10.1 Continued

Description of the landscape	Nest type	Higher predation in small than in large habitat patches (range)	Higher predation close to habitat edge than inside the patch (range)	Source, study site
Coniferous forest –farmland edge	Dummy eggs in old woodpecker holes	–	Yes 0–>200 m	Johnsson (1993) South-central Sweden
Coniferous forest –farmland edge	Real hole-nesting birds	–	No 0–>200 m	Johnsson, Nilsson and Tjernberg (1993) South-central Sweden
Deciduous forest patches surrounded by farmland	Open dummy nests above ground	No 1–200 ha	No 0–125 m	Nour, Matthysen and Dhondt (1993) Belgium

Predation rate measured in forest landscapes, mainly in mature forest neighboring clear-cuts or young forest stands, but also in young stands neighboring mature forest

Mature aspen forest–young forest edge. Mature forest dominates the landscape	Dummy nests on ground	–	No 0–50 m	Yahner and Wright (1985) Pennsylvania, USA
Coniferous forest –clear-cut edge. Mature forest dominates the landscape	Open dummy nests both on ground and above ground	–	No, opposite 0–160 m (inside mature forest)	Ratti and Reese (1988)[a] Idaho, USA
Coniferous forest –clear-cut edge. Mature forest dominates the landcape	Open dummy nests both on ground and above ground	–	No 0–160 m (on clear-cuts)	Ratti and Reese (1988)[a] Idaho, USA
Mature forest–plantation edge	Open dummy nests above ground	Yes 1.1–190 ha	Yes 5–>500 m	Gibbs (1991)[a] Costa Rica

Table 10.1 Continued

Description of the landscape	Nest type	Higher predation in small than in large habitat patches (range)	Higher predation close to habitat edge than inside the patch (range)	Source, study site
Coniferous forest patches neighboring clear-cut (13% clear-cut, 58% mature forest)	Dummy nests on ground	No 5–25 ha	No, opposite 0–100 m	Storch (1991) Southern Germany
Coniferous forest –clear-cut edge (7% clear-cut, 88% mature forest)	Dummy nests both on ground and above ground	No 9–203 ha (in mature forest)	No 0–300 m (in mature forest)	Rudnicky and Hunter (1993)[a] Maine, USA
Coniferous forest –clear-cut edge (7% clear-cut, 88% mature forest)	Dummy nests both on ground and above ground	No 2–107 ha (on clear-cuts)	No 0–300 m (on clear-cuts)	Rudnicky and Hunter (1993)[a] Maine, USA
Forest patches surrounded by clear-cut	Dummy nests on ground	No 3–50 ha	No 0–400 m	Huhta (*in manus*) Northern Finland

Predation rate measured in open habitats neighboring different kinds of habitat

Description of the landscape	Nest type	Higher predation in small than in large habitat patches (range)	Higher predation close to habitat edge than inside the patch (range)	Source, study site
Grassland surrounded by farmland. Farmland dominated the landscape	Real duck nests	No 12–54 ha	–	Duebbert and Lokemoen (1976) South Dakota, USA
Grassland surrounded by farmland. Farmland dominated the landscape	Real duck nests	No	–	Gatti (1987) Wisconsin, USA
Farmland–forest edge (58% farmland)	Dummy nests on ground	–	No 0–400 m	Andrén and Angelstam (1988)[a] South-central Sweden

Table 10.1 Continued

Description of the landscape	Nest type	Higher predation in small than in large habitat patches (range)	Higher predation close to habitat edge than inside the patch (range)	Source, study site
Moorland surrounding forest (90% moorland)	Dummy nests on ground	–	No 100–1300 m	Avery, Winder and Egan (1989) Northern Scotland
Farmland–forest edge (98% farmland)	Dummy nests on ground	–	Yes 0–200 m	Møller (1989)[a] Denmark
Tall grass prairie –forest edge. Farmland and prairie dominated the landscape	Real nests	Yes 16–486 ha	Yes 0–>45 m	Johnson and Temple (1990) Minnesota, USA
Grassland surrounded by farmland. Farmland dominated the landscape	Real duck nests	No 100–204 ha	–	Clark, Nudds and Bailey (1991) Saskatchewan, Canada
Farmland–forest edge. Farmland dominated the landscape	Curlew nests	–	No 0–>100 m	Berg (1992) South-central Sweden
Bogs surrounded by forest. Forest dominated the landscape	Dummy nests on ground	No 25–>100 ha	No 50–250 m	Berg, Nilsson and Boström (1992) Southern, south-central and northern Sweden
Farmland surrounded by forest. Forest dominated the landscape	Real nests on ground	–	No 0–400 m	Vickery, Hunter and Wells (1992) Maine, USA
Meadows surrounded by forest. Forest dominated the landscape	Dummy nests on ground	–	No	Esler and Grand (1993) Alaska, USA

Table 10.1 Continued

Description of the landscape	Nest type	Higher predation in small than in large habitat patches (range)	Higher predation close to habitat edge than inside the patch (range)	Source, study site
Predation rate measured in other kinds of habitats				
Chaparral surrounded by urban habitat. Urban habitat dominated the landscape	Open dummy nests above ground	No 4.8–4500 ha	–	Langen, Bolger and Case (1991) California, USA
Predation rate measured on real islands				
One island and mainland	Open dummy nests both on ground and above ground	Yes 14.8 km² versus mainland	–	Loiselle and Hoppes (1983) Panama
Archipelago in a lake	Real thrush nests and old ones with dummy eggs	No 0.03–2.2 ha	–	Nilsson *et al.* (1985) Southern Sweden
One island and mainland	Open dummy nests above ground	No, opposite 8.5 km² versus mainland	–	George (1987) California, USA

[a] These studies occur twice, because they have measured predation rate in relation to edge and patch size in different habitats.

The studies in Table 10.1 are ordered according to patch/matrix characteristics. The first group of studies was performed in forest patches neighboring farmland; sometimes the landscape was dominated by farmland and sometimes by forest. The second group is studies from forest landscapes, i.e. landscape mosaics with forest stands of different ages. In the third group predation rate was measured in open habitats, i.e. two-dimensional habitat, surrounded by different kinds of habitat. The surrounding habitats might be, for example, other open habitats or forest. Thus, this group is more heterogeneous than the former groups. Finally,

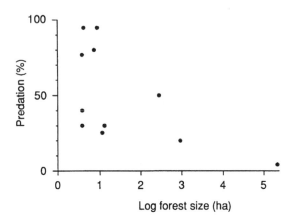

Figure 10.1 Predation rate on dummy nests in relation to forest fragment size. (Redrawn from Wilcove, 1985a.)

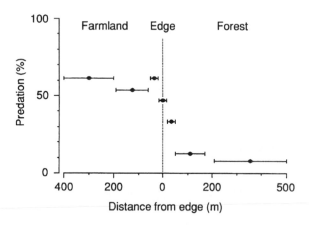

Figure 10.2 Predation rate on dummy nest in relation to distance from farmland–forest edge. Note that the predation rate was related to distance to edge in the forest but not in the farmland. (Redrawn after Andrén and Angelstam, 1988.)

I have included one study from chaparral habitat surrounded by urban habitat and three studies from real islands. The studies from real islands could be viewed as habitat patches where predation rate is only influenced by the predators restricted to that habitat patch, i.e. there is no influence of the surroundings.

Twenty-two studies have measured the predation rate inside forest neighboring farmland or other open habitats (two-dimensional habitats), excluding clear-cuts. Four of them failed to find any edge and/or size effect on predation rate, i.e. 82% of the studies found an effect (Figure 10.3). In two of the studies that failed to find any edge effect (Johnsson, Nilsson and Tjernberg, 1993; Nour, Matthysen and Dhondt, 1993) the major predators were forest-living species – red squirrel (*Sciurus vulgaris*) and pine marten (*Martes martes*), respectively. The two other studies (Angelstam, 1986; Santos and Telleria, 1992) were performed in landscapes where forest dominated, i.e. farmland patches were surrounded by forest. Both Angelstam (1986) and Santos and Telleria (1992) argued that the absence of edge-related increase in predation rate could be due to the landscape composition. Predators living in the small farmland patches had a low impact on the predation rate in the surrounding forest. However, comparing studies from landscape dominated by farmland with those dominated by forest does not reveal any large difference due to landscape composition. Fourteen studies reported the landscape composition. All seven studies from landscapes dominated by farmland (100%) and five of seven studies from forest-dominated landscapes (71%) found that the predation rate in the forest was related to distance from forest–farmland edge or to forest patch size.

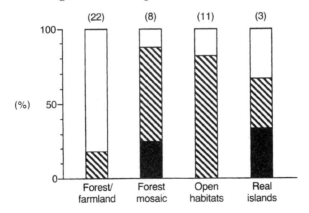

Figure 10.3 Proportion of studies that found a raised predation rate close to habitat edges or in small habitat patches (open part of the bars); that failed to find any effect (hatched part of the bars); and that found a lower predation in habitat edges or in small patches (black part of the bar). Sample sizes are shown above each bar. Data from Table 10.1.

Six studies have been performed in forest landscape, i.e. a landscape mosaic of forest patches of different age. Two of the studies have measured predation rate on both sides of the habitat edge. i.e. a total sample of eight. In this kind of landscape, only the study by Gibbs (1991) showed an effect of edge or patch size, i.e. 12% of the studies in this landscape (Figure 10.3). This study differs from the other ones because it was the only study not performed in the northern parts of the world. It was performed in Costa Rica. There could be a geographical difference in how predators use the landscape mosaic. Two studies (Ratti and Reese, 1988; Storch, 1991) actually found that the predaton rate was lower at the edge than inside the old forest. Yahner, Morrell and Rachael (1989) tested the effect of contrast, i.e. the degree of difference between adjacent habitats, on predation rate in a forest landscape. They found no difference in predation rate on dummy nests placed in edges between mature forest on one side and 2- and 12-year-old forest on the other, respectively. In another study, Yahner and Scott (1988) found the predation rate in old forest to be higher in landscapes with a high proportion of clear-cut areas than in a landscape with no clear-cuts. This difference might indicate an effect at a regional level. The total density of predators might increase with increasing proportion of clear-cut in the landscape, but the predators are neither restricted to a certain habitat nor attracted by habitat edges – they use the entire landscape mosaic. All of the studies from forest mosaics have used dummy nests. Therefore, studies of predation rate on real nests in relation to patch size and distance to habitat edge is really needed in this landscape type.

Only two of the 11 studies that measured the predation rate in open habitats, e.g. farmland, grassland, bogs or moorland, found any effect of patch size or distance from habitat edge, i.e. 18% of the studies in this landscape (Figure 10.3). However, this group was more heterogeneous than the other groups. For example, three studies were from grasslands surrounded by farmland, i.e. the surroundings were also open habitat, and all of them failed to find any effect of patch size on predation rate. Five studies were from farmland neighboring forest; two of these (40%) found an effect of distance to forest edge on predation rate.

One study was from chaparral habitat surrounded by urban areas and in this study there was no effect of patch size on predation rate (Langen, Bolger and Case, 1991). Finally, three studies have measured predation rate on real islands. Loiselle and Hoppes (1983) found a higher pre-dation rate on an island (Barro Colorado Island) than on the neighboring mainland, probably due to an increased number of small predators on the island (Karr, 1982). Nilsson et al. (1985) found no differences in predation rate between islands and the mainland. In contrast, George (1987) found the predation rate on dummy nests on one isolated island (2.5 km from the mainland and with an area of 8.5 km²) to be lower than on the neighboring mainland. Furthermore, the density of avian nest predators was higher on the mainland, and the density of black-throated sparrow (*Amphispiza bilineata*) lower. Thus, the results of the three studies from real islands were different from one another (Figure 10.3).

The shape of a habitat fragment might also influence the predation rate within it, because the proportion of edge is higher in a long and narrow habitat fragment than in a circular fragment. Unfortunately, only one study reports the effect of shape. Wilcove (1985a) found that the predation rate in a long narrow forest fragment was higher than in similar sized, but more circular, habitat fragments. The long narrow habitat fragment was easily penetrated by nest predators from the adjacent habitat.

Angelstam (1986) suggested that the intensity of the edge-related predation rate could be explained by the productivity difference between the habitat patch and the surrounding habitat. A more productive habitat (e.g. farmland) should support more predators than a less productive one (e.g. forest). The high density of predators in the productive habitat should increase the predation rate on prey close to the edge in the less productive one. This idea was partly supported by this review, because edge-related increase in predation was most commonly found inside forests surround by farmland, i.e. a landscape mosaic with large differences in productivity. In forest landscape, i.e. with small differences in productivity between the patches, edge-related increase in predation rate was uncommon. However, as edge-related increase in predation rate can be caused by other factors than predators living in the surrounding habitat and penetrating into small habitat patches, Angelstam's (1986) suggestion cannot explain all predation–habitat edge relationships.

To conclude, it seems that an edge-related increase in predation is most commonly found inside forests surrounded by farmland and is rarely found in forest mosaics. The results from the group of open habitats is unclear, mainly because it is a more heterogeneous group than the other ones. In a recent review on nest predation in relation to habitat edges, Paton (1994) found when reanalysing published data that the effect of edge usually occurred within 50m of an edge. He also found that edge effects were as likely to occur in unforested habitats as in forest habitats, whereas I concluded that edge effect mainly occurred in forest patches surrounded by farmland, although the discrepancy between the reviews is mainly due to differences in the classification of habitat edges. However, independent of these uncertainties, one can conclude that a habitat edge does not necessarily imply an edge-related increase in predation pressure on birds' nests (Figure 10.3).

10.3 THE USE OF DUMMY NESTS IN PREDATION EXPERIMENTS

Edge effects on predation rate have often been studied using dummy nests. However, the use of dummy nests have been criticized by some authors (e.g. Storaas, 1988; Willebrand and Marcström, 1988), and the results obtained from studies using dummy nests should be used with caution (Martin, 1987; Yahner and Voytko, 1989; Roper, 1992). However, one may use dummy nests for very different reasons and the experimental error might have different importance. For example, one may use dummy nests to get the actual predation rate on real nests. If that is the case one must mimic exactly the real nests, the size and color of eggs, the placement of the nest, etc. Three studies that have compared predation rate of real nests and dummy nests found different effects. Martin (1987) and Roper (1992) found that predation rate was lower for dummy nests than for real nests, while Götmark, Neergaard and Åhlund, (1990) found no difference in predation rate between real nests and dummy nests. However, the most common purpose for using dummy nests has been to get an index of predation rate.

Two studies have compared patterns in predation rate on real grouse nests with dummy nests (hen's eggs put on the ground) in different years (Storaas, 1988; Willebrand and Marcström, 1988). Both of these studies found that the changes between years in predation rate were different for real and dummy nests. Sonerud (1993) found that the temporal pattern of predation on dummy nests mimicked the predation pattern on real Tengmalm's Owl (*Aegolius funereus*) nests, but the change was not significant for dummy nests while it was for real nests. Thus, there might be a difference between artificial and real nests.

However, the most common use of dummy nests has been to compare predation rate in different habitats. Some studies have measured patterns in predation rate on both real nests and dummy nests. Nilsson *et al.* (1985) found no differences in predation rate between small islands and the mainland for both real thrush nests and open dummy nests. Similarly, Møller (1988, 1989, 1991) found the same predation pattern for both real blackbird nests and open dummy nests, namely that the predation rate was higher in small forest patches and close to forest–farmland edges. On the other hand, the predation rate on real nests (jackdaw, *Corvus monedula*; stock dove, *Columba oenas*; black woodpecker, *Dryocopus martius*; goldeneye, *Bucephala clangula*) in old black woodpecker nests was not related to forest edge (Johnsson, Nilsson and Tjernberg, 1993), while the predation on dummy eggs put in similar old black woodpecker nests was higher close to forest–farmland edge (Johnsson, 1993)

Out of the 22 studies that have measured predation rate on nests in forest patches in a farmland landscape (Table 10.1), eight out of nine studies of real nests (89%) and 11 out of 14 studies using dummy nests (78%) found an effect of forest patch size or distance to forest–farmland edge on predation rate. In another habitat (open fields or grassland), five studies of real nests and four studies using dummy nests did not find any effect of patch size or distance to habitat edge, while one study (Johnson and Temple, 1990) on real nests (17%) and one study (Møller, 1989) using dummy nests (20%) found an effect of distance to forest edge. Thus there does not seem to be any consistent difference between real nests and dummy nests with respect to predation rate in relation to size of habitat patches and distance to habitat edge, although the predation pattern differed between landscape mosaics.

Dummy nests have also been used to measure the effect of density and dispersion pattern (clumped, random or over-dispersed) on predation rate (Andrén, 1991). Four out of five studies on real nests did not show any effect of inter-nest distance (Blancher and Robertson, 1985; Galbraith, 1988; Andrén, 1991; Schieck and Hannon, 1993; but see Krebs, 1971) and the studies using densities of dummy nests that were within the normal range of densities did not find any effect of inter-nest distance on predation rate (Loman and Göransson, 1978; Boag, Reebs and Schroeder, 1984; O'Reilly and Hannon, 1989). Thus, the predation pattern, effects of inter-nest distance and nest density seem to agree between real nests and dummy nests.

Another use of dummy nests is to get an idea of which predators may prey upon nests. Angelstam (1986) and Andrén (1992) used track boards around dummy nests to find out what predator species had preyed upon ground nests. Both of these studies found that about 80% of the nests were taken by birds (corvids). In the same kind of landscape (boreal forest), Willebrand and Marcström (1988) found that only about 30% of real nests, with a dummy egg, were taken by birds. The type of dummy used by Angelstam (1986) and Andrén (1992) was very open – two eggs placed on a 0.4×0.4 m board. Thus, it is likely that this type of dummy nest overestimates the importance of avian predators. However, within the group of avian predators (corvids), the dummy nests might be used as an index of the probability of different corvids robbing real nests in different habitats and landscapes (Andrén, 1992).

To conclude, dummy nests can be used as an index of predation rate in some situations, although the results should be interpreted with caution. There is certainly a need for more studies on real nests, especially comparing real nests and dummy nests in the same study.

10.4 HABITAT SELECTION OF PREDATORS

To understand why there is sometimes an edge-related increase in predation rate on birds' nests, it is important to establish which predators have preyed upon the nests and the habitat selection of these nest predators. Very few studies have actually studied the habitat selection of the predator in relation to habitat edge (e.g. Widén, 1989), and most of them are based on indications from tracks (e.g. Bider, 1968; Forsyth and Smith, 1973), observations of predators (e.g. Brittingham and Temple, 1983), track boards (e.g. Angelstam, 1986; Andrén, 1992) or bill markings left in dummy eggs (e.g. Møller, 1989).

If predators are attracted by habitat edges, because of the density of birds, then the relative use of habitat edges by predators and the relative predation rate should be related to differences in prey density in the edge and other parts of the habitat. One of the few studies of habitat selection in relation to habitat edges actually found that goshawks (*Accipiter gentilis*) avoided forest edges and preferred hunting inside large patches of old forest (Widén, 1989). Although the goshawk preys upon birds and mammals and is not a potential nest predator, it could respond to the increased density of adult birds in forest edges (Hansson, 1983; Helle, 1983).

Habitat edges with a higher density of birds are found in different landscape mosaics. Gates and Gysel (1978) and Chasko and Gates (1982) found that both density of birds and density of nests increase from the interior parts of the forest to the forest–field edges. Both also found that predation rate on nests was higher in the habitat edge than in the interior parts of the forest. In forest landscapes, Strelke and Dickson (1980), Hansson (1983) and Helle (1983) found that the density of birds was higher in forest–clear-cut edges than inside the forest. However, Helle (1983) also found that the amount of shrub layer in the forest–clear-cut edge explained variation in bird density. In forest edges with a well-developed shrub layer, the bird density was twice as high as in edges with a poor shrub layer. The density in the latter edge type was not different from the density in the interior parts of the forest. Also, Morgan and Gates (1982) found that forest edges with a well-developed shrub layer had a higher density than open forest edges. Although increased density of breeding birds is reported from forest landscape, most of the studies of predation failed to find any edge-related increase in predation rate. Thus, it is unlikely that differences in bird density in habitat edge as compared to deeper inside a habitat patch, can explain why an edge-related increase in predation rate is found mainly in forest neighboring farmland, and not in other landscape types.

Another explanation for the raised predation rate close to habitat edges is that predators move along habitat edges. The predators are not primarily looking for food, but as they spend more time close to the habitat edge than in other parts of the habitat patches, the risk of predation is raised. Bider (1968) and Forsyth and Smith (1973) found that many potential nest predators were more active close to forest edges and to some extent moved along the habitat edge. There is also evidence that predation on birds' nests is incidental, i.e. the predators are basically looking for other food items and find birds' nests just by chance (Angelstam, 1986; Vickery, Hunter and Wells, 1992; but see Andrén, 1992). Thus, a high density of birds and nests may not be the only reason why predators are more active in habitat edges. Two studies found that predation rate on nests in forest decreases with distance from the habitat edge of very long narrow habitat elements, i.e. tranmission-lines (Chasko and Gates, 1982) and fields along gravel roads (Burkey, 1993). Forest habitat dominated the landscape in both these studies. The raised predation rate close to the habitat edge could probably be due to predators moving along the habitat edge. Whether raised predation in edge is due to an attraction and active search for nests or movements along edges and incidental predation on nests can be tested by comparing nest predation in habitat edges with higher nest density than other parts of the habitat with habitat edges with no increased nest density. Attraction of predators to edges predicts that there should only be raised predation in edges that have higher densities of nests than deeper inside the habitat patch, while movement of predators along edge predicts raised predation independent of nest density.

Predators living in the matrix and penetrating into other habitat patches have a particularly large potential to raise the predation pressure in small habitat patches. For example, the density of farmland predators per unit forest area increases very rapidly as the forest is fragmented and interspersed with farmland. However, a high density of predators in farmland does not *per se* imply a heavy predation pressure; they must also be shown to prey upon nests inside forest fragments, i.e. they must be habitat generalists. On the other hand, if the predators are habitat generalists that penetrate into habitat patches, their activity can cover considerable distances. Using dummy nests, Andrén (1992) showed that the hooded crow (*Corvus corone*), an open habitat bird, was the most important corvid for causing the increased predation pressure close to forest–farmland edges. Similarly, Møller (1989) found that hooded crows preyed upon nests inside the forest, as well as along the farmland–forest edge. Furthermore, the density of hooded crows was higher in landscapes with a mixture of farmland and forest than in landscapes dominated by either farmland or forest, indicating its use of both farmland and forest (Figure 10.4; Andrén, 1992). This was also the case for the American crow (*Corvus brachyrhynchos*), where both nest density and reproductive success were higher in a mixed farmland–forest landscape than in a pure farmland landscape (Ignatiuk and Clark, 1991). Angelstam (1986) and Angelstam and Andrén (in preparation) demonstrated that the badger (*Meles meles*) is also an important predator that may cause the edge-related increase in predation pressure close to farmland–forest edges. In North America, raccoon (*Procyon lotor*) and blue jay (*Cyanocitta cristata*) may be the most important species causing the inverse relationship between nest predation and forest patch size, as shown by dummy nest experiments (Wilcove, 1985b; PhD thesis, cited in Terborgh, 1989). Brittingham and Temple (1983) and Rothstein, Verner and Stevens (1984) found that the brown-headed cowbird (*Molothrus ater*) was more active close to forest edges than inside the forest. Its feeding activities and social behavior are restricted to open habitats, and the proportion of nests parasitized by brown-headed cowbirds was higher close to the edge than inside the forest.

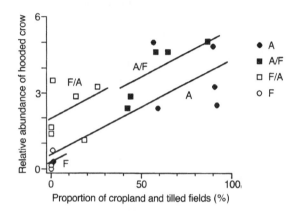

Figure 10.4 Relative abundance of hooded crow in relation to proportion of cropland and tilled fields in 4k m² large study plots. A, A/F, F/A and F represent four different landscape types: A – a landscape dominated by agricultural land (10% forest); A/F – agricultural land with some forest (40% forest); F/A – forest area with some agricultural land (7% agricultural land); and F – almost pure forest landscape (2% agricultural land) (From Andrén, 1992, reproduced with permission.)

On the other hand, if the raised predation in habitat edges is not caused by predators entering from the surroundings, but by predators living in the habitat patch, then the patch has to fulfil the area requirements of the predators. Since predators generally have larger area requirements than prey (Schoener, 1968), it is unlikely that predators restricted to only one particular habitat fragment can cause the increased predation rate in a small fragment. One may expect instead that as the fragment size falls below the area requirements of the predators, the predation rate should decline. However, if several small fragments together can support the predators, i.e. the predators view the landscape as a fine-grained environment (Levins, 1968), then there might not be any change in predation rate. Real islands could be viewed as habitat patches where predation rate is only influenced by the predators restricted to that habitat patch, i.e. there is no influence of the surroundings. Therefore, as the island size is not fulfilled for the predators but for the prey, the predation rate should decline. Accordingly, the predation rate should be negatively related to island size. However, the three studies from real islands show all possible effects of size on predation rate. Loiselle and Hoppes (1983) found a positive relationship between island size and predation rate, Nilsson *et al.* (1985) found no relationship and George (1987) found a negative relationship.

In forest–farmland mosaics, Andrén (1992) found that the impact of jay (*Garrulus glandarius*) and raven (*Corvus corax*), two forest-living corvids, as nest predators was lower as the forest fragments became smaller. In a landscape with only small forest fragments these two predators were not found at all. Therefore, it is likely that the high density of farmland predators, which to some extent also exploit forest habitat, may explain why edge-related increases in predation rate are mainly found in the forest fragments neighboring farmland. Based on predator activity, track boards or other identification of predators, nine studies (Brittingham and Temple, 1983; Wilcove, 1985a; Wilcove, McLellan and Dobson, 1986; Andrén and Angelstam, 1988; Møller, 1988, 1989, 1991; Temple and Cary, 1988; Andrén 1992) concluded that raised predation rate close to forest–farmland edges was due to predators living in farmland and then penetrating the forest. Wilcove, McLellan and Dobson (1986) and Andrén and Angelstam (1988) reported that the raised predation pressure on nests extended about 200–300 m into the forest. On the other hand, an increased density of birds and nests is usually found very close to the forest edge, i.e. < 50m (Gates and Gysel, 1978; Hansson, 1983; Helle, 1983). Thus, it is unlikely that raised density of nests in habitat edge can explain the increased predation rate far away from habitat edges.

In open, two-dimensional habitat, like farmland, several studies have reported increased predation on nests close to trees or other perches suitable for avian predators (Loman and Göransson, 1978; Erikstad, Blom and Myrberget, 1982; Sullivan and Dinsmore, 1990; Berg, Lindberg and Källebrink, 1992). These perches could be single trees, groups of trees or forest edges. However, according to Table 10.1, only two of 11 studies found an effect of distance to habitat edge or patch size on nest predation rate. Thus, there seems to be a contradiction between these two observations. Three of the studies in Table 10.1 in open habitat were surrounded by other open habitat. Thus, the habitat edges were not perchsites for avian predators. All three of them failed to find any patch size-related predation. (Duebbert and Lokemoen, 1976; Gatti, 1987; Clark, Nudds and Bailey, 1991). In one study that failed to find any edge-related increase in predation rate, the main predator was striped skunk (*Mephitis mephitis*; Vickery, Hunter and Wells, 1992). If forest edges are perches for avian predators one should not expect the predation to be related to distance to edge if the main predator is a mammal. Møller (1989) reported that farmland-living birds avoided nesting close to forest edges and that predation rate on dummy nests placed in farmland was higher close to the edge. Similarly, Berg (1992) found that curlew (*Numenius arquata*) nested far away from forest edges, but predation rate was not related to distance to forest edge. Thus, the curlew probably avoided dangerous areas. Thus, it seems unclear how farmland-living birds respond to forest–farmland edges. There might be an increased predation risk close to forest edges in farmland and birds might avoid nesting close to the edge. Thus, this needs to be studied, both on real nests and on dummy nests.

On the basis of these data, one may distinguish three major types of predators depending on their habitat use. The first is predators whose activity is mainly restricted to one type of habitat, i.e. habitat specialists, for which the landscape is divided (Addicott *et al.*, 1987). Examples include the magpie (*Pica pica*) and the jackdaw (*Corvus monedula*), which prey upon nests mainly in farmland (Møller, 1989; Andrén, 1992). Other examples of habitat specialist are the jay (Møller, 1989; Andrén, 1992), pine marten (Storch, Lindström and de Jounge, 1990; Johnsson, Nilsson and Tjernberg, 1993) and the red squirrel (Nour, Matthysen and Dhondt,

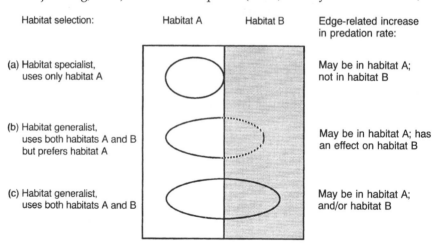

Figure 10.5 Different types of predators depending on their activity in different habitats and their potential effect on edge-related predation rates. (a) Habitat specialist, e.g. jackdaw and magpie using farmland in a farmland–forest mosaic; European jay, pine marten and red squirrel using forest in a forest–farmland mosaic. (b) Habitat generalist that prefers one habitat, but also uses the other one, e.g. hooded crow, badger, blue jay and raccoon preferring farmland but also using forest in a farmland–forest mosaic. (c) Habitat generalist that uses both habitats approximately to the same extent, e.g. raven, blue jay and American crow in a forest mosaic with patches of forest stand of different age; hooded crow, American crow, badger, striped skunk, raccoon and red fox within a farmland mosaic with different kinds of habitat patches such as meadows, tilled fields, etc.

1993), which mainly prey upon nests inside forest patches. This group of predators may cause an edge-related increase in predation rate if they are attracted by the habitat edge or if they use habitat edges as traveling lines, but the effect will be only close to the habitat edge within the habitat they prefer (Figure 10.5).

The second type includes predators whose activity is not restricted to only one habitat type, i.e. species for which the landscape is heterogeneously undivided (Addicott et al., 1987). Such predators may prefer certain habitats but also frequent others; thus, they are habitat generalists. Hooded crows and badgers both showed a preference for farmland but also used the forest to some extent (Angelstam, 1986; Andrén, 1992; Angelstam and Andrén, in preparation). The same is probably true for the raccoon and blue jay (Wilcove, 1985b; PhD thesis, cited in Terborgh, 1989), as well as for the nest-parasitic brown-headed cowbird (Brittingham and Temple, 1983; Rothstein, Verner and Stevens, 1984). This kind of predator will cause an edge-related increase in the habitat type it penetrates, but might also cause an increased predation rate close to the edge in the preferred habitat, if it is attracted by the habitat edge (Figure 10.5). However, predators penetrating into habitat patches may have an effect on prey considerable distances from the habitat edge.

Finally, some habitat generalists may use several habitats in the landscape mosaic without a strong preference for a certain habitat. The raven might be an example, as it was the most important corvid nest predator in a landscape dominated by forest, but within this kind of landscape its impact as a nest predator was about the same in mature forest and on clear-cuts (Andrén, 1992; Angelstam and Andrén, in preparation). Furthermore, since the density of raven in boreal parts of Sweden was three pairs per 100 km^2 (Angelstam, Lindström and Widén 1984), the species obviously needs very large home ranges (3000 ha) compared to the typical grain size of the managed forest (10 ha). Thus, from the raven's perspective the landscape mosaic is fine-grained, and the raven may use the entire landscape mosaic. Similarly, Yahner and Wright (1985) and Yahner and Scott (1988) failed to find any edge-related increase of predation rates but instead found an effect of the degree of forest fragmentation. The main predators in their study were blue jay and American crow. Both species increased in density with increasing degree of forest fragmentation (Yahner and Scott, 1988). This was interpreted as an effect at the regional level, where predator density depends on the landscape mosaic but the predators exploit the entire landscape mosaic. Some predators living in the farmland probably also exploit most habitats in the landscape mosaic. Examples of important nest predators within the farmland mosaic are hooded crow and badger, both of which use the entire farmland (Møller, 1989; Andrén, 1992; Angelstam and Andrén, in preparation). Examples of nests predators from North America that probably use most habitats in a farmland landscape, are the striped skunk (*Mephitis mephitis*), red fox (*Vulpes vulpes*), raccoon and American crow (Duebbert and Lokemoen, 1976; Gatti, 1987; Crabtree and Wolfe, 1988; Johnson, Sargeant and Greenwood, 1989; Clark, Nudds and Bailey, 1991; Vickery, Hunter and Wells, 1992). Although the predators use the entire landscape mosaic, the nest predation can be edge-related, if this type of predator is attracted by the habitat edge (Figure 10.5).

In a mixed farmland–forest landscape, the blue jay was suggested to be one of the most important predators for the increased predation rate close to the forest edge (Wilcove, 1985b; PhD thesis, cited in Terborgh, 1989), whereas in a forest landscape, where the blue jay was one of the major predators, no edge-related elevation of predation rate was recorded (Yahner and Scott, 1988). Thus, the same predator species play different roles in different landscapes, and have greatly different effects on prey populations.

10.5 EFFECT ON PREY OF EDGE-RELATED INCREASE IN PREDATION PRESSURE

The most obvious effect on prey of an increased nest predation rate close to habitat edges is that prey should avoid nesting close to edges. However, as the cause of raised predation in edges might be different, prey could respond differently. If predators are attracted to an edge because of the high density of birds and nests, the prey themselves could also be attracted, because of the high density of prey for them. Helle and Muona (1985) found that the density index of several abundant invertebrate groups was higher in forest edges than inside the forest. This could explain the increased density of birds and nests in habitat edges (Gates and Gysel, 1978; Chasko and Gates, 1982; Hansson, 1983; Helle, 1983). Thus, birds might be attracted to habitat edges and locate their nests there, although the breeding success is lower in habitat edges. Therefore, Gates and Gysel (1978) suggested that artificial habitat edges were ecological traps, by concentrating nests and thereby increasing density-dependent mortality.

On the other hand, there are examples where the birds seem to avoid areas with higher risk of predation. Møller (1989) found that birds with open nests tended to nest at some distance from edges, whereas species with partially covered nests frequently nested close to woodland edges. He also reported that predation rate was higher in the forest–farmland edge than inside the forest. Similarly, Berg (1992) found that curlew avoided nesting close to forest edges. Another example of nest site selection that could be explained by minimizing the predation risk was studied by Storch (1991). However, she found the opposite pattern; capercaillie hens (*Tetrao urogallus*) preferred to nest in the edge zone between mature forest and clear-cuts, where nest predation on dummy nests was lower than inside the mature forest stands. An alternative explanation for a preference to locate nests in the edge zone by capercaillie hens, might be that the snow melts earlier in the edge zone between mature forest and clear-cut than inside the mature forest.

An inevitable effect of an edge-related increase in predation rate is that predation rates will be higher in small fragments compared to large ones, as the proportion of edge zone is inversely related to fragment size (Levenson, 1981). Increased predation pressure in small habitat fragments has been suggested to explain why certain species are missing from small forest fragments, which otherwise seem to fulfil the habitat and area requirements for the species (Whitcomb et al., 1981; Wilcove, 1985a; Terborgh, 1989; Askins, Lynch and Greenberg, 1990). The susceptibility to predation differs between nest types and species; for example, open nests, especially if close to the ground, are more vulnerable to predation than nests in tree holes (Martin, 1988a,b; Møller, 1991). Long-distance migrating birds are claimed to have been severely affected by forest fragmentation in eastern USA. Several of these species have only single broods and nest in open nests close to the ground or in the trees, whereas few of the residents or short-distance migrants place their nests on the ground. Therefore, differences in vulnerability to nest predation may explain the disproportional disappearance of long-distance migrants (Whitcomb et al., 1981; Terborgh, 1989; Askins, Lynch and Greenberg, 1990; but see Greenwood, 1990, for alternative explanations). Corresponding differences in vulnerability to nest predation were argued to have resulted in a lower proportion of open-nesting birds than hole-nesting birds in small as compared to large forest fragments (Martin, 1988c; Møller, 1988).

Temple and Cary (1988) modeled the effect of edge-related increase in nest predation on population density of birds in landscapes with different proportions of suitable habitat. Their model suggested a significant effect on prey population density only in landscapes with highly fragmented habitat, because the proportion of poor territories, i.e. territories close to habitat edge, increased very rapidly as the size of patches declined. Temple (1986) compared two different models to predict the presence and abundance of forest birds in forest patches of different sizes. The first model was based on the total area of the forest patch, while the other only used the core-area of the forest patch, i.e. the area of forest more than 100 m from forest edges. Both models were fairly good predictors of the presence and abundance of birds, but the core-area model was consistently better than the total-area model.

Berg, Nilsson and Boström (1992) tested a suggestion, advanced by Boström and Nilsson (1983), that a higher predation pressure close to the forest edge might explain why bogs with relatively little edge zone between bog and forest (circular bogs) had higher bird density than bogs with relatively much edge zone (long narrow bogs). Since Berg, Nilsson and Boström (1992) failed to find any difference in predation rate between dummy nests placed in the center of the bog and those placed in the edge zone, they concluded that nest predation rate was approximately similar on bogs regardless of shape. However, predation on adult birds in relation to the forest edge might explain the pattern, but has not been tested (Berg, Nilsson and Boström, 1992).

The suggestion that species with open nests, especially close to the ground, should be more sensitive to increased density of predators assumes that these nests are more vulnerable to nest predation than other nests. However, the sensitivity of different nest types depends on the type of predators involved. Clark and Nudds (1991) reviewed studies that have examined the importance of concealment. If birds were the major predators, concealment was found to be important for nesting success in 76% of the studies (16 out of 21). On the other hand, if mammals alone or birds and mammals together were the major predators, concealment was found to be important for nesting success in only 31% of the studies (9 out of 29).

10.6 WHEN SHOULD PREDATION NOT BE RELATED TO EDGE?

So far I have discussed possible explanations for raised predation in habitat edges. However, it is equally important to be able to predict when predation rate should not be related to habitat edge. The first explanation, that predators are attracted to edges because of high prey density in edges, predicts that there should not be any raised predation in the habitat edge if the density of prey is the same in the habitat edge as in other parts of the habitat. The second explanation, that predators move along habitat edges, does not give any clear prediction about the absence of edge-related increase in predation rate, because all habitat edges are potential traveling lines. Finally, if the raised predation rate in habitat edges is due to predators entering from the surrounding habitat, there should not be any edge-related increase in predation rate if the predators are habitat specialists or if the density of predators is about equal in the habitats on both sides on the edge.

10.7 SUMMARY

The effect of habitat edge and patch size on rates of nest predation depends on the predator species involved and especially how they use the landscape mosaic. Edge-related increase in predation seems to be most commonly found inside forests surrounded by farmland and was rarely found in forest mosaics. The effect in open habitat seems to be unclear, mainly because it is a more heterogeneous group of habitat edges than the other ones. Thus, a habitat edge does not necessarily generate an edge-related increase in predation pressure. However, the same type of habitat patch can have different qualities, owing to differences in the surrounding habitats. Therefore, one must also take into account the properties of neighboring habitats to understand the abundance and distribution of species in habitat patches. It is not enough to just to consider the size and internal habitat quality of the patch, as has been done in most studies of effects of habitat fragmentation. Even prey species that experience the landscape as a mosaic of isolated habitat patches, i.e. a divided landscape, might be influenced by predators entering from the surroundings.

ACKNOWLEDGEMENTS

Per Angelstam, Bob Clark, Lennart Hansson, Przemek Majewski, Sven G. Nilsson, Tarja Oksanen, Jørund Rolstad, Jon Swenson and Staffan Ulfstrand (in alphabetical order) gave valuable comments on the manuscript. My work was funded by the Swedish Environmental Protection Agency and the private foundations Olle och Signhild Engkvists stiftelser.

REFERENCES

Addicott, J.F., Aho, J.M., Antolin, M.F. *et al.* (1987) Ecological neighbourhoods: scaling environmental patterns. *Oikos*, **49**, 340–6.

Andrén, H. (1991) Predation: an over-rated factor for over-dispersion of birds' nests? *Anim. Behav.*, **41**, 1063–9.

Andrén, H. (1992) Corvid density and nest predation in relation to forest fragmentation. A landscape perspective. *Ecology*, **73**, 794–804.

Andrén, H. and Angelstam, P. (1988) Elevated predation rates as an edge effect in habitat islands: experimental evidence. *Ecology*, **69**, 544–7.

Angelstam, P. (1986) Predation on ground-nesting birds' nests in relation to predator density and habitat edge. *Oikos*, **47**, 365–73.

Angelstam, P. (1992) Conservation of communities – the importance of edges, surroundings and landscape mosaic structure, in *Ecological Principles of Nature Conservation* (ed. L. Hansson), Elsevier Applied Science, London, pp. 9–70.

Angelstam, P., Lindström, E. and Widén, P. (1984) Role of predation in short-term population fluctuations of some birds and mammals in Fennoscandia. *Oecologia*, **62**, 199–208.

Askins, R.A., Lynch, J.F. and Greenburg, R. (1990) Population declines in migratory birds in eastern North America. *Current Ornith.*, **7**, 1–57.

Avery, M.I., Winder, F.L.R., and Egan, V.M. (1989) Predation on artificial nests adjacent to forestry plantations in northern Scotland. *Oikos*, **55**, 321–3.

Berg, Å. (1992) Factors affecting nest-site choice and reproductive success of curlews *Numenius arquata* on farmland. *Ibis*, **134**, 44–51.

Berg, Å., Nilsson, S.G. and Boström, U. (1992) Predation on artificial wader nests on large and small bogs along a south–north gradient. *Ornis Scand.* **23**, 13–16.

Berg, Å., Lindberg. T. and Källebrink, K.G. (1992) Hatching success of lapwings on farmland: differences between habitats and colonies of different sizes. *J. Anim. Ecol.*, **61**, 469–76.

Bider, J.R. (1968) Animal activity in uncontrolled terrestrial communities as determined by a sand transect technique. *Ecol. Monogr.*, **38**, 269–308.

Björklund, M. (1990) Nest failure in the Scarlet Rosefinch *Carpodacus erythrinus*. *Ibis*, **132**, 613–17.

Blancher, P.J. and Robertson, R.J. (1985) Predation in relation to spacing of kingbird nests. *Auk*, **102**, 654–8.

Boag, D.A., Reebs, S.G. and Schroeder, M.A. (1984) Egg loss among spruce grouse inhabiting lodgepole pine forests. *Can. J. Zool.*, **62**, 1034–7.

Boström, U. and Nilsson, S.G. (1983) Latitudinal gradients and local variations in species richness and structure of bird communities on raised peat-bogs in Sweden. *Ornis Scand.*, **14**, 213–26.

Brittingham, M.C. and Temple, S.A. (1983) Have cowbirds caused forest songbirds to decline? *Bioscience*, **33**, 31–5.

Burkey, T.V. (1993) Edge effects in seed and egg predation at two neotropical rainforest sites. *Biol. Cons.*, **66**, 139–43.

Chasko, G.G. and Gates, J.E. (1982) Avian habitat suitability along a transmission-line corridor in an oak–hickory forest region. *Wildl. Monogr.*, **82**, 1–41.

Clark, R.G. and Nudds, T.D. (1991) Habitat patch size and duck nesting success: the crucial experiments have not been performed. *Wildl. Soc. Bull.*, **19**, 534–43.

Clark, R.G., Nudds, T.D. and Bailey, R.O. (1991) Populations and nesting success of upland-nesting ducks in relation to cover establishment. *Can. Wildl. Serv. Prog. Notes 193*.

Crabtree, R.L. and Wolfe, M.L. (1988) Effects of alternate prey on skunk predation of waterfowl nests. *Wildl. Soc. Bull.*, **16**, 163–9.

Duebbert, H.F. and Lokemoen, J.T. (1976) Duck nesting in fields of undisturbed grass–legume cover. *J. Wildl. Manage.*, **40**, 39–49.

Erikstad, K.E., Blom, R. and Myrberget, S. (1982) Territorial hooded crows as predators on willow ptarmigan nests. *J. Wildl. Manage.*, **46**, 109–14.

Esler, D. and Grand, J.B. (1993) Factors influencing depredation of artifical duck nests. *J. Wildl. Manage.*, **57**, 244–8.

Forman, R.T.T. (1981) Interactions among landscape elements: a core of landscape ecology, in *Perspective in Landscape Ecology* (eds S.P. Tjallingii and A.A. de Veer), Centre for Agricultural Publication and Documentation, Wageningen, pp. 35–48.

Forman, R.T.T. and Godron, M. (1981) *Landscape Ecology*, Wiley, New York.

Forsyth, D.J. and Smith, D.A. (1973) Temporal variability in home ranges of eastern chipmunks (*Tamias striatus*) in southeastern Ontario woodlots. *Am. Midl. Nat.*, **90**, 107–17.

Galbraith, H. (1988) Effects of agriculture on the breeding ecology of lapwings *Vanellus vanellus*. *J. Appl. Ecol.*, **25**, 487–503.

Gates, J.E. and Gysel, L.W. (1978) Avian nest dispersion and fledging outcome in field–forest edges. *Ecology*, **59**, 871–83.

Gatti, R.C. (1987) *Duck Production: The Wisconsin Picture*. Wisc. Dep. Nat. Res., Bur. Res., Res. Manage., Madison, WI, Findings 1.

George, T.L. (1987) Greater land bird densities in island vs. mainland: relation to nest predation. *Ecology*, **68**, 1393–400.

Gibbs, J.P. (1991) Avian nest predation in tropical wet forest: an experimental study. *Oikos*, **60**, 155–61.

Götmark, F., Neergaard, R. and Åhlund, M. (1990) Predation of artificial and real arctic loon nests in Sweden. *J. Wildl. Manage.*, **54**, 429–32.

Greenwood, J.J.D. (1990) What the little birds tell us. *Nature*, **343**, 22–3.

Hansson, L. (1983) Bird numbers across edges between mature conifer forest and clearcuts in Central Sweden. *Ornis Scand.*, **14**, 97–104.

Harris, L.D. (1988) Edge effects and conservation of biotic diversity. *Cons. Biol.*, **2**, 330–2.

Helle, P. (1983) Bird communities in open ground–climax forest edges in northeastern Finland. *Oulanka Reports*, **3**, 39–46.

Helle, P. and Muona, J. (1985) Invertebrate numbers in edges between clear-fellings and mature forests in northern Finland. *Silva Fennica*, **19**, 281–94.

Huhta, E. Ground nest predation in a fragmented boreal forest landscape (in press).

Ignatiuk, J.B. and Clark, R.G. (1991) Breeding biology of American crows in Saskatchewan parkland habitat. *Can. J. Zool.*, **69**, 168–75.

Janzen, D.H. (1986) The external threats, in *Conservation Biology. The Science of Scarcity and Diversity* (ed. M.E. Soulé), Sinauer, Sunderland, MA, pp. 286–303.

Johnson, R.G. and Temple, S.A. (1990) Nest predation and brood parasitism of tallgrass prairie birds. *J. Wildl. Manage.*, **54**, 106–11.

Johnson, D.H., Sargeant, A.B. and Greenwood, R.J. (1989) Importance of individual species of predators on nesting success of ducks in the Canadian Prairie Pothole Region. *Can. J. Zool.*, **67**, 291–7.

Johnsson, K. (1993) Nest site quality in old black woodpecker holes – a predation experiment, in The black woodpecker *Dryocopus martius* as a keystone species in forest. PhD thesis, Report 24, Department of Wildlife Ecology, Swedish University of Agricultural Sciences, Uppsala, Sweden.

Johnsson, K., Nilsson, S.G. and Tjernberg, M. (1993) Increased nest predation on hole-nesting birds at forest edges? in The black woodpecker *Dryocopus martius* as a keystone species in forest. PhD thesis, Report 24, Department of Wildlife Ecology, Swedish University of Agricultural Sciences, Uppsala, Sweden.

Karr, J. (1982) Avian extinctions on Barro Colorado Island, Panama: a reassessment. *Am. Nat.*, **119**, 220–39.

Krebs, J.R. (1971) Territory and breeding density in the great tit (*Parus major* L.). *Ecology*, **52**, 2–22.

Kuitunen, M. and Helle, P. (1988) Relationship of the common treecreeper *Certhia familiaris* to edge and forest fragmentation. *Ornis Fenn.*, **65**, 150–5.

Langen, T.A., Bolger, D.T. and Case, T.J. (1991) Predation on artificial bird nests in chaparral fragments. *Oecologia*, **86**, 395–401.

Levenson, J.B. (1981) Woodlots as biogeographic islands in southeastern Wisconsin, in *Forest Island Dynamics in Man-dominated Landscapes* (eds R.L. Burgess and D.M. Sharpe), Ecol. Stud. 41, Springer-Verlag, pp. 13–39.

Levins, R. (1968) *Evolution in Changing Environments*, Princeton University Press, Princeton, NJ.

Loiselle, B.A, and Hoppes, W.G. (1983) Nest predation in insular and mainland lowland rainforest in Panama. *Condor*, **85**, 93–5.

Loman, J. and Göransson, G. (1978) Egg shell dumps and crow (*Corvus cornix*) predation on simulated birds' nests. *Oikos*, **30**, 461–6.

Martin, T.E. (1987) Artificial nest experiments: effects of nest appearance and type of predators. *Condor*, **89**, 925–8.

Martin, T.E. (1988a) Processes organizing open-nesting bird assemblages: competition or predation? *Evol. Ecol.*, **2**, 37–50.

Martin, T.E. (1988b) On the advantage of being different: Nest predation and the coexistence of bird species. *Proc. Natl Acad. Sci. USA*, **85**, 2196–9.

Martin, T.E. (1988c) Habitat and area effects on forest bird assemblages: Is nest predation an influence? *Ecology*, **69**, 74–84.

Møller, A.P. (1988) Nest predation and nest site choice in passerine birds in habitat patches of different size: a study of magpies and blackbirds. *Oikos*, **53**, 215–21.

Møller, A.P. (1989) Nest site selection across field–woodland ecotones: the effect of nest predation. *Oikos*, **56**, 240–6.

Møller, A.P. (1991) Clutch size, nest predation, and distribution of avian unequal competitors in a patchy environment. *Ecology*, **72**, 1336–49.

Morgan, K.A. and Gates, J.E. (1982) Bird population patterns in forest edge and strip vegetation at Remington farms, Maryland. *J. Wildl. Manage.*, **46**, 933–44.

Nilsson, S.G., Björkman, C., Forslund, P. and Höglund, J. (1985) Egg predation in forest bird communities on islands and mainland. *Oecologia (Berlin)*, **66**, 511–15.

Nour, N., Matthysen, E. and Dhondt, A.A. (1993) Artificial nest predation and habitat fragmentation: different trends in bird and mammal predators. *Ecography*, **16**, 111–16.

Odum, E.P. (1971) *Fundamentals of Ecology*, Saunders, Philadelphia, PA.

O'Reilly, P. and Hannon, S.J. (1989) Predation of simulated willow ptarmigan nests: the influence of density and cover on spatial and temporal patterns of predation. *Can. J. Zool.*, **67**, 1263–7.

Paton, P.W. (1994) The effect of edge on avian nest success: How strong is the evidence? *Conservation Biology*, **8**, 17–26.

Ratti, J.T. and Reese, K.P. (1988) Preliminary test of the ecological trap hypothesis. *J. Wildl. Manage.*, **52**, 484–91.

Roper, J.J. (1992) Nest predation experiments with quail eggs: too much to swallow? *Oikos*, **65**, 528–30.

Rothstein, S.I., Verner, J. and Stevens, E. (1984) Radio-tracking confirms a unique diurnal pattern of spatial occurrence in the parasitic brown-headed cowbird. *Ecology*, **65**, 77–88.

Rudnicky, T.C. and Hunter, M.L. (1993) Avian nest predation in clearcuts, forests, and edges in a forest-dominated landscape. *J. Wildl. Manage.*, **57**, 358–64.

Sandström, U. (1991) Enhanced predation rates on cavity bird nests at deciduous forest edges – an experimental study. *Ornis Fenn.*, **68**, 93–8.

Santos, T. and Telleria, J.L. (1992) Edge effects on nest predation in Mediterranean fragmented forest. *Biol. Conserv.*, **60**, 1–5.

Schiech, J.O. and Hannon, S.J. (1993) Clutch predation, cover, and the overdispersion of nests of the willow ptarmigan. *Ecology*, **74**, 743–50.

Schoener, T.W. (1968) Sizes of feeding territories among birds. *Ecology*, **49**, 123–41.

Small, M.F. and Hunter, M.L. (1988) Forest fragmentation and avian nest predation in forested landscapes. *Oecologia*, **76**, 62–4.

Sonerud, G. (1993) Reduced predation by nest box relocation: differential effect on Tengmalm's Owl and artificial nests. *Ornis Scand.*, **24**, 249–53.

Stamps, J.A., Buechner, M. and Krishnam, V.V. (1987) The effect of edge permeability and habitat geometry on emigration from patches of habitat. *Am. Nat.*, **129**, 533–52.

Storaas, T. (1988) A comparison of losses in artificial and naturally occurring capercaillie nests. *J. Wildl. Manage.*, **52**, 123–6.

Storch, I. (1991) Habitat fragmentation, nest site selection, and nest predation risk in Capercaillie. *Ornis. Scand.*, **22**, 213–17.

Storch, I., Lindström, E. and de Jounge, J. (1990) Diet and habitat selection of the pine marten in relation to competition with the red fox. *Acta Theriol.*, **35**, 311–20.

Strelke, W.K. and Dickson, J.G. (1980) Effects of forest clear-cut edge on breeding birds in east Texas. *J. Wildl. Manage.*, **44**, 559–67.

Sullivan, B.D. and Dinsmore, J.J. (1990) Factors affecting egg predation by American crows. *J. Wildl. Manage.*, **54**, 433–7.

Telleria, J.L. and Santos, T. (1992) Spatiotemporal patterns of egg predation in forest islands: an experimental approach. *Biol. Conserv.*, **62**, 29–33.

Temple, S.A. (1986) Predicting impacts of habitat fragmentation on forest birds: a comparison of two models, in *Wildlife 2000. Modeling Habitat Relationships of Terrestrial Vertebrates* (eds J. Verner, M.L. Morrison and C.J. Ralph), The University of Wisconsin Press, Madison, WI, pp. 301–4.

Temple, S.A. and Cary, J.R. (1988) Modeling dynamics of habitat-interior bird populations in fragmented landscapes. *Conserv. Biol.*, **2**, 340–7.

Terborgh, J. (1989) *Where Have All the Birds Gone?* Princeton University Press, Princeton, NJ.

Vickery, P.D., Hunter, M.L., Jr and Wells, J.V. (1992) Evidence of incidental nest predation and its effects on nests of threatened grassland birds. *Oikos*, **63**, 281–8.

Whitcomb, R.F., Robbins, C.S., Lunch, J.F. *et al.* (1981) Effects of forest fragmentation on avifauna of eastern deciduous forest, in *Forest Island Dynamics in Man-dominated Landscapes* (eds R.L. Burgess and D.M. Sharpe), Ecol. Stud. 41, Springer-Verlag, pp. 125–205.

Widén, P. (1989) The hunting habitats of goshawks *Accipiter gentilis* in boreal forests of central Sweden. *Ibis*, **131**, 205–13.

Wiens, J., Crawford, C. and Gosz, J. (1985) Boundary dynamics: a conceptual framework for studying landscape ecosystems. *Oikos*, **45**, 421–7.

Wilcove, D.S. (1985a) Nest predation in forest tracts and the decline of migratory songbirds. *Ecology*, **66**, 1211–14.

Wilcove, D.S. (1985b) Forest fragmentation and the decline of migratory songbirds. PhD thesis, Princeton University, Princeton, NJ.

Wilcove, D.S., McLellan, C.H. and Dobson, A.P. (1986) Habitat fragmentation in the temperate zone, in *Conservation Biology. The Science of Scarcity and Diversity* (ed. M.E. Soulé), Sinauer Associates, Sunderland, MA, pp. 237–56.

Willebrand, T. and Marcström, V. (1988) On the danger of using dummy nests to study predation. *Auk*, **105**, 378–9.

Yahner, R.H. (1988) Changes in wildlife communities near edges. *Conserv. Biol.*, **2**, 333–9.

Yahner, R.H. and Scott, D.P. (1988) Effects of forest fragmentation on depredation of artificial nests. *J. Wildl. Manage.*, **52**, 158–61.

Yahner, R.H. and Voytko, R.A. (1989) Effects of nest-site selection on depredation of artificial nests. *J. Wildl. Manage.*, **53**, 21–5.

Yahner, R.H. and Wright, A.L. (1985) Depredation on artificial ground nests: effects of edge and plot age. *J. Wildl. Manage.*, **49**, 508–13.

Yahner, R.H., Morell, T.E. and Rachael, J.S. (1989) Effects of edge contrast on depredation of artificial nests. *J. Wildl. Manage.*, **53**, 1135–8.

The plant–pollinator landscape   11

Judith L. Bronstein

11.1 INTRODUCTION

Food resources are patchily distributed for most animals, regulating their feeding behavior, population dynamics, and ultimately their evolution. Other chapters in this book have addressed the effect of mosaic landscapes on predators (Chapter 10), competitors (Chapter 9) and herbivores (Chapter 2). Foraging decisions of mutualistic animals such as pollinators and seed dispersers are also made within patchy environments. Interestingly, however, mutualisms have almost never been studied from a landscape perspective. In general, issues of ecological scale and spatiotemporal heterogeneity have not yet become an important concern in the study of mutualism; most research in this field still focuses on describing the natural history of particular interactions at specific places and times (Bronstein, 1994).

Superficially, the plant–pollinator landscape might be thought to resemble the landscape in which plants and herbivores interact. In both cases, herbivorous animals (most often, insects) forage on patchily distributed plants, forcing them to decide when to move between patches and when to switch foods. However, three important differences may cause the dynamics of these two systems to differ greatly. First, most pollinators probably can exploit a greater diversity of plants than can (insect) herbivores. On the other hand, pollinator resources are located within flowers, which in many species are present for a very brief period relative to foragers' lifespans; this should lead to distinct problems of food limitation in certain situations. Finally, plants are under selection to be 'access-

Mosaic Landscapes and Ecological Processes
Edited by Lennart Hansson, Lenore Fahrig and Gray Merriam.
Published in 1995 by Chapman & Hall, London. ISBN 0 412 45460 2.

ible' in space and time to their mutualistic pollinators, whereas they are more likely to evolve ways to 'escape' in space and time from herbivores. The landscapes of pollinators and herbivores have thus been shaped by very different selection pressures.

I will argue in this chapter that the landscape in which pollinators forage emerges from an interacting set of plant and pollinator attributes. Foremost among these are the timing, duration and intensity of flowering in plants, and the search behaviors and diet breadth of their pollinators: flowering phenology sets the distance that pollinators would have to travel between patches of a given species to obtain food, while the mode of search and dietary specificity determines the likelihood that pollinators can and will make that journey. I will describe first the critical plant traits and then the critical pollinator traits, considering how they vary both within and among species. These sets of traits do not vary independently, allowing us to identify several 'plant–pollinator landscapes' (i.e. combinations of plant and pollinator traits) and the types of organisms that occupy them. For each of these landscapes, I will offer hypotheses on how pollinators are able to persist on their floral resources. We will see that certain problems of persistence may explain the relative rarity of some of these landscapes.

At present, we know remarkably little about how pollinators respond to spatial and temporal variation in their floral resources. Mutualisms have almost always been studied from the perspective of one of the two partners only (Cushman and Beattie, 1991; Bronstein, 1994). Although plant phenological variation exists at every spatial and temporal scale imaginable, little is yet known about how foraging pollinators respond; conversely, while we know that pollinators commonly move among resource patches, we rarely know the degree of phenological difference among those patches. Furthermore, plant–pollinator interactions have probably been studied at an inappropriate scale, or an insufficient range of scales, to be able to interpret them easily from a landscape perspective. For example, while extremely local, short-term foraging decisions of certain pollinators have been extensively studied, the total distances traveled by these animals in their lifetime, or even in a single foraging bout, are still poorly known. Phenological variation in a given plant species could be critical in pollinator dynamics: for example, migrating pollinators might track temporal gradients of flowering of preferred species. Alternatively, the temporal and spatial scales over which floral availability varies may simply be too different from the scale at which pollinators choose their resources to matter. Thus, the central question I will attempt to answer in this chapter is: does phenological variation within plant species help to explain the ability of pollinating animals to persist within a given landscape?

11.2 COMPONENTS OF THE PLANT–POLLINATOR LANDSCAPE: PLANTS IN FLOWER

Phenological patterns exist at a wide range of spatial and temporal scales. Individuals in a population might flower simultaneously; alternatively, plants may flower at somewhat different times or for varying lengths of time, either within or across populations. Quantitative information on these phenomena is integral to building up a picture of the plant–pollinator landscape. In this section I briefly summarize the origins of phenological differences within and among plant species, then review some of the most common flowering patterns.

11.2.1 ORIGINS OF PHENOLOGICAL DIFFERENCES

Plants that exploit animals as pollen vectors face a dilemma: pollinators need to be attracted to the flowers, yet they must be forced to leave at some point if the plant is going to outcross. A large suite of plant traits, many of them phenological, have evolved that increase the likelihood that this will occur (Bawa, 1983, 1990; Waser, 1983a, b; Rathcke and Lacey, 1985; Zimmerman, 1988). In the neotropics, for example, moth-pollinated plants flower primarily in the wet season when moths are most abundant (Frankie, 1975), whereas the flowering peak for many deciduous-forest woody plants is in the dry season when bee abundance peaks (Janzen et al., 1982). Whether natural selection has shaped phenology any more precisely than this has been debated extensively (Waser, 1983a; Rathcke and Lacey, 1985; Ollerton and Lack, 1992). There is good evidence in a few cases that flowering times of sympatric species that compete for pollinators have shifted apart (McNeilly and Antonovics, 1968; Waser, 1978, 1983a; Campbell and Motten, 1985). Community-wide flowering sequences often appear regular and orderly, prompting arguments that selection to reduce competition for pollinators has been pervasive (e.g. Frankie, 1975; Heinrich, 1976a; Stiles, 1977). However, many of these community patterns have proven difficult to distinguish from randomly generated sequences (Poole and Rathcke, 1979; Rathcke, 1984; but see Stiles, 1979; Cole, 1981).

Certain flowering patterns are difficult to explain through natural selection. Selection may simply not act very strongly on the precise timing of flowering, allowing chance variations among individuals, populations and species to persist (Ollerton and Lack, 1992). Alternatively, various constraints may limit the set of possible phenologies. Phylogenetic constraints on flowering time may be important: certain plant families and genera are well known to have characteristic flowering seasons (Kochmer and Handel, 1986). The timing of flowering may also be closely linked with other traits and life history events, raising the possibility that flowering time could shift due to selection on a correlated trait. For example, selection could be acting to fine-tune 'fruiting' phenology to correspond to seed disperser availability, optimal seed germination time, or an absence of seed predators (Augspurger, 1981; Rathcke and Lacey, 1985; Primack, 1987; Newstrom *et al.*, 1993). Complex physiological relationships may also exist between timing of flowering and vegetative growth. In many tropical trees, the timing of water stress regulates both processes (e.g. Reich and Borchert, 1982). Evidence for such tight correlations is still limited, however, (Kelly, 1992; Newstrom *et al.*, 1993). Flowering time in some plants is known to change rapidly under selection, sometimes dramatically so (Waser, 1983a).

While biologists generally focus on phenological differences among plant species, marked variability can be found within species as well. The onset of flowering may be triggered by either endogenous or exogenous cues; only three such exogenous factors – photoperiod, temperature and onset of rainfall – have been identified, although these may interact in complex ways (Murfet, 1977; Rathcke and Lacey, 1985; Bernier, 1988). Variation in exogenous factors is likely to result in a certain degree of intraspecific flowering time variation over space and across time. Geographic differences in flowering time have long been known for certain temperate plant species; in fact, quantitative models that predict the degree of variation surprisingly well, usually on the basis of cumulative degree-days, have been in use for hundreds of years (e.g. Réaumur, 1735). Plants usually flower predictably later at progressively higher altitudes (Macior, 1977; Arroyo, 1990) and more extreme latitudes (Hodgkinson and Quinn, 1978; Reader, 1983; but see Ray and Alexander, 1966). Longitudinal flowering patterns also exist: 'Hopkins' bioclimatic law' (Hopkins, 1938) predicts that in North America, fall flowering for many plants species will occur four days later for each 5° longitude west. Geographic variation in moisture may also predict differences in flowering time (Burk, 1966). Experiments have shown many of these geographic differences in flowering time to be genetically based (e.g. Vance and Kucera, 1960; Burk, 1966; Ray and Alexander, 1966; Hodgkinson and Quinn, 1978; McIntyre and Best, 1978; Reader, 1983).

Local variation in flowering time within plant populations can be striking as well. Jackson (1966) found that individual herbs at one Indiana woodland site varied in date of first flowering by as much as 11 days, correlated with microsite differences in air temperature. Tepedino and Stanton (1980) have documented a range of four to five weeks in date of first bloom for most species on the Great Plains shortgrass prairie, a habitat characterized by great spatial and temporal variation in rainfall. In tropical trees, varying degress of water stress over space can explain some local variation in flowering time and intensity (Augspurger, 1981;

Reich and Borchert, 1982). Spatial and temporal variation in resource availability may also help explain local differences in flowering time, and certainly in flowering intensity (Rathcke and Lacey, 1985).

It is clear that the proximate and ultimate causes of flowering phenology are tightly entangled and easily misunderstood. For example, geographic variation in flowering time may be 'explained' by clines in proximate flowering cues, yet may clearly be adaptive: geographical shifts may, for instance, result in flowering after the risk of night freezes has passed (Vance and Kucera, 1960; Ray and Alexander, 1966; Hodgkinson and Quinn, 1978; Waser, 1983a). Hereafter, I will shift my focus away from the causes of phenological variation, and concentrate on documenting common phenological patterns within species and their ecological and biogeographical correlates.

11.2.2 PATTERNS OF PHENOLOGICAL SYNCHRONY WITHIN SPECIES

The degree of synchrony within and among patches of a given plant species plays a major role in determining whether a pollinator can persist on that resource. Although phenological records exist for a remarkable number of species, their value is surprisingly limited for approaching this issue. Observers usually record average flowering dates for species at a given site, rather than the range of flowering times of individuals or patches of individuals; variation within individuals or patches is even more rarely noted. Furthermore, studies are usually extremely local and of relatively short duration (1–3 years). These limitations are compounded by the fact that there have been no agreed-upon quantitative measures of phenological phenomena. Newstrom *et al.* (1991,1993), however, have recently suggested some very useful schemes to quantify phenological variation. In their system, a phenological pattern at a given level such as the individual or population is described by six parameters: flowering frequency, duration, amplitude, date, synchrony and regularity. Adoption of such a scheme should make rigorous comparisons possible at a variety of scales and impose some order upon what can be a confusing diversity of possible phenologies.

Even without a rigorous classification scheme, some phenological patterns can easily be recognized. As a rule, flowering synchrony and the length of the floral period are correlated: in species with short blooming seasons, individuals within a population tend to start and stop at the same time, whereas in species with more prolonged flowering, there is less among-individual synchrony (Augspurger, 1983). This phenomenon is illustrated in Figure 11.1, using phenological data from three Panamanian shrubs studied by Augspurger (1983).

Certain phenologies fairly logically accompany certain plant breeding systems (Bawa, 1983; Wyatt, 1983). In dioecious species, males commonly

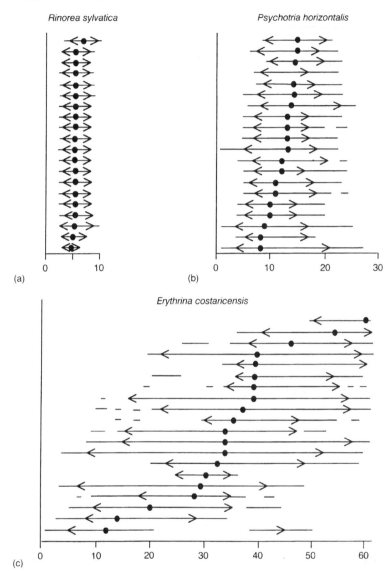

Figure 11.1 Flowering time of three shrub species of semi-deciduous lowland forest of Panama: species with briefer flowering periods show greater intrapopulation synchrony. Each horizontal line represents the total duration of flowering of each of 20 individuals, listed in order of date the median flower was open (represented by the dot). A broken line indicates discontinuous flower production. The brackets enclose the period around the median flowering date during which 90% of an individual's flowers were open. (a) *Rinorea sylvatica* (Seem.) O. Kunze (Violaceae). (b) *Psychotria horizontalis* Sw. (Rubiaceae). (c) *Erythrina costaricensis* Micheli var. panamensis (Standl) comb. nov. (Fabaceae). (Redrawn from Augspurger, 1983.)

flower first and produce more floral rewards, so visitors have usually picked up pollen by the time they reach the females (Bullock and Bawa, 1981; Bullock, Beach and Bawa, 1983). Male and female flowers on hermaphroditic plants may open at different times, increasing the likelihood that pollen will move between rather than within individuals (Bawa, 1977).

In addition, certain phenological patterns predominate in certain habitats and biogeographical regions. The greatest phenological diversity by far is found in the tropical rainforest, where continual warmth and moisture may permit reproduction year-round. Flowering phenology has been studied most intensively in the wet forest at the La Selva Biological Station, Costa Rica, although some comparable data are available for other regions (e.g. Putz, 1979; Augspurger, 1983; Ashton, Givnish and Appanah, 1988; Heideman, 1989; Bullock and Solis-Magallanes, 1990; Kinnaird, 1992). At La Selva, four broad species-level patterns have been recognized (Newstrom, Frankie and Baker, 1991; see also Frankie, Baker and Opler, 1974; Gentry, 1974; Opler, Frankie and Baker, 1980).

(1) The most common phenological pattern is subannual flowering (Figure 11.2(a)). These species flower multiple times each year, with highly variable flowering amplitudes and durations. Well-studied examples include *Guarea rhopalocarpa* (Bullock, Beach and Bawa, 1983) and *Piper arieianum* (Marquis, 1988).

(2) In annual flowering (e.g. *Dipteryx panamensis*), individuals undergo one major flowering episode each year; the duration of flowering and tendency to skip years vary widely among species (Figure 11.2(b)).

(3) Supra-annual species such as *Andira inermis* and *Posoqueria grandiflora* flower at multiyear intervals (Figure 11.2(c)).

(4) The rarest pattern is continual flowering (Figure 11.2(d)), in which individuals flower (with some amplitude variations) all year for many years, with only brief interruptions (e.g. *Hamelia patens*).

Frankie and his collaborators (Frankie, Baker and Opler, 1974; Opler, Frankie and Baker, 1980) have contrasted phenological patterns between La Selva and the seasonally dry deciduous forest at La Pacifica about 200 km away. Phenologies at the dry site are much more seasonal: flowering is more synchronous within species, fewer species show extended blooming periods, and few flower more than once each year. Year-to-year variation in flowering is much more noticeable at the dry site. Species present at both wet La Selva and dry La Pacifica often show flowering patterns typical of the site, rather than a constant phenotype (Newstrom *et al.*, 1993).

Most temperate-zone species seem to fit loosely into Newstrom *et al.*'s 'annual' flowering pattern (Figure 11.2(b)), due to strong environmental

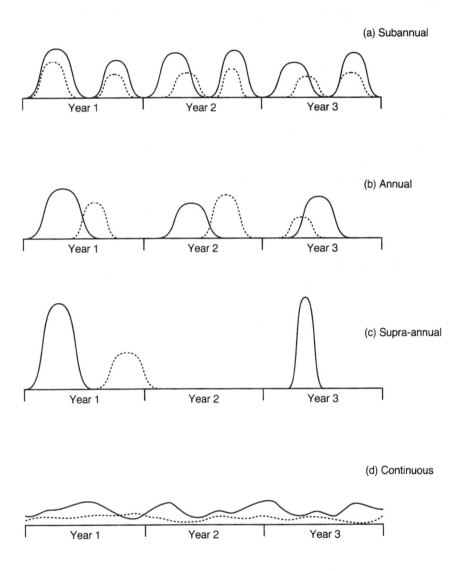

Figure 11.2 Four species-level flowering patterns documented in the wet forest of La Selva, Costa Rica (Newstrom, Frankie and Baker, 1991). Two hypothetical species (solid and dotted lines) are illustrated for each pattern. (a) Subannual flowering: each species flowers multiple times each year. This pattern was shown by 55% of species surveyed at La Selva. (b) Annual flowering: each species flowers once per year (29%). (c) Supra-annual flowering: each species flowers at multiyear intervals (9%). (d) Continuous flowering: each species flowers continuously with rare interruptions (7%).

constraints on the length of the flowering season. Bawa (1990) recognizes four general differences between pollination in forests of the north temperate zone and the wet tropics: temperate-zone flowering is largely confined to late spring and summer, individual flowers last many days longer, pollination by wind rather than animals is more common, and adaptations for pollination by vertebrates are much rarer.

It is important to note that none of the factors I have discussed is independent of the others, or independent of pollinator identity. For example, certain pollinator taxa are only present in the temperate zone or the tropics, or in certain strata within the forest (Bawa *et al.*, 1985; Bawa, 1990). Breeding systems are differentially associated with certain habitats (Bawa, Perry and Beach, 1985) and pollinators (Wyatt, 1983). Furthermore, phylogenetic relationships may account for more of the phenological similarities and differences among species than is currently recognized.

11.3 COMPONENTS OF THE PLANT–POLLINATOR LANDSCAPE: POLLINATORS IN MOTION

How pollinators forage is of course not strictly related to the phenology of their food plants. The value of a food patch to a given pollinator (and even what could be said to constitute a patch in the first place) depends on that visitor's energetic requirements, not on the absolute quantity of food present. It is the animal's perception of patch structure, not our own, that is of relevance.

Optimal foraging theory predicts that if organisms behave so as to enhance their food intake per unit time spent foraging, they should change foraging tactics in response to changes in food availability (Pyke, Pulliam and Charnov, 1977). We therefore would expect pollinators to alter their search behavior and possibly diet breadth in response to changes in floral abundance. Such decision-making by pollinators has been studied extensively, although almost always within plants or single patches (e.g. Pyke, 1978a, b, 1980; Heinrich, 1979; Hodges, 1985; Harder, 1988; Pleasants, 1989).

11.3.1 SEARCHING BEHAVIOR

The tracking of food resources drives much of the daily and seasonal movement of nectar-feeding animals (Karr, 1990; Fleming, 1992). Movements are generally spatially restricted when foraging is locally profitable; regardless of the pollinator taxon, most flights are made between neighboring flowers (Waddington, 1983). Visits become increasingly wide-ranging as local reward levels decline (Levin and Kerster, 1969). The likelihood that pollinators will discover phenologically different patches, and thus whether they could ever track a single resource over space and time, is a function of the spatio-temporal scale on which they search relative to the scale at which plant phenology varies.

Insects are by far the most common animal pollinators, and among insect pollinators, the bees are by far the dominant taxon (Barth, 1991). Bees are 'central place foragers', returning to their nest after each foraging trip with food to provision the young. Foraging behavior has been primarily studied in bumblebees and honeybees, and almost exclusively in agricultural settings or low-diversity natural habitats (Roubik, 1989). Different species and sexes vary, however, in mean and maximum flight distances from the nest; the extent to which floral resources are clumped or dispersed also influences foraging behavior. Of two Costa Rican *Trigona* species studied by Johnson and Hubbell (1975), one forages in large groups, visiting large, dense *Cassia* patches, while the other forages individually or in smaller groups, visiting widely spaced or isolated plants; it appears to be competitively excluded from large patches. Patch-size specialization has also been documented in some temperate-zone bees (Sih and Baltus, 1987; Sowig, 1989). Certain large tropical bees 'trapline', or visit a series of isolated plants or patches on a surprisingly long circuit in a predictable order (Janzen, 1971). In the temperate zone, the forest-dwelling honeybees studied by Visscher and Seeley (1982) did 95% of their foraging within a 6 km radius of the nest, much further than suggested by studies of the same species in agricultural settings. On a daily basis, however, they show site fidelity to patches only a few tens of meters wide (Visscher and Seeley, 1982). Roubik (1989) has pointed out that knowing how far bees can potentially fly does not reveal what they usually do, which is probably of more relevance for understanding their interactions with plants. It is unlikely that most bees ever travel more than a moderate fraction of their potential flight distance. Even the far-ranging euglossine bees show local site fidelity under most circumstances (Ackerman *et al.*, 1982; Armbruster, 1993).

Although Lepidoptera are the second most important group of insect pollinators, their pollination behavior is still poorly known. Some butterfly species move constantly, whereas others stay extremely localized; the causes of both local and long-distance movements have been heavily debated (Baker, 1989; Ehrlich, 1989). Flower-visiting in butterflies appears to be closely linked with mating and oviposition behavior, with females often foraging on flowers in the immediate vicinity of the plant species on which they prefer to lay eggs (Ehrlich, 1989). Some tropical butterflies, including *Heliconius* species, show traplining behavior reminiscent of euglossine bees (Gilbert, 1975).

Pollination by vertebrates is fairly uncommon in the temperate zone but quite important in the tropics (Bawa, 1990). Megachiroptera bats of tropical Africa and Asia forage opportunistically, often concentrating their attention on a single tree or a few local trees. The Microchiroptera of the American and Australian tropics are more agile and use sonar, allowing them to adopt traplining behavior (Frankie and Baker, 1974). Larger species fly longer distances between plants (Heithaus, Fleming and Opler, 1975), with the largest able to fly up to 100 km a night in search of flowers (Fleming, 1992). Bats can thus avoid local food shortages by seeking out distant, scattered food resources. Important nectarivorous birds include Australian honeyeaters, African sunbirds, Hawaiian honey-creepers, and New World hummingbirds, although only the last of these have received extensive attention in the context of plant–pollinator inter-actions (Feinsinger, 1976, 1980; Wolf, Stiles and Hainsworth, 1976; Stiles, 1980, 1985). Traplining hummingbirds are relatively large, powerful fliers; they specialize on patchily distributed but rich flowers. Other species are chronic wanderers ('nomads') that follow a series of resource flushes from community to community. Finally, small, agile territorialists feed upon and aggressively defend rich floral patches. Carpenter (1987) has shown that resource defense by the territorial species is particularly intense when regional food levels are low, even when local food levels are high. This observation suggests that territorial hummingbirds, like the trapliners and nomads, have the ability to assess floral resources over a broad spatial scale.

It is clear that, as a rule, vertebrate pollinators regularly assess their foraging options at a broader spatial scale than most invertebrates. Taxo-nomic differences in foraging distances have been thought to relate in part simply to body size: larger foragers with higher metabolic require-ments tend to fly longer distances. However, Waser (1982) found that this size rule did not hold for insects and hummingbirds visiting three flowering plant species. He suggests that long flights between flowers may in some cases be attributable to territorial defense and mate search rather than to foraging decisions.

Flower-feeding animals that migrate (hummingbirds, some bats and some insects, particularly Lepidoptera) face a second set of foraging decisions: they must choose not only what to feed upon at each stop, but, at a broader scale, the migration route itself. These routes evidently follow 'nectar pathways' (Fleming, 1992) formed by a predictably shifting regional array of food resources. Along the route, migrants encounter a series of plant communities differing in peak flowering times; as we have seen, flowering times of the individual plant species they encounter may also differ geographically.

All nectar-feeding birds in the north temperate zone are latitudinal migrants. The best-studied of these is the Rufous hummingbird (*Selasphorus rufus*), whose migration from north-western North America to southern Mexico is the longest of any North American hummingbird (Grant and Grant, 1967). This route clearly follows the blooming sequence of hummingbird-specialized plants. From its winter range in Mexico the Rufous hummingbird migrates north along the coast of the Pacific Ocean, where lowland flowers are blooming in response to winter rains. In late summer it begins to move south, visiting flowers blooming copiously in mountain meadows of the inland ranges. Both temperate and tropical hummingbirds also undergo altitudinal migrations, a relatively less studied phenomenon. California species that breed in the lowlands ascend to higher mountain regions to feed on late summer-flowering plants before beginning their migration south (Grant and Grant, 1967). Costa Rican hummingbirds move to the lowlands between April and August when flower abundance is high, then return upslope to feed for the rest of the year (Stiles, 1980).

Much less is known about long-distance migrations in other pollinator taxa. Impressive latitudinal migrations have been documented for two genera of nectar-feeding bat species in the Sonoran Desert of North America (Fleming, 1992). Certain butterflies undergo spectacular but still poorly understood long-distance migrations (Baker, 1989; Ehrlich, 1989). Many Monarch butterfly populations move each year between overwintering grounds in Mexico and North American breeding sites. Even supposedly nonmigratory populations, however, often travel long distances: Costa Rican Monarchs move back and forth between seasonally dry Pacific habitats and evergreen Atlantic habitats (Haber, 1993), while California individuals travel between coastal and mountain regions (Nagano *et al.*, 1993).

11.3.2 FAITHFULNESS TO INDIVIDUAL PLANT SPECIES

Pollinators move among local food patches when the one they are in falls in relative profitability, but do not necessarily stay faithful to the same food species when they move. If patches of a single species differ phenologically, pollinators can either choose to track this temporal variation across space, or to stay local but switch foods. In part, it is the degree of faithfulness to an individual plant species that will determine whether intraspecific phenological variation plays a role in explaining how populations of a given pollinator species can persist.

Vertebrate pollinators seem to forage opportunistically, moving in a single foraging bout among flowers of the set of local plant species that rewards them about equivalently. For example, Rufous hummingbirds defend dense floral patches during stops along their migratory route; different species within these patches are visited roughly in proportion to their abundance (Kodric-Brown and Brown, 1978). Seven Costa Rican bat species studied by Heithaus, Fleming and Opler (1975) are similarly opportunistic. Nectarivorous birds and bats may specialize on single food resources only at times of year when little else is available to them (e.g. Waser, 1979).

Most insect pollinators use more limited subsets of available floral resources than do vertebrates. Certain generalist bees show constancy, the tendency for individuals to forage for a time within, rather than among, plant species even when many species are available (Waser, 1986). Within a preferred meadow, for instance, bumblebees will fly over many species in full flower in order to reach the one on which they are currently 'majoring' (Heinrich, 1976b). If the peak blooming season of a preferred plant species is longer than the lifetime of a given bumblebee worker, she will specialize on that resource for her entire life. On the other hand, the selectivity of a constant bee is reduced or obliterated when food availability is low (Heinrich, 1975). As central-place foragers with strong site-fidelity, these bees prefer to switch foods rather than to attempt to track the current preferred one across space.

Some insect pollinators show less behavioral flexibility. Many semisocial and solitary bees in the temperate zone are oligolectic, collecting pollen from only one or a few flowering species (e.g. Linsley, 1958). However, even these species can broaden their diets somewhat during times of local food shortage (Cruden, 1972). More extreme specificity is largely a tropical phenomenon, and includes many of the most spectacular cases known of plant/animal coevolution, such as the interaction between figs and their species-specific wasp pollinators (section 11.4.2). It is among truly obligate pollinators that we can expect the closest tracking of plant phenology over space and time.

11.4 PLANT–POLLINATOR LANDSCAPES

The resource landscape in which pollinators live is thus determined by a combination of plant attributes (e.g. timing and synchrony of flowering) and pollinator attributes (including their ability to fly among patches and their dietary flexibility). While many possible combinations of plant and pollinator traits can be imagined, some combinations are clearly more common in nature than others. Below, I sketch out five very general types of plant–pollinator landscapes. The first two are occupied by highly specialized pollinators and (respectively) synchronously flowering and asynchronously flowering plants. The third and fourth landscapes are occupied by relatively generalized pollinators and (respectively) synchronously flowering and asynchronously flowering plants, while the fifth is occupied by generalist pollinators that migrate. In the following sections I describe each of these hypothetical landscapes, offer possible real-life examples, and consider some of the problems that a pollinator might face in each case.

11.4.1. SPECIALIZED POLLINATORS AND SYNCHRONOUSLY FLOWERING PLANTS

Many plants, particularly those in seasonal environments, have very brief flowering periods, in part due to constraints set by climatic conditions suitable for flowering. Extremely few of these species have species-specific relationships with pollinators. We can easily recognize two reasons why this should be the case. First, the specialist pollinator would face a total food shortage for most of the year. This should restrict such specificity to animals that could either track their resource by migrating, or else not feed for much of the year (e.g. via hibernation or diapause). Second, extremely close phenological synchrony would seem to be necessary between the mutualists. Flowering time commonly shifts from year to year with ambient conditions, sometimes dramatically; it is unlikely that pollinator timing can always shift to exactly the same extent. If the pollinator arrived before flowering began, it would starve, whereas if flowering began first, the plant would fail reproductively. If pollinators completely missed the flowering period even for one year, it could result in extinction of the local pollinator population. A high failure rate would seem likely in these systems, even if selection for phenological matching were intense.

There is at least one good example, however, that fits into this plant–pollinator landscape: the interaction between yuccas and their obligate yucca moth pollinators (Addicott, Bronstein and Kjellberg, 1990; Powell, 1992). Yuccas (*Yucca* spp., Agavaceae) are native to most of North and Central America, occurring primarily in relatively arid, highly seasonal environments. The roughly 40 yucca species are pollinated exclusively and obligately by moths in two closely related genera, *Tegeticula* and *Parategeticula* (Prodoxidae); the interaction is probably species-specific or nearly so (Miles, 1983; Addicott, unpublished).

In late spring, newly eclosed adult yucca moths leave the soil, fly to yucca inflorescences, and mate there. Female moths then actively gather pollen. They then fly to a flower on another plant, oviposit into the pistil, and actively transfer pollen to its receptive stigma. Adult yucca moths live only a few days; they do not feed as adults. Their offspring develop within the maturing fruit, consuming several seeds apiece. In late summer, larvae drop to the soil and enter diapause, which lasts at least until the next spring. Thus, yucca moths only feed during a fairly brief portion of their life cycle, permitting their exclusive association with a seasonally flowering and fruiting plant.

The extent to which spring emergence of yucca moths is well coordinated with yucca flowering is unclear. The onset of yucca flowering does shift somewhat from year to year (Aker, 1982) as well as with elevation (Webber, 1953). Individuals in some yucca populations do not flower every year, leading to massive annual variations in flower abundance (Kingsolver, 1984; Powell, 1984; Fuller, 1990). Unfortunately, comparable information on spatial and temporal variation in yucca moth populations is not yet available. Casual observations on several species, however, suggest that flowering does not closely match the period of moth availability (Powell and Mackie, 1966; Aker, 1982; Kingsolver, 1984; Powell, 1984). In the best-studied case, Aker (1982) found great variation in flowering onset and duration within a *Yucca whipplei* population, with moths consistently emerging well after the onset of flowering.

It is still an open question how this interaction can persist (Addicott, Bronstein and Kjellberg, 1990). Three intriguing possibilities deserve further attention. First, we need data on the proximate cues for yucca flowering and yucca moth emergence. These cues may be either the same or very closely linked, decreasing the likelihood of a severe phenological mismatch. Second, we have no information on how far yucca moths can migrate in search of open flowers, and very little data on phenological differences among yucca or yucca moth populations. Many obligate insect pollinators show remarkable abilities to find isolated plants, usually via olfactory cues (Williams, 1983; Ware *et al.*, 1993). Finally, some data suggest that certain yucca moths can remain in diapause for at least several seasons (Fuller, 1990). If different cohorts of a single genotype emerge in different years, the system as a whole should be more resilient to year-to-year phenological and climatic variation.

The yucca–yucca moth interaction shows an exceptional degree of species-specificity. However, this type of landscape may also exist for pollinator species constrained to be relatively specialized ecologically, due to an absence of alternative resources when these animals are present. A good example may be insects that pollinate early-spring wildflowers in temperate-zone deciduous forests (Schemske, 1977; Schemske *et al.*, 1978). These plants must flower during the brief period after temperatures are warm enough for pollinator flight, but before the forest canopy has closed and light levels have fallen. In the Illinois forest studied by Schemske *et al.* (1978), this period shifts 8–20 days from year to year. The abundance of pollinators (flies and several families of bees, including some specialists) shifts among years as well, but not closely matched with flowering time. Consequently, reproductive success of these plants is frequently pollinator-limited. No information is yet available on the consequences for the pollinators.

11.4.2. SPECIALIZED POLLINATORS AND ASYNCHRONOUSLY FLOWERING PLANTS

Not surprisingly, obligate plant–pollinator relationships are more common in relatively aseasonal environments, where individuals can flower for prolonged periods and/or populations can remain in flower for much or all of the year. Even these situations, however, pose difficulties for pollinator persistence. Individuals of a given species are often widely dispersed, particularly in highly diverse tropical forests (Hubbell and Foster, 1986); how can specialized pollinators locate these small, scattered patches? Secondly, the fact that these species flower for a prolonged period does not preclude the possibility that temporal gaps in flower availability will arise. How can specialist pollinators survive these periods?

The system that best exemplifies this landscape is the relationship between monoecious fig species and their pollinator wasps (Janzen, 1979; Wiebes, 1979; Bronstein, 1992). There are about 700 species of fig (*Ficus*, Moraceae), each pollinated by its own species of fig wasp (Agaonidae). Like yucca moths, the minute (1–3 mm) female fig wasps both actively pollinate and deposit their seed-eating offspring within the flowers of their host plant. Unlike yucca moths, however, fig wasps have no

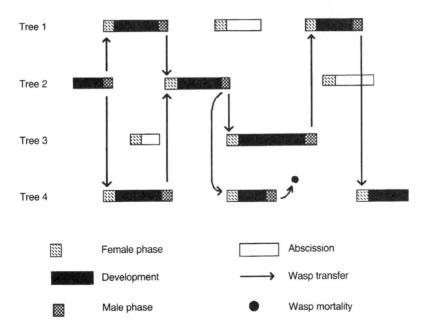

Figure 11.3 Hypothetical flowering sequence of a four-tree population of monoecious fig trees. The 'female phase' is the brief period when pollinator wasps are attracted to receptive female florets within the inflorescences; fruit development is initiated only if wasps arrive. The 'male phase' is the brief period when anthers mature, female wasps collect pollen, and depart in search of another female-phase fig. Note that if any of these trees were absent, the pollinator population would become locally extinct during the time period shown.

diapause. When they have reached adulthood, female offspring collect pollen from their natal fig and then depart in search of a female-phase inflorescence in which to lay their eggs. In most fig species, trees show great within-crown synchrony, but flower out of synchrony with other trees in the population. Thus, the departing wasps must locate another tree that happens to be in the correct phenological stage. They live a few days at most and do not feed as adults (like yucca moths), so there is a very short period during which they can succeed.

Some of the consequences of this unique natural history emerge when we examine Figure 11.3, which illustrates pollinator movements within a hypothetical four-tree population of monoecious figs. Population-level flowering asynchrony is clearly essential for the wasps to have any chance of reaching another tree, and thus for both trees and wasps to be able to reproduce. In fact, such asynchrony must extend year-round: if there is even a brief gap in the flowering sequence, pollen-laden wasps should go locally extinct. (For example, imagine that Tree 1 in Figure 11.3 skipped its last flowering episode; the wasp population would go extinct about three-quarters of the way through the period shown.) The more trees there are, the more likely it is that wasps departing at any time of year will be able to find a flowering neighbor.

In order to determine how large a fig population must be for a fig wasp population to persist successfully, Bronstein *et al.* (1990) devised a simulation model of this interaction. We randomly generated flowering sequences for individual trees, using field data on durations of flowering and intervals between flowering episodes of one well-studied African fig species, *Ficus natalensis*. After each new tree was added, we asked: given an initial input of wasps, will wasps still be present in the system four years later? The simulation ended after the first tree was added that permitted the wasps to persist. Notice that this simple model includes no spatial structuring: we defined a fig population as the number of trees within an area in which wasp transfer was possible, providing that wasps were available at the right time.

In 100 simulations, we found that a median of 95 *F. natalensis* trees (range 65–294) was necessary to allow a pollinator population to persist for four years. (This critical population size would have been even larger had we required the pollinators to persist for longer periods.) These results indicate that long-term pollinator persistence should be problematic even in very large fig populations. The model also proved to be useful for identifying some of the wasp and fig traits that are most critical for allowing wasp persistence in relatively small fig populations; many of these have been given minimal study to date, although they are important to document if we wish to examine fig wasp persistence in nature. Notable among these traits are the length of fig sexual phases, i.e. the time over which a fig attracted or released wasps (longer sexual phases increased pollinator persistence) and the lifespan of adult wasps. The model also suggests that the range limits of this largely tropical interaction may be set by climatic conditions that force a gap in the fig flowering sequence.

While the dilemmas of an obligate pollinator associated with an asynchronously flowering plant are perhaps most striking in the fig–fig wasp mutualism, parallels probably exist in less-studied interactions. The understorey herb *Dieffenbachia longispatha* (Araceae), for example, is pollinated by two genera of scarab beetles in the rainforest of Costa Rica (Young, 1988). Plants are widely scattered; they flower asynchronously from March to September, with individual inflorescences receptive for a single day and a several-day interval between flowering of different inflorescences on an individual. A very low density of inflorescences is therefore available for pollinators at any one time. Beetles fly an average of at least 83 m between inflorescences on successive nights (Young, 1988), evidently using relatively specific volatile odors to locate them (Schatz, G., personal communication). While it has some obvious parallels to the fig–fig wasp mutualism, we might predict that this system will show more resilience to temporal and spatial variation because it is somewhat less species-specific.

11.4.3 GENERALIZED POLLINATORS AND SYNCHRONOUSLY FLOWERING PLANTS

The combination of generalized pollinators and synchronously flowering plants is probably very common in nature. If we interpret 'synchrony' loosely, it probably includes most temperate animal-pollinated plants; if we interpret 'generalization' loosely, it includes animals that forage relatively indiscriminately, those with broad diets but temporal specialization or constancy, as well as those with more restricted diets that can be expanded somewhat when resources are scarce.

Associations between relatively generalized pollinators and plants resemble mosaics at most spatial and temporal scales. For example, Horvitz and Schemske (1990) have shown that the set of pollinators visiting *Calathea ovandensis*, a neotropical herb, shifts over short distances and among years. At a global scale, many introduced plants are effectively pollinated and fed upon by native pollinators (Valentine, 1978); conversely, introduced honeybees are now very effective mutualists of native floras throughout the world. Therefore, persistence of generalist pollinators and plants should be facilitated because both partners have fallback options in times of scarcity of any one of their partners. Plants pollinated by generalists do not necessarily experience high reproductive success, however. Not all visitors to *Calathea ovandensis*, for example, are good pollinators; butterflies are less effective than bees (Schemske and

Horvitz, 1984) and the most effective bee pollinator is only occasionally abundant (Horvitz and Schemske, 1990). Furthermore, plants that share pollinators with simultaneously blooming species may be forced to compete for pollinator attention (Rathcke, 1983). Plants that can attract the attention of the greatest number and variety of pollinators are often those displaying huge numbers of flowers for a brief period. However, while the number of visits that these mass-blooming species experience may be high, the quality of those visits may be low. For instance, Frankie, Opler and Bawa (1976) recorded 70 bee species on *Andira inermis*, a mass-flowering tree of the neotropical dry forest, but identified only eight of those species that ever moved between trees. Within-plant synchrony and intensity of flowering may thus reflect selection for a display large enough to attract visitors, but small enough that the visitors must depart before long in search of more food (Carpenter, 1976; Stephenson, 1982; Augspurger, 1983; Frankie and Haber, 1983).

Mass-blooming plants are highly desirable resources that are often defended by competitively dominant pollinators (Johnson and Hubbell, 1975; Kodric-Brown and Brown, 1978; Roubik, 1989). There are risks associated with such resource defense, however, including costs of competitive resource depletion, aggression and territorial defense. Johnson and Hubbell (1975) have argued that these trade-offs can lead to 'density specialization', in which one species controls the high-density resource while others profitably harvest more dispersed resources. These alternative foraging strategies have been particularly well-studied within certain neotropical hummingbird communities. Research by Feinsinger (1976, 1980), Wolf, Stiles and Hainsworth (1976), and others indicates that many short- and long-distance movements of hummingbirds between patches of flowers are related to intense interspecific competition. Hierarchies can be recognized of which species are most likely to depart the patch temporarily as aggression increases and resource levels fall. Aggressive territorialists tend to remain faithful to one patch. In contrast, local populations of more nomadic species fluctuate in size as some individuals invade nearby areas to exploit local resource flushes, then retreat again when resources in the new patch decline and competition from its residents increases. When patch resources are sufficient, 'fugitive species' (Chapter 9), inferior competitors that can efficiently use leftover resources, migrate in from further away. Studies of island hummingbird communities show that the number of species exhibiting each foraging strategy, as well as the distance each travels in search of a new patch, can be related to the regional abundance and dispersion of floral resources (Feinsinger, 1980; Feinsinger, Wolfe and Swarm, 1982). It would be fascinating to model these phenomena from a landscape ecology approach, e.g. using source–sink models (Pulliam, 1988; Danielson, 1991; see also Levin, 1974). Such an effort would augment our minimal ability to predict how pollinator population dynamics will be influenced by alterations to the landscape.

11.4.4 GENERALIST POLLINATORS AND ASYNCHRONOUSLY FLOWERING PLANTS

The combination of generalist pollinators and asynchronously flowering plants is relatively common in the tropics, although some examples can also be found in more seasonal environments. One well-studied temperate example involves lavender (*Lavandula latifolia*), an evergreen shrub that flowers from mid-July through October in southern Spain, where it is visited by at least 70 species of bees, flies, butterflies and moths (Herrera, 1987, 1988). There is pronounced seasonal turnover in pollinator species, in part because the flowering period lasts longer than many of these insects' lifetimes. Variation in the pollinator assemblage is also striking within a population across years, as well as in space (Herrera, 1988).

Longer-lived pollinator individuals can become fairly dependent on species that flower reliably for prolonged periods. Widely dispersed tropical plants that bear few flowers per individual cannot compete for the aggressive, territorial pollinators that mass-flowering species attract. These species nevertheless provide spatially predictable resources, often leading competitively inferior but strong-flying pollinators to incorporate them into their traplining routes. Trapliners and their preferred plant species tend to have fairly specialized relationships compared to the associations that revolve around more synchronously flowering plants. For example, most traplining hummingbirds have relatively long bills, matching the long corolla tubes of their preferred flowers; there are relatively few of these flowers per plant, but each holds a high volume of nectar. In contrast, territorial hummingbirds at the same sites are short-billed. The plants that they patrol generally have copious numbers of small, nectar-poor flowers, or else have fewer flowers but grow patchily, producing the same effect.

There is much that we do not yet know about these interactions. Do traplines consists of single or multiple plant species? Do traplining pollinators track the flowering of individual plant species, to the point where they will stay faithful to that species even if they have to shift their foraging route? What factors induce a trapliner to switch to a new resource, or to drop an individual plant from its route? Do trapliners opportunistically exploit mass-flowering species in addition to their usual resources? Answers to these questions would augment our minimal understanding of this plant–pollinator landscape.

11.4.5 GENERALIST MIGRATORY POLLINATORS

As I have discussed, generalist pollinators usually switch to another local plant species when the one they are using falls in profitability, rather than trying to track their current resource into distant regions in which it is still in full flower. Long-distance resource tracking may actually occur, however, in generalist pollinators that migrate. Migrations commonly follow nectar corridors formed by predictable regional gradients in flowering time (Fleming, 1992). We do not yet know whether individual animals track the flowering of individual plant species on these migrations, as opposed to a broader set of acceptable resources. While most migrating pollinators are rather opportunistic in their feeding choices, under some circumstances so few flowering species are available along the route that they might in fact track certain critical ones.

Waser (1979) has studied the interaction between ocotillo (*Fouquieria splendens*), a desert shrub, and several species of migratory hummingbirds. In the south-western United States, the onset of flowering corresponds to a temporally predictable three to four week pulse of northward migration in the hummingbirds. At Waser's study site in Arizona, ocotillo is almost the only hummingbird-adapted flower during that season, and hummingbirds feed at it almost exclusively. Waser (1979) argues that when migratory pollinators appear in a brief seasonal pulse and visit a single species, selection should foster a close phenological match between them. The real test of Waser's hypothesis would be to see whether, at a geographic scale, ocotillo flowering shifts predictably in parallel with the date of hummingbird arrival.

Fleming (1992, unpublished) has been studying the same question from a pollinator's perspective, looking at the sequence of resources that the Lesser long-nosed bat (*Leptonycteris curasoae*) relies upon during its latitudinal migrations. Carbon-isotope analysis suggests that they visit flowering paniculate agaves (C_4) on the fall voyage from the Sonoran Desert to tropical southern Mexico, and columnar cacti (C_3) on the trip back north. It is still an open question, however, whether or not these nectar corridors are formed by single plant species (Fleming, T., personal communication).

The evolutionary significance of phenological matching, whether between flowering plants and migrating pollinators (discussed here) or between fruiting plants and migrating seed dispersers (Snow, 1971; Thompson and Willson, 1979; Skeate, 1987), must be interpreted with caution. Natural selection may have fine-tuned plant phenology to match the availability of animal mutualists (Waser, 1979; Fleming, 1992). Alternatively, selection may have shaped the migration route to match a pre-existing nectar corridor, or the choice of route may not be genetically based at all. Regardless of what is cause and what is consequence, however, geographical variation in flowering time of individual species may now critically aid in nectarivores' survival during migrations. To test whether this is the case, two types of data will be necessary. First, we need community-wide descriptions of the resources available for pollinators at a range of locations along the migratory route, as well as detailed phenological information about those resources. This will reveal whether animals even have the option of tracking individual plant species over a geographic gradient of flowering time. Second, and more problematically, we must have data on what an individual animal, rather than the species as a whole, feeds upon at each location.

11.5 CONCLUSIONS

I have argued in this chapter that the ability of pollinating animals to persist within a given landscape is a function of an interacting set of critical plant and pollinator traits. We have seen that some fairly simple predictions about persistence can be made, in part because these traits do not vary independently. Some correlations between plant and pollinator traits are probably the result of coevolution, particularly in obligate interactions. Others may have resulted from unilateral evolution (e.g. plants that may have evolved to flower at the time when migrating pollinators pass through) or behavioral plasticity (e.g. competitively inferior pollinators excluded from all but relatively scattered floral resources). This issue, however, is complicated (and made more interesting) by the fact that the plant and pollinator traits of interest can vary greatly over time and space, even for a given species. Flowering phenology varies at spatial scales ranging from the microhabitat to the geographic region; temporal variation in some of the exogenous cues for flowering can result in yearly shifts in the onset of flowering at each location. Similarly, pollinator phenology and abundance is known to shift in time as well as over rather fine spatial scales, related to factors as diverse as climatic variability, the presence of competitors and availability of nest sites.

Are foraging decisions of pollinators made at the same scale that flowering varies, such that intraspecific phenological variation could help to explain the ability of pollinating animals to persist? The answer appears to be a qualified yes for many highly specialized plant–pollinator interactions. Selection to co-ordinate timing between the partners must be very strong in these cases; depending on the interaction, extreme flowering synchrony (e.g. for yuccas) or asynchrony (for figs) may be crucial to producing such co-ordination. However, misses in time seem to be common anyway, suggesting that spatio-temporal unpredictability precludes the possibility of fine-tuning phenology past a certain point.

These apparently inevitable phenological mismatches may be one of many factors contributing to the rarity of specialized modes of pollination. For facultative plant–pollinator interactions, it is less likely that phenological variation in any one plant species plays a major role in pollinator persistence. Most generalist pollinators appear to make their foraging decisions at a spatial scale much finer than the one at which phenology usually varies. When local flowering of a preferred species wanes, these foragers are more likely to switch to a different local species than to search out phenologically retarded patches of their current favorite. One exception may be generalist pollinators that migrate, because these individuals move across a landscape where differences in flowering time may occur at the correct spatial and temporal scale to permit tracking. In general, studying local and regional flowering sequences of the entire set of acceptable plant species, rather than the phenology of any one of them, may be the most useful approach for understanding how generalist pollinators can persist.

While issues of scale are attracting increasing attention in most areas of ecological research (e.g. Wiens, 1989), the study of plant–pollinator interactions has remained an exception. Pollinator foraging decisions have habitually been studied within single plants or single patches, but we do not yet have a solid grasp of whether this in fact is the scale at which individuals perceive their resource environment. There is some evidence that both bees and birds keep track of floral availability at a regional scale and incorporate this information into local foraging decisions (Visscher and Seeley, 1982; Carpenter, 1987). Furthermore, understanding the extent to which phenological variation shapes the landscape has been clouded by lack of precision about the scale at which such variation occurs (i.e. within individual, within population, or among populations). It is entirely possible that many plant and pollinator traits have been misunderstood because they have been studied at an inappropriate spatial or temporal scale. For example, arguments that community-level flowering sequences are orderly and shaped by selection rarely take into account the often dramatic variation in overlap in flowering periods from year to year (e.g. Schemske et al., 1978). Studies at inappropriate scales will also lead us to misinterpret the resilience of particular interactions. The flexible opportunism of hummingbirds increases the likelihood that they will find acceptable food resources at a given stop on their migration route, but they nevertheless starve to death in huge numbers when blooming is reduced over a significant portion of their migration route (Kodric-Brown and Brown, 1978). In contrast, the surprising resilience of the highly specialized yucca–yucca moth and fig–fig wasp interactions can probably only be understood with better knowledge of the flight ranges of these small insects; they are almost sure to be much larger than currently believed.

Better knowledge of plant–pollinator landscapes will surely aid efforts to make more informed conservation decisions. For example, pollinator conservation efforts probably must operate on a regional or continental scale for migrants but on a much finer scale for relatively sedentary generalists; to conserve specialized pollinators like fig wasps, the focus must be on preserving sufficiently large populations of their obligate host species. More extensive field studies will generate more quantitative data on foraging ranges and the set of acceptable resources used by individual pollinators in their lifetimes. These data will be critical for developing predictive models of minimum areas and population sizes necessary to conserve pollinators, as well as the plants that they benefit.

11.6 SUMMARY

Pollinators, like herbivores, feed on patchily distributed plants, forcing them to decide when to move between patches and when to switch plant species. The 'plant–pollinator landscape' is shaped by an interacting set of plant and pollinator attributes: flowering phenology sets the distance that pollinators need to travel between patches of a given species to obtain sufficient food, while the pollinators' search capacities and dietary specificity determines the likelihood that they can and will make that journey. These sets of traits do not vary independently, allowing us to identify several characteristic landscapes and the types of organisms that occupy them. This chapter has described five such landscapes, attempting in each case to determine whether phenological variation within plant species helps to explain the ability of pollinators to persist. The first two landscapes are occupied by highly specialized pollinators and (respectively) synchronously flowering and asynchronously flowering plants. The first landscape is the rarest, probably related to the likelihood of a costly phenological mismatch between plant and pollinator; the second may be more common, although only one example (the fig–fig wasp interaction) has been studied in depth. Intraspecific phenological variation is likely to have critical consequences for pollinator persistence in these two landscapes. The third and fourth landscapes are occupied by relatively generalized pollinators and (respectively) synchronously flowering and asynchronously flowering plants. The former is probably the most common landscape, encompassing most temperate plant–pollinator interactions. Pollinators in these landscapes are less likely to be strongly influenced by phenological variation in any one of their resource plants, due to their ability to switch foods easily while remaining local. The fifth landscape is occupied by generalist pollinators that may migrate. Little is known about resource use of individual migrants, but it is possible that at least some of them rely on geographical gradients in flowering time of certain preferred species. Studying plant–pollinator interactions with a landscape approach holds great promise from both a basic and applied perspective, although such research is currently hampered by incomplete information about the temporal and spatial scales at which critical plant and animal traits vary.

ACKNOWLEDGEMENTS

I thank the many colleagues who supplied me with literature on this diverse topic and who commented on the manuscript, particularly Bruce Milne, Ted Fleming, Francisco Ornelas, Brent Danielson, Linda Newstrom, Carol Augspurger, Sarah Richardson, Amy Faivre, Cecelia Smith and Sandy Adondakis. Carol Augspurger generously gave permission to reprint the data shown in Figure 11.1. The fig studies described here were supported by a NATO postdoctoral fellowship and an NSF/ROW grant (BSR 90–07492). Finally, special thanks to Beverly Rathcke for awakening and supporting my interest in plant–pollinator interactions, phenology and coevolution.

REFERENCES

Ackerman, J. D., Mesler, M. R., Lu, K. L. and Montalvo, A. M. (1982) Food foraging behavior of the male Euglossini (Hymenoptera: Apidae): vagabonds or trapliners? *Biotropica*, **14**, 241–8.

Addicott, J. F., Bronstein, J. and Kjellberg, F. (1990) Evolution of mutualistic life-cycles: yucca moths and fig wasps, in *Insect Life Cycles: Genetics, Evolution and Co-ordination* (ed. F. Gilbert), Springer-Verlag, London, pp. 143–61.

Aker, C. L. (1982) Spatial and temporal dispersion patterns of pollinators and their relationship to the flowering strategy of *Yucca whipplei*. *Oecologia*, **54**, 243–52.

Armbruster, W. S. (1993) Within-habitat heterogeneity in baiting samples of male euglossine bees: possible causes and implications. *Biotropica*, **25**, 122–8.

Arroyo, J. (1990) Spatial variation of flowering phenology in the mediterranean shrublands of southern Spain. *Isr. J. Bot.*, **39**, 249–62.

Ashton, P. S., Givnish, T. J., and Appanah, S. (1988) Staggered flowering in the Dipterocarpaceae: new insights into floral induction and the evolution of mast fruiting in the aseasonal tropics. *Am. Nat.*. **132**, 44–66.

Augspurger, C. K. (1981) Reproductive synchrony of tropical plants: experimental effects of pollinators and seed predators on *Hybanthus prunifolius* (Violaceae). *Ecology*, **62**, 775–88.

Augspurger, C. K. (1983) Phenology, flowering asynchrony, and fruit set of six neotropical shrubs. *Biotropica*, **15**, 257–67.

Baker, R. R. (1989) The dilemma: when and how to go or stay, in *The*

Biology of Butterflies (eds R. I. Vane-Wright and P. R. Akery), Princeton University Press, Princeton, NJ, pp. 279–96.

Barth, F. G. (1991) *Insects and Flowers: The Biology of a Partnership*, Princeton University Press, Princeton, NJ.

Bawa, K. S. (1977) The reproductive biology of *Cupania guatemalensis* Radlk. (Sapindaceae). *Evolution*, **31**, 52–63.

Bawa, K. S. (1983) Patterns of flowering in tropical plants, in *Handbook of Experimental Pollination Biology* (eds C. E. Jones and R. J. Little), Van Nostrand Reinhold, NY, pp. 394–410.

Bawa, K. S. (1990) Plant–pollinator interactions in tropical rain forests. *Ann. Rev. Ecol. Syst.*, **21**, 399–422.

Bawa, K. S., Bullock, S. H., Perry, D. R. *et al.* (1985) Reproductive biology of tropical lowland rain forest trees. II. Pollination systems. *Am. J. Bot.*, **72**, 346–56.

Bawa, K. S., Perry, D. R. and Beach, J. H. (1985) Reproductive biology of tropical lowland rain forest trees. I. Sexual systems and incompatibility mechanisms. *Am. J. Bot.*, **72**, 331–45.

Bernier, G. (1988) The control of floral evocation and morphogensis. *Ann. Rev. Plant Physiol. Plant Mol. Biol.*, **39**, 175–219.

Bronstein, J. L. (1992) Seed predators as mutualists: ecology and evolution of the fig–pollinator interaction, in *Insect–Plant Interactions*, Vol. IV (ed. E. Bernays), CRC Press, Boca Raton, pp. 1–44.

Bronstein, J. L. (1994) Our current understanding of mutualism. *Q. Rev. Biol.*, **69**, 31–51.

Bronstein, J. L., Gouyon, P.-H. Gliddon, C. *et al.* (1990) Ecological consequences of flowering asynchrony in monoecious figs: a simulation study. *Ecology*, **71**, 2145–56.

Bullock, S. H. and Bawa, K. S. (1981) Sexual dimorphism and the annual flowering pattern in *Jacaratia dolichaula* (D. Smith) Woodson (Caricaceae) in a Costa Rican rain forest. *Ecology*, **62**, 1494–504.

Bullock, S. H. and Solis-Magallanes, J. A. (1990) Phenology of canopy trees of a tropical deciduous forest in Mexico. *Biotropica*, **22**, 22–35.

Bullock, S. H., Beach, J. H. and Bawa, K. S. (1983) Episodic flowering and sexual dimorphism in *Guarea rhopalocarpa* Radlk. (Meliaceae) in a Costa Rican rain forest. *Ecology*, **64**, 851–62.

Burk, C. J. (1966) Rainfall periodicity as a major factor in the formation of flowering races of camphorweed (*Heterotheca subaxillaris*). *Am. J. Bot.*, **53**, 933–6.

Campbell, D. R. and Motten, A. F. (1985) The mechanism of competition for pollination between two forest herbs. *Ecology*, **66**, 554–63.

Carpenter, F. L. (1976) Plant–pollinator interactions in Hawaii: pollination energetics of *Metrosideros collina* (Myrtaceae). *Ecology*, **57**, 1125–44.

Carpenter, F. L. (1987) Food abundance and territoriality: to defend or not to defend? *Am. Zool.*, **27**, 387–99.

Cole, B. J. (1981) Overlap, regularity, and flowering phenologies. *Am. Nat.*, **117**, 993–7.

Cruden, R. W. (1972) Pollination biology of *Nemophila menziesii* (Hydrophyllaceae) with comments on the evolution of oligolectic bees. *Evolution*, **26**, 373–89.

Cushman, J. H. and Beattie, A. J. (1991) Mutualisms: assessing the benefits to hosts and visitors. *Trends Ecol. Evol.*, **6**, 193–5.

Danielson, B. J. (1991) Communities in a landscape: the influence of habitat heterogeneity on the interactions between species. *Am. Nat.*, **138**, 1105–20.

Ehrlich, P. R. (1989) The structure and dynamics of butterfly populations, in *The Biology of Butterflies* (eds R. I. Vane-Wright and P. R. Ackery), Princeton University Press, Princeton, NJ, pp. 25–40.

Feinsinger, P. (1976) Organization of a tropical guild of nectarivorous birds. *Ecol. Monogr.*, **46**, 257–91.

Feinsinger, P. (1980) Asynchronous migration patterns and coexistence of tropical hummingbirds, in *Migrant Birds in the Neotropics* (eds A. Keast and E. S. Morton), Smithsonian Institution Press, Washington, DC, pp. 411–19.

Feinsinger, P., Wolfe, J. A. and Swarm, L. A. (1982) Island ecology: reduced hummingbird diversity and the pollination biology of plants, Trinidad and Tobago, West Indies. *Ecology*, **63**, 494–506.

Fleming, T. H. (1992) How do fruit- and nectar-feeding birds and mammals track their food resources? in *Effects of Resource Distribution on Animal–Plant Interactions* (eds M. D. Hunter, T. Ohgashi and P. W. Price), Academic Press, NY, pp. 355–91.

Frankie, G. W. (1975) Tropical forest phenology and pollinator–plant coevolution, in *Coevolution of Animals and Plants* (eds L. E. Gilbert and P. H. Raven) University of Texas Press, Austin, TX, pp. 192–209.

Frankie, G. W. and Baker, H. G. (1974) The importance of pollinator behavior in the reproductive biology of tropical trees. *An. Inst. Biol. Univ. Autón Mexico 45, Ser. Botánica*, **1**, 1–10.

Frankie, G. W. and Haber, W. A. (1983) Why bees move among mass-flowering neotropical trees, in *Handbook of Experimental Pollination Biology* (eds. C. E. Jones and R. J. Little), Van Nostrand Reinhold, NY, pp. 360–72.

Frankie, G. W., Baker, H. G. and Opler, P. A. (1974) Comparative phenological studies of trees in tropical wet and dry forests in the lowlands of Costa Rica. *J. Ecol.*, **62**, 881–919.

Frankie, G. W., Opler, P. A. and Bawa, K. S. (1976) Foraging behaviour of solitary bees: implications for outcrossing of a neotropical forest tree species. *J. Ecol.*, **64**, 1049–57.

Fuller, O. S. (1990) Factors influencing the balance of cooperation and conflict between the yucca moth, *Tegeticula yuccasella* and its mutualist, *Yucca glauca*. PhD dissertation, Univ. New Mexico.

Gentry, A. H. (1974) Flowering phenology and diversity in tropical Bignoniaceae. *Biotropica*, **6**, 64–8.

Gilbert, L. E. (1975) Ecological consequences of a coevolved mutualism between butterflies and plants, in *Coevolution of Animals and Plants* (eds L. E. Gilbert and P. H. Raven), University of Texas Press, Austin, TX, pp. 210–40.

Grant, K. A. and Grant, V. (1967) Effects of hummingbird migration on plant speciation in the California flora. *Evolution*, **21**, 457–65.

Haber, W. A. (1993) Seasonal migration of monarchs and other butterflies in Costa Rica, in *Biology and Conservation of the Monarch Butterfly* (eds S. B. Malcolm and M. P. Zalucki), Nat. Hist. Museum of Los Angeles County, Los Angeles, CA, pp. 201–8.

Harder, L. D. (1988) Choice of individual flowers by bumblebees: interaction of morphology, time and energy. *Behaviour*, **104**, 60–77.

Heideman, P. D. (1989) Temporal and spatial variation in the phenology of flowering and fruiting in a tropical rainforest. *J. Ecol.*, **77**, 1059–79.

Heinrich, B. (1975) Energetics of pollination. *Ann. Rev. Ecol. Syst.*, **6**, 139–70.

Heinrich, B. (1976a) Flowering phenologies: bog, woodland, and disturbed habitats. *Ecology*, **57**, 890–9.

Heinrich, B. (1976b) The foraging specializations of individual bumblebees. *Ecol. Monogr.*, **46**, 105–28.

Heinrich, B. (1979) Resource heterogeneity and patterns of movement in foraging bumblebees. *Oecologia*, **40**, 235–45.

Heithaus, E. R., Fleming, T. H. and Opler, P. A. (1975) Foraging patterns and resource utilization in seven species of bats in a seasonal tropical forest. *Ecology*, **56**, 841–54.

Herrera, C. M. (1987) Components of pollinator 'quality': comparative analysis of a diverse insect assemblage. *Oikos*, **50**, 79–90.

Herrera, C. M. (1988) Variation in mutualisms: the spatiotemporal mosaic of a pollinator assemblage. *Biol. J. Linn. Soc.*, **35**, 95–125.

Hodges, C. M. (1985) Bumble bee foraging: the threshold departure rule. *Ecology*, **66**, 179–87.

Hodgkinson, K. C., and Quinn, J. A. (1978) Environmental and genetic control of reproduction in *Danthonia caespitosa* populations. *Aust. J. Bot.*, **26**, 351–64.

Hopkins, A. D. (1938) Bioclimatics: a science of life and climatic relations. *US Dept Agric. Misc. Publ. 280*.

Horvitz. C. C., and Schemske, D. W. (1990) Spatiotemporal variation in insect mutualists of a neotropical herb. *Ecology*, **71**, 1085–97.

Hubbell, S. P. and Foster, R. B. (1986) Commonness and rarity in a neotropical forest: implications for tropical tree conservation, in *Conser-*

vation Biology (ed. M. E. Soulé), Sinauer Associates, Sunderland, MA, pp. 205–31.

Jackson, M. T. (1966) Effects of microclimate on spring flowering phenology. *Ecology*, **47**, 407–15.

Janzen, D. H. (1971) Euglossine bees as long-distance pollinators of tropical plants. *Science*, **17**, 203–5.

Janzen, D. H. (1979) How to be a fig. *Ann. Rev. Ecol. Syst.*, **10**, 13–51.

Janzen, D. H., De Vries, P. J., Higgins, M. I. and Kimsey, L. S. (1982) Seasonal and site variation in Costa Rican euglossine bees at chemical baits in lowland deciduous and evergreen forests. *Ecology*, **63**, 66–74.

Johnson, L. K. and Hubbell, S. P. (1975) Contrasting foraging and coexistence of two bee species on a single resource. *Ecology*, **56**, 1398–406.

Karr, J. R. (1990) Interactions between forest birds and their habitats: a comparative synthesis, in *Biogeography and Ecology of Forest Bird Communities*, SPB Academic Publishing, The Hague, The Netherlands, pp. 379–86.

Kelly, C. A. (1992) Reproductive phenologies in *Lobelia inflata* (Lobeliaceae) and their environmental control. *Am. J. Bot.*, **79**, 1126–33.

Kingsolver, R. W. (1984) Population biology of a mutualistic interaction: *Yucca glauca* and *Tegeticula yuccasella*. PhD dissertation, Univ. Kansas.

Kinnaird, M. (1992) Phenology of flowering and fruiting of an East African riverine forest ecosystem. *Biotropica*, **24**, 87–194.

Kochmer, J. P. and Handel, S. N. (1986) Constraints and competition in the evolution of flowering phenology. *Ecol. Monogr.*, **56**, 303–25.

Kodric-Brown, A. and Brown, J. H. (1978) Influence of economics, interspecific competition, and sexual dimorphism on territoriality of migrant Rufous Hummingbirds. *Ecology*, **59**, 285–96.

Levin, D. A., and Kerster, H. W. (1969) The dependence of bee-mediated pollen and gene dispersal upon plant density. *Evolution*, **23**, 560–71.

Levin, S. A. (1974) Dispersion and population interactions. *Am. Nat.*, **168**, 207–28.

Linsley, E. G. (1958) The ecology of solitary bees. *Hilgardia*, **27**, 453–599.

Macior, L. W. (1977) The pollination ecology of *Pedicularis* (Scrophulariaceae) in the Sierra Nevada of California. *Bull. Torrey Bot. Club*, **104**, 148–54.

Marquis, R. J. (1988) Phenological variation in the neotropical understory shrub *Piper arieianum*: causes and consequences. *Ecology*, **69**, 1552–5.

McIntyre, G. I. and Best, K. F. (1978) Studies on the flowering of *Thlaspi arvense* L. IV. Genetic and ecological differences between early- and late-flowering strains. *Bot. Gaz.*, **139**, 190–5.

McNeilly, T. and Antonovics, J. (1968) Evolution in closely adjacent plant populations. IV. Barriers to gene flow. *Heredity*, **23**, 205–18.

Miles, N. J. (1983) Variation and host specificity in the yucca moth,

Tegeticula yuccasella (Incurvariidae): a morphometric approach. *J. Lepid. Soc.*, **37**, 207–16.

Murfet, I. C. (1977) Environmental interaction and the genetics of flowering. *Ann. Rev. Plant Physiol.*, **28**, 253–78.

Nagano, C. D., Sakai, W. H., Malcolm, S. B. *et al.* (1993) Spring migration of monarch butterflies in California, in *Biology and Conservation of the Monarch Butterfly* (eds S. B. Malcolm and M. P. Zalucki), Nat. Hist. Museum of Los Angeles County, Los Angeles, CA, pp 219–32.

Newstrom, L. E., Frankie, G. W. and Baker, H. G. (1991) Survey of long-term flowering patterns in lowland tropical rain forest trees at La Selva, Costa Rica, in *L'arbre, Biologie et Developpement* (ed. C. Edelin), Naturalia Monspeliensia, no. hors série A7, Montpellier, France, pp. 345–66.

Newstrom, L. E., Frankie, G. W., Baker, H. G. and Colwell, R. K. (1993) Diversity of flowering patterns at La Selva, in *La Selva: Ecology and Natural History of a Lowland Tropical Rainforest* (eds L. A. McDade, K. S. Bawa, G. S. Hartshorn and H. A. Hespenheide), University of Chicago Press, Chicago, IL, pp. 142–60.

Ollerton, J., and Lack, A. J. (1992) Flowering phenology: an example of relaxation of natural selection? *Trends Ecol. Evol.*, **7**, 274–6.

Opler, P. A., Frankie, G. W. and Baker, H. G. (1980) Comparative phenological studies of treelet and shrub species in tropical wet and dry forests in the lowlands of Costa Rica. *J. Ecol.*, **68**, 167–88.

Pleasants, J. M. (1989) Optimal foraging by nectarivores: a test of the marginal value theorem. *Am. Nat.*, **134**, 51–71.

Poole, R. W. and Rathcke, B. J. (1979) Regularity, randomness, and aggregation in flowering phenologies. *Science*, **203**, 470–1.

Powell, J. A. (1984) Biological interrelationships of moths and *Yucca schottii*. *Univ. Calif. Berkeley, Publ. Entomol.*, **100**, 1–93.

Powell, J. A. (1992) Interrelationships of yuccas and yucca moths. *Trends Ecol. Evol.*, **7**, 10–15.

Powell, J. A. and Mackie, R. A. (1966) Biological relationships of moths and *Yucca whipplei* (Lepidoptera: Gelechiidae, Blastobasidae, Prodoxidae). *Univ. Calif. Berkeley, Publ. Entomol.*, **42**, 1–59.

Primack, R. B. (1987) Relationships among flowers, fruits, and seeds. *Ann. Rev. Ecol. Syst.*, **18**, 409–30.

Pulliam, H. R. (1988) Sources, sinks, and population regulation. *Am. Nat.*, **132**, 652–61.

Putz, F. E. (1979) Aseasonality in Malaysian tree phenology. *Malay. For.*, **42**, 1–24.

Pyke, G. H. (1978a) Optimal foraging in hummingbirds: testing the marginal value theorem. *Am. Zool.*, **18**, 739–52.

Pyke, G. H. (1978b) Optimal foraging: movement patterns of bumblebees between inflorescences. *Theor. Pop. Biol.*, **13**, 72–98.

Pyke, G. H. (1980) Optimal foraging in bumblebees: calculation of net rate of energy intake and optimal patch choice. *Theor. Pop. Biol.*, **17**, 232–46.

Pyke, G. H., Pulliam, H. R. and Charnov, E. L. (1977) Optimal foraging: a selective review of theory and test. *Q. Rev. Biol.*, **52**, 137–54.

Rathcke, B. (1983) Competition and facilitation among plants for pollination, in *Pollination Biology* (ed. L. Real), Academic Press, New York, pp. 305–29.

Rathcke, B. J. (1984) Patterns of flowering phenologies: testability and causal inference using a random model, in *Ecological Communities: Conceptual Issues and the Evidence* (eds D. R. Strong, D. Simberloff, L. G. Abele and A. B. Thistle), Princeton University Press, Princeton, NJ, pp. 383–93.

Rathcke, B. and Lacey, E. P. (1985) Phenological patterns of terrestrial plants. *Ann. Rev. Ecol. Syst.*, **16**, 179–214.

Ray, P. M. and Alexander, W. E. (1966) Photoperiodic adaptation to latitude in *Xanthium strumarium*. *Am. J. Bot.*, **53**, 806–16.

Reader, R. J. (1983) Using heat sum models to account for geographic variation in the floral phenology of two ericaceous shrubs. *J. Biogeogr.*, **10**, 47–64.

Réaumur, R. A. F. de (1735) Observation du thérmomètre, faites à Paris pendant l'anneé 1735, comparées avec celles qui ont été faites sous la ligne, à l'Isle de France, à Alger et quelques-unes de nos isles de l'Amerique. *Acad. des Sci. Paris*, **1735**, 545.

Reich, P. B. and Borchert, R. (1982) Phenology and ecophysiology of the tropical tree, *Tabebuia neochrysantha* (Bignoniaceae). *Ecology*, **63**, 294–9.

Roubik, D. W. (1989) *Ecology and Natural History of Tropical Bees*, Cambridge University Press, Cambridge.

Schemske, D. W. (1977) Flowering phenology and seed set in *Claytonia virginica*. *Bull. Torrey Bot. Club*, **104**, 254–63.

Schemske, D. W. and Horvitz, C. C. (1984) Variation among floral visitors in pollination ability: a precondition for mutualism specialization. *Science*, **225**, 519–21.

Schemske, D. W., Willson, M. F., Melampy, M. N. *et al.* (1978) Flowering ecology of some spring woodland herbs. *Ecology*, **59**, 351–66.

Sih, A. and Baltus, M. S. (1987) Patch size, pollinator behavior, and pollinator limitation in catnip. *Ecology*, **68**, 1679–90.

Skeate, S. T. (1987) Interactions between birds and fruits in a northern Florida hammock community. *Ecology*, **68**, 297–309.

Snow, D. W. (1971) Evolutionary aspects of fruit-eating by birds. *Ibis*, **113**, 194–202.

Sowig, P. (1989) Effects of flowering plant's patch size on species composition of pollinator communities, foraging strategies, and resource partitioning in bumblebees (Hymenoptera: Apidae). *Oecologia*, **78**, 550–8.

Stephenson, A. G. (1982) When does outcrossing occur in a mass-flowering plant? *Evolution*, **36**, 762–7.

Stiles, F. G. (1977) Coadapted competitors: the flowering seasons of hummingbird-pollinated plants in a tropical forest. *Science*, **198**, 1177–8.

Stiles, F. G. (1979) Reply to Poole and Rathcke. *Science*, **203**, 471.

Stiles, F. G. (1980) The annual cycle in a tropical wet forest hummingbird community. *Ibis*, **122**, 322–43.

Stiles, F. G. (1985) Seasonal patterns and coevolution in the hummingbird-flower community of a Costa Rican subtropical forest, in *Neotropical Ornithology, Ornithology Monographs 36* (eds P. A. Buckley, M. S. Foster, E. S. Morton *et al.*), Am. Ornith. Union, Washington, DC, pp. 757–87.

Tepedino, W. J. and Stanton, N. L. (1980) Spatio-temporal variation on phenology and abundance of floral resources on shortgrass prairie. *Great Basin Nat.*, **40**, 197–215.

Thompson, J. N. and Willson, M. F. (1979) Evolution of temperate fruit-bird interactions: phenological strategies. *Evolution*, **33**, 973–82.

Valentine, D. H. (1978) The pollination of introduced species, with special reference to the British Isles and the genus *Impatiens*, in *The Pollination of Flowers by Insects* (ed. A. J. Richards), Academic Press, London, pp. 117–23.

Vance, B. D. and Kucera, C. L. (1960) Flowering variations in *Eupatorium rugosum*. *Ecology*, **41** 340–5.

Visscher, P. K. and Seeley, T. D. (1982) Foraging strategy of honeybee colonies in a temperate deciduous forest. *Ecology*, **63**, 1790–801.

Waddington, K. D. (1983) Foraging behavior of pollinators, in *Pollination Biology* (ed. L. Real), Academic Press, NY, pp. 213–39.

Ware, A. B., Kaye, P. T., Compton, S. G. and van Noort, S. (1993) Fig volatiles: their role in attracting pollinators and maintaining pollinator specificity. *Plant Syst. Evol.*, **186**, 147–56.

Waser, N. M. (1978) Competition for hummingbird pollination and sequential flowering in two Colorado wildflowers. *Ecology*, **59**, 934–44.

Waser, N. M. (1979) Pollinator availability as a determinant of flowering time in ocotillo (*Fouquieria splendens*). *Oecologia*, **39**, 107–21.

Waser, N. M. (1982) A comparison of distances flown by different visitors to flowers of the same species. *Oecologia*, **55**, 251–7.

Waser, N. M. (1983a) Competition for pollination and floral character differences among sympatric plant species: a review of the evidence, in *Handbook of Experimental Pollination Biology* (eds C. E. Jones and R. J. Little), Van Nostrand Reinhold, NY, pp. 277–93.

Waser, N. M. (1983b) The adaptive nature of floral traits: ideas and evidence, in *Pollination Biology* (ed. L. Real), Academic Press, New York, pp. 242–5.

Waser, N. M. (1986) Flower constancy: definition, cause and measurement. *Am. Nat.*, **127**, 593–603.

Webber, J. M. (1953) Yuccas of the southwest. *US Dept Agric. Monogr. 17*.

Wiebes, J. T. (1979) Co-evolution of figs and their insect pollinators. *Ann. Rev. Ecol. Syst.*, **10**, 1–12.

Wiens, J. A. (1989) Spatial scaling in ecology. *Funct. Ecol.*, **3**, 385–97.

Williams, N. H. (1983) Floral fragrances as cues in animal behavior, in *Handbook of Experimental Pollination Biology*, (eds C. E. Jones and R. J. Little), Van Nostrand Reinhold, NY, pp. 50–72.

Wolf, L. L., Stiles, F. G. and Hainsworth, F. R. (1976) Ecological organization of a tropical highland hummingbird community. *J. Anim. Ecol.*, **45**, 349–79.

Wyatt, R. (1983) Pollinator–plant interactions and the evolution of breeding systems, in *Pollination Biology* (ed. L. Real), Academic Press, New York, pp. 51–96.

Young, H. J. (1988) Neighborhood size in a beetle pollinated tropical aroid: effects of low density and asynchronous flowering. *Oecologia*, **76**, 461–6.

Zimmerman, M. (1988) Nectar production, flowering phenology, and strategies for pollination, in *Plant Reproductive Ecology* (eds J. Lovett Doust and L. Lovett Doust), Oxford University Press, NY, pp. 157–78.

Part Five

Implications for Conservation

Basic studies of plant and animal ecology may be performed within relatively homogeneous plots. However, studies on small homogeneous plots give results with unknown relevance for larger areas. When applying ecological principles to practical problems such as pest management, game harvesting or conservation, they have to be put into the context of real landscapes, which include a mix of habitats of differing qualities and some non-habitat. Among the various subdivisions of applied ecology, biodiversity conservation has been at the focus recently. Within this subject, landscape effects are especially important; many endangered animal and plant species occur as fragmented populations dependent on interchange of individuals or genes, or as 'island' populations dependent on dispersal from some remote pool of surplus production.

In this section Susan Harrison and Lenore Fahrig first provide some general background, including relevant theory, for effects of spatial and temporal variability on fragmented or subdivided populations. Predictions from this theory are related to dynamics and persistence (or extinction) in some well-studied (meta)populations of endangered or rare species. The authors offer some conclusions regarding the testability and applicability of general landscape theory in conservation research. In the final chapter, Graham Arnold examines the actual use of landscape approaches in practical conservation and planning programmes. His analysis produces suggestions for applications of landscape ecology in future large-scale planning, including geographic information systems and simulation models or expert systems.

Landscape pattern and population conservation

12

Susan Harrison and Lenore Fahrig

12.1 INTRODUCTION

Current estimates of species extinctions indicate a sharp increase in extinction rate over the past two decades due to human-caused changes in species' habitat (Groombridge, 1992). The most noticeable and probably most important change is the reduction of the amount of habitat available for many species due primarily to expansion of agriculture and increased deforestation. However, in addition to the amount of habitat available, the spatial and spatio-temporal pattern of the habitat can have important implications for population survival. In this chapter we first briefly review the theoretical literature exploring the relationships between landscape pattern and population survival. We then present several case studies from the empirical literature that illustrate these relationships. Finally we review areas that are unresolved and where future research should be directed.

12.2 REVIEW OF THEORY

Landscape pattern refers to the distribution of habitat and resources in the landscape. Although habitat types used by a species can be many and varied, the distribution and availability of breeding habitat is of primary importance for long-term population survival. In this chapter we restrict our definition of landscape pattern to the distribution of

Mosaic Landscapes and Ecological Processes.
Edited by Lennart Hansson, Lenore Fahrig and Gray Merriam.
Published in 1995 by Chapman & Hall, London. ISBN 0 412 45460 2

breeding habitat in the landscape; our use of the term 'habitat' refers to breeding habitat only. Landscape pattern can be divided into spatial pattern and spatio-temporal pattern (Fahrig and Merriam, 1994). In the former the landscape pattern is static, at least on a time-scale relevant to the dynamics of a species of interest. In the latter the landscape spatial pattern changes over time for one or both of two possible reasons: either the habitat itself is ephemeral (e.g. the habitat is patches of annual plants), or disturbances are common and widespread in the habitat.

In the context of single-species conservation, the spatial and temporal scales of landscape pattern are best viewed in relation to the inherent spatial and temporal scales of the species. For example, patch size and inter-patch distance are considered relative to the dispersal distance of the organism, and disturbance rate and patch lifespan are considered relative to the lifespan of the organism.

12.2.1 LANDSCAPE SPATIAL PATTERN

The main components of landscape spatial pattern are

- amount of habitat in the landscape
- mean size of habitat patches
- mean inter-patch distance
- variance in patch sizes
- variance in inter-patch distances
- landscape connectivity

Note that, since we are discussing breeding habitat, patch size refers to the number of breeding sites in the patch. If patches vary in quality, a high-quality patch of the same physical size as the lower-quality patches would be a 'larger' patch.

A non-resolvable difficulty in theory relating landscape pattern to population survival is that, within a given hypothetical landscape, habitat amount, mean patch size and mean inter-patch distance may not be varied independently (Fahrig, 1992). For example, for a given amount of habitat in an area, larger patches imply larger inter-patch distances. For a given patch size, larger inter-patch distances imply less habitat overall in the landscape (Figure 12.1). Alternatively, in order to maintain patch size and habitat amount constant while increasing inter-patch distance, the total landscape under consideration must increase. Therefore, when interpreting results from a theoretical study it is important to be aware of the components of habitat pattern that vary in addition to (or as a consequence of) the components that are the focus of the study.

The relationship between habitat amount and population survival is straightforward. As the amount of habitat decreases, the population size decreases. In a stochastic model this leads to an increasing probability

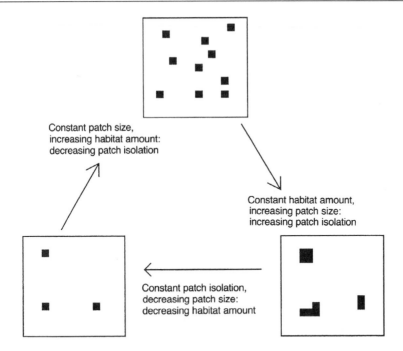

Constant patch size,
increasing habitat amount:
decreasing patch isolation

Constant habitat amount,
increasing patch size:
increasing patch isolation

Constant patch isolation,
decreasing patch size:
decreasing habitat amount

Figure 12.1 Relationship between habitat amount, mean patch size and mean patch isolation in a landscape of fixed size. If one of the three parameters is held constant and another changed, the third must change.

of regional population extinction (e.g. Stacey and Taper, 1992). Because of this clear and strong effect of habitat amount, many theoretical studies examining effects of habitat spatial pattern hold habitat amount constant and vary other aspects of landscape pattern.

Habitat clumping (i.e. increasing patch size) increases the probability of population survival (Herben, Rydin and Söderström, 1991; Adler and Nürnberger, 1994). In Adler and Nürnberger's study the size of the landscape decreases with increasing habitat clumping. However, Fahrig (in preparation) has found that this result holds even when the landscape size is constant and when the dispersal range of the organism is very limited. This suggests that, for a constant amount of habitat in a landscape of a certain size, the positive effects of increasing patch size outweigh the associated negative effects of increasing patch isolation (i.e. decreasing colonization probability).

The importance of high variance in patch sizes has been suggested in many theoretical studies. The early theory of island biogeography (MacArthur and Wilson, 1967) is an extreme example of this in which one 'patch', the mainland, is assumed to be infinitely large. This means that regional survival is ensured. The source–sink model (Pulliam, 1988; Danielson, 1992) is a less extreme version of inter-patch size variation, in which some patches (probably large, high-quality, stable ones) are sources of colonists for other patches (probably small, low-quality, frequently disturbed ones) which are sinks. The positive effect of increasing variance in patch size on regional survival is related to the result for habitat clumping above. In both cases the point is that whenever there is one or a few large patches in which population survival is virtually

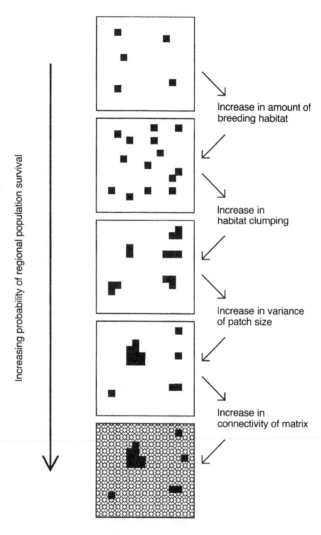

Figure 12.2 Effects of landscape spatial pattern on regional population survival, derived from general modeling studies.

ensured, these patch populations can act as colonists for other patches, and survival of the population on the landscape is ensured (Harrison, 1991; Wootton and Bell, 1992).

The final aspect of landscape spatial pattern that has a large effect on population survival is the nature of the inter-patch area or 'matrix'. The nature of the matrix determines the 'connectivity' of the landscape, or the ease with which individuals can move about within it (Taylor *et al.*, 1993). If the inter-patch area is inhospitable to moving individuals, rates of recolonization are reduced (Hansson, 1991). The effects of landscape spatial pattern on regional population survival are illustrated in Figure 12.2.

12.2.2 LANDSCAPE SPATIO-TEMPORAL PATTERN

Theoretical studies examining the effects of landscape spatial pattern, when that pattern changes over time, have generally found that the rate of change in landscape pattern is far more important than the spatial pattern itself in affecting population survival. Aspects of spatio-temporal pattern are, for disturbances,

- disturbance rate
- disturbance size
- temporal correlation in disturbances

and for ephemeral patches,

- rate of patch formation
- patch lifespan

(Herben, Rydin and Söderström, 1991).

Fahrig (1991) examined the importance of landscape spatial pattern in the presence of disturbance. Importance of spatial pattern was measured as the degree to which dispersal distance of the organism affects population survival. In an area of continuous habitat in which disturbances occur, when disturbance rate is low to moderate (less than half of the area in a disturbed state at any one time), spatial pattern of the habitat is not important for population survival. This is because disturbed areas are quickly recolonized from neighboring habitat. However, if disturbances are very large, the regional survival probability declines, since the central areas of the disturbances are not likely to be colonized before another disturbance occurs (Coffin and Lauenroth, 1989; Fahrig, 1991).

In an area of patchy non-ephemeral habitat, the higher the disturbance rate, the less important is the effect of habitat spatial pattern on population survival (Fahrig, in preparation). However, the effect of spatial pattern does remain detectable even at high disturbance rates. The effect of spatial pattern only disappears completely when the disturbance rate is so high that regional persistence is unlikely. In addition, the degree of synchrony of disturbances among habitat patches may affect survival rate (Harrison and Quinn, 1989; Gilpin, 1990). The more synchronous the disturbances, the less likely that there will be colonists available to recolonize disturbed sites during a time of low disturbance probability.

In ephemeral habitats, the 'birth' rate of new habitat patches is most important for population survival. There are two effects of patch birth rate. First, the lower the birth rate, the less habitat is available at any one time, therefore the smaller the regional population size and therefore the higher the extinction probability. Second, since habitat is ephemeral the residents of any patch must move to a newly formed patch before the one they are in 'dies'. The lower the patch birth rate, the less likely they are to successfully move to a new patch. Similarly, the shorter the lifespan of each patch, the greater is this imperative for individuals to move to other patches. Shorter patch lifespan therefore leads to lower population size even when patches are 'born' more often such that the overall amount of habitat at any one time is constant (Fahrig, 1992). This effect of patch lifespan far outweighs the effect of habitat spatial pattern on population survival. Notice that the case of extremely high disturbance rates in continuous habitat (above) is essentially the same as that of a landscape of ephemeral patches of short lifespan. The effects of landscape spatio-temporal pattern on regional population survival are illustrated in Figure 12.3.

12.3 CASE STUDIES

12.3.1 PATCHY NON-EPHEMERAL HABITATS

Interacting effects of habitat amount, patch size and patch isolation: forest birds

As temperate-zone forests become increasingly fragmented, the viability of remnant bird populations has become an increasing focus of concern. Habitat spatial pattern is clearly important; many studies have found that species composition and rates of local extinction are affected by patch size, isolation and the presence or absence of corridors (reviews in Opdam, 1990; Rolstad, 1991). The European nuthatch (*Sitta europaea*) is a typical example. Its local extinction rate on woodlots was negatively correlated with patch area, and colonization rate was negatively correlated with patch isolation (Verboom *et al.*, 1991). For species such as this, with relatively low rates of dispersal in comparison to inter-patch distance, there is a danger of population collapse if patches become fewer and more isolated. Note, however, that increasing patch isolation alone will not necessarily increase regional extinction probability. Theory predicts that increasing isolation will only have a negative effect on survival

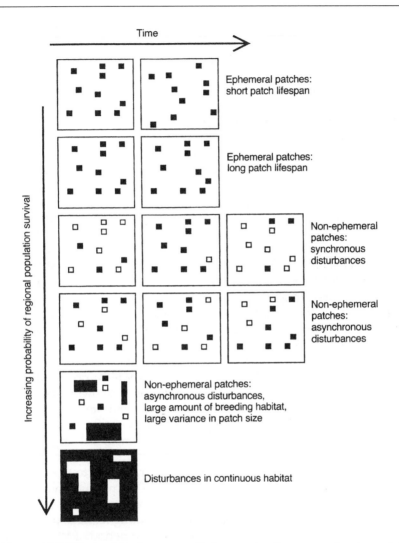

Figure 12.3 Effects of landscape spatio-temporal pattern on regional population survival, derived from general modeling studies.

if the overall amount of habitat in the landscape declines. If, instead, the amount of habitat remains constant but patch sizes increase, regional survival is predicted to be enhanced. Therefore, for making predictions about regional survival, habitat amount, patch size and patch isolation must be considered simultaneously.

One of the most ambitious attempts to deal with the spatial aspects of population viability concerns the northern spotted owl (*Strix occidentalis caurina*) in the north-western US. Less than 100 years ago, habitat of this species was an unbroken band of old-growth coniferous forest. Today less than 20% of its habitat remains, and it is highly fragmented by timber clear-cuts. The owl's philopatric behavior, and its requirement for pair territories of 1000–25 000 ha, makes it perhaps the most vulnerable of the species endemic to the north-west's old-growth forests. In addition to being a US endangered species, the owl is also a 'Management Indicator Species' for the US Forest Service, proprietor of 80% of the remaining old-growth forest in the owl's range.

With a few exceptions, such as the Olympic National Forest, few remaining fragments are large enough to support a viable population of the owl. Thus the focus has been on creating a network of forest fragments, close enough to one another to permit mutual exchange and recolonization. A team of agency scientists address this problem using a spatial simulation model of the owl population (Lamberson et al., 1992; McKelvey, Noon and Lamberson, 1992). Like previous models of the owl and other spatially distributed populations (Lande, 1987, 1988; Doak, 1989), this model predicted the existence of thresholds for regional collapse. When forest fragments become too sparse and isolated, dispersing juvenile owls cannot find territories at a sufficient rate to balance the deaths of adult territory-holders, and the population quickly becomes extinct. Some evidence suggests the real owl population may already be at or near this threshold (Lande, 1988; Doak, 1989).

The owl population model used by the agency team predicted that a network of forest patches each large enough for > 20 owl pairs, and no more than 12 miles (19 km) from another patch, could maintain a viable owl population. This plan was adopted by the Forest Service. However, the plan was later blocked in court by environmentalist litigants who argued, among other things, that habitat dynamics had not been taken into account. The plan called for many small fragments to be cut, while other areas were to be allowed to regrow in order to produce the network of larger patches. Based on theoretical predictions, increasing the size of patches should be the most effective way of ensuring regional population survival. However, the plan involves a transition period of 50–150 years of net habitat loss followed by regrowth. The serious risk posed by this habitat loss to the regional survival of the owl was not addressed by the agency plan (Harrison, Stahl and Doak, 1993).

In some cases fragmentation has resulted in inter-patch distances much greater than a given species' dispersal ability. For example, populations of red-cockaded woodpeckers in remnant patches of pine forest in the US south-east (Walters, 1991) and amphibians in the same coniferous forest inhabited by the spotted owl (Welsh, 1990) may be so isolated from each other that there is essentially no interchange between patches. Therefore, patch spacing and corridors are probably not relevant to their survival. In this case the survival of each local population depends primarily on the patch size, and the survival of the regional population depends on the existence of one or more patches large enough to effectively ensure population survival there.

12.3.2 PATCHY, DISTURBED HABITATS

(a) Effect of variance in patch size: bay checkerspot butterfly

In central California, the bay checkerspot butterfly (*Euphydryas editha bayensis*) occupies patches of grassland on serpentine soil, which support its larval host plants. In dry years these plants senesce early, causing high larval mortality. Hence checkerspot populations appear to fluctuate and go extinct in response to yearly climatic patterns (Ehrlich *et al.*, 1980; Murphy and Erlich, 1980). At the same time, the butterfly's capacity to disperse to and colonize new patches is extremely limited (Harrison, Murphy and Ehrlich, 1988; Harrison, 1989). Regional persistence in this species appears to depend not on recolonization *per se*, but on the existence of a large mainland population. In 1987, one colony of about 10^6 adults was surrounded by nine much smaller (10^1–10^2) colonies, all within 4.5 km of the large colony. At distances of 5–20 km from the large colony there were 18 suitable patches of habitat, none of them occupied by the butterfly (Harrison, Murphy and Ehrlich, 1988).

The butterfly's regional distribution is thus highly sensitive to the spatial pattern of habitat, but its regional persistence is probably not. Schoener and Spiller (1987) found similar patterns in spider metapopulations on islands. Frequent population turnover occurred on small islands, but populations on large islands were highly persistent.

(b) Importance of habitat clumping: amphibians in freshwater ponds

At the northern limit of its range, the European pool frog (*Rana lessonae*) is found in shallow freshwater ponds at the edge of the Baltic Sea. Large ponds seldom contain frog populations, since they are inhabited by the predatory pike (*Esox lucius*). Within small ponds, frog populations may go extinct during cold winters that cause reproductive failure. Extinctions are also caused by the disappearance of ponds, due either to natural succession or to draining for forestry. Sjögren (1991) compared the distribution of frogs across 60 ponds to that recorded by another researcher in 1962, and found an average extinction rate of 2% per year. However, extinctions seldom occurred in ponds < 1 km from other occupied ponds, suggesting that the dispersal of frogs over such distances is sufficient to rescue declining populations. In this system, where all patches are small, theory predicts that regional survival is enhanced by close-spaced clumping of ponds. This effectively increases patch size.

Strikingly similar patterns are seen in other freshwater amphibians. Gill (1978) found that the newt (*Notopthalmus viridescens*) dispersed readily among neighboring ponds, and in each year a few ponds supplied the majority of recruits to all other ponds. Sinsch (1992) likewise found that natterjack toads (*Bufo calamita*) move frequently enough among ponds to compensate for local reproductive failures. Both *Bufo calamita* (Sinsch, 1992) and the wood frog *Rana sylvatica* (Berven and Grudzien, 1990) show patterns of genetic variation that indicate movement is frequent among nearby (< 1 km) ponds, but declines rapidly at greater distances. Because habitat (pond) quality is variable in space and time, movement among ponds is essential to the regional persistence of the species, and clusters of nearby ponds are necessary to support viable regional populations.

12.3.3. EPHEMERAL HABITATS

Importance of patch birth rate and patch lifespan: Butterflies and other insects

Many insects colonize patches of their food resource, and maintain local populations that last from one to several generations. These dynamics are exemplified by recent studies of milkweed beetles on patches of their host plant (McCauley, 1991) and forked fungus beetles on rotting logs (Whitlock, 1992). In both these cases, the founder effects associated with colonization created spatial genetic structure in the metapopulation (McCauley, 1991; Whitlock, 1992). However, insects immigrated to and emigrated from patches fairly frequently, suggesting that metapopulation persistence was not highly sensitive to habitat spatial pattern, as predicted by theory.

Several well-studied butterfly species in Great Britain occupy spatially and temporally variable habitats (e.g. Warren, Thomas and Thomas, 1984; Thomas and Harrison, 1992; Thomas, Thomas and Warren, 1992; Thomas and Jones, 1993). For example, in some regions the silver-studded blue butterfly (*Plebejus argus*) requires grassland 3–5 cm in height, and successional overgrowth may lead to the extinction of local populations. In other regions, *P. argus* is found in patches of heathland recently (< 10 years) affected by fire and other disturbance. The extinction and colonization rates of *P. argus* are higher in heathland than in grassland, because of more rapid disturbance and succession. But within either biotope, the butterfly is relatively good at tracking its shifting habitat mosaic. Only at distances of > 1 km from occupied patches are suitable patches likely to remain unoccupied for a significant period of time (Thomas and Harrison, 1992).

For *P. argus* and many similar species, regional population dynamics are essentially controlled by the rates and patterns of habitat processes, i.e. disturbance and succession (Thomas, 1994). The key to conservation of these species is the maintenance of a habitat mosaic in which the required successional stage(s) is always present. This may involve very active interventive management and/or the preservation of relatively large areas of biotope (patch plus matrix). Once the temporal continuity of the habitat mosaic is broken and the species becomes regionally extinct, natural long-distance recolonization may be extremely slow. For example, the skipper (*Hesperia comma*) may take decades to recolonize areas of chalk grassland from which it went extinct in the mid–1950s, when the destruction of rabbits by myxomatosis caused overgrowth of the habitat (Thomas and Jones, 1993).

In the US, the Karner blue butterfly (*Lycaeides melissa samuelis*) is an even more extreme example of an ephemeral habitat species. Its host plant, *Lupinus perennis*, flourishes in recently burned patches of pine barren habitat, found on sandy soils in the north-eastern US. The butterfly cannot tolerate fire in any of its life stages, and thus must continually track a shifting mosaic of burned and unburned patches. Habitat loss combined with fire suppression now threatens this species; Givnish, Menges and Schweitzer (1988) describe an attempt to devise an appropriate management regime for the butterfly on surviving fragments of pine barrens.

12.4 FUTURE DIRECTIONS

From a practical standpoint, the most pressing question remaining to be answered is: to what extent can alteration of landscape pattern compensate for loss of habitat? In other words, can we maintain high population survival probability while reducing the amount of habitat available, by carefully selecting the sizes and spatial locations of the remaining habitat fragments? Unfortunately it will be difficult or impossible to answer this question using the empirical data available. In most cases, habitat fragmentation involves simultaneously loss of habitat, reduction of patch sizes and increased patch isolation, which means that it is not possible to estimate the separate effects of these factors on population density and survival probability. Conservation programs will therefore depend to an increasing extent on spatially explicit population models which integrate local population dynamics and dispersal with landscape spatial and temporal pattern (e.g. Burgman, Ferson and Akçakaya, 1993). Although such detailed modeling is possible for some organisms (e.g. Fahrig and Merriam, 1985; Pulliam, Dunning and Liu, 1992), in most cases the information required is not available. Even for the well-studied spotted owl, the exemplary modeling effort by US Forest Service scientists (McKelvey, Noon and Lamberson, 1992; Lamberson *et al.*, 1992) was found to have significant shortcomings (Harrison, Stahl and Doak, 1993).

An additional important research question is therefore: under what circumstances (for what kinds of species in what kinds of landscapes) is spatially explicit modeling necessary for predicting the effects of habitat fragmentation on population survival?

Future research will also need to address the question: how can we predict the effects of alteration of habitat pattern on species diversity in the landscape? This is a difficult problem since the pattern of any one landscape is particular to each species living there. Species differ both in the kinds of habitat they use in the landscape and the spatial and temporal scales at which they interact with the landscape. For example, endangered amphibians of old-growth conifer forest in the north-western US occur as isolated independent remnant populations (Welsh, 1990) whereas birds using these same forest patches occur as more interconnected subpopulations. It appears unlikely that predictions of effects on diversity can be made using a composite of many spatially explicit population models. A more promising approach will be a statistical modeling approach in which the actual diversity in landscapes of differing spatial patterns are measured and compared to develop general relationships between species diversity and landscape pattern as viewed from the human perspective.

12.5 SUMMARY

Theory relating population survival to landscape spatial pattern is reviewed. Important generalizations are:

1. as the amount of habitat decreases the probability of regional population survival decreases;
2. for the same total amount of habitat, increased habitat clumping (i.e. increasing patch size) increases the probability of population survival, and this positive effect of increasing patch size outweighs the negative effect of increasing inter-patch distance;
3. increasing inter-patch variance in patch size increases the probability of regional survival;
4. when the landscape pattern is dynamic (patches are ephemeral or disturbance rate is high), landscape spatial pattern is relatively unimportant; and
5. when habitat is ephemeral, regional population survival increases with increasing patch lifespan.

Case studies illustrating these effects in real populations are given. However, we conclude that empirical studies cannot be used to rigorously test the theoretical predictions because human activities usually alter several aspects of landscape pattern simultaneously. Important questions for future research are

1. To what extent can careful planning of landscape pattern compensate for loss of habitat?
2. Under what circumstances (for what kinds of species in what kinds of landscapes) is spatially explicit modeling necessary for predicting the effects of habitat fragmentation on population survival?
3. How can we make predictions of effects of alteration of landscape pattern on species diversity?

REFERENCES

Adler, F. R. and Nürnberger, B. (1994) Persistence in patchy irregular landscapes. *Theor. Pop. Biol.*, **45**, 41–75.

Berven, K. A. and Grudzien, T. A. (1990), Dispersal in the wood frog (*Rana sylvatica*): implications for genetic population structure. *Evolution*, **44**, 2047–56.

Burgman, M. A., Ferson, S. and Akçakaya, H. R. (1993) *Risk Assessment in Conservation Biology*, Chapman and Hall, New York.

Coffin, D. P. and Lauenroth, W. K. (1989) Disturbances and gap dynamics in a semiarid grassland: a landscape-level approach. *Landsc. Ecol.*, **3**, 19–27.

Danielson, B. J. (1992) Habitat selection, interspecific interactions and landscape composition. *Evol. Ecol.*, **6**, 399–411.

Doak, D. (1989) Spotted owls and old growth logging in the Pacific Northwest. *Conserv. Biol.*, **3**, 389–96.

Ehrlich, P. R., Murphy, D. D., Singer, M. C. *et al.* (1980) Extinction, reduction, stability and increase: the responses of checkerspot butterfly (*Euphydryas*) populations to the California drought. *Oecologia*, **46**, 101–5.

Fahrig, L. (1991) Simulation methods for developing general landscape-level hypotheses of single species dynamics, in *Quantitative Methods in Landscape Ecology*, (eds M. G. Turner and R. H. Gardner), Springer Verlag, New York, pp. 417–42.

Fahrig, L. (1992) Relative importance of spatial and temporal scales in a patchy environment. *Theor. Pop. Biol.*, **41**, 300–14.

Fahrig, L. and Merriam, G. (1985) Habitat patch conectivity and population survival. *Ecology*, **66**, 1762–8.

Fahrig, L. and Merriam, G. (1994) Conservation of fragmented populations. *Conserv. Biol.*, **8**, 50–59.

Gill, D. E. (1978) The metapopulation of the red-spotted newt, *Notopthalmus viridescens* (Rafinesque). *Ecol. Monog.*, **48**, 145–66.

Gilpin, M. E. (1990) Extinction of finite metapopulations in correlated environments, in *Living in a Patchy Environment*, (eds B. Shorrocks and I. R. Swingland), Oxford Scientific Publications, Oxford, UK, pp. 177–86.

Givnish, T. J., Menges, E. S. and Schweitzer, D. F. (1988) Minimum area requirements for the Karner blue butterfly: an assessment. Report to the City of Albany, New York.

Groombridge, B. (ed.) (1992) *Global Biodiversity: Status of the Earth's Living Resources*, Chapman and Hall, London and New York.

Hansson, L. (1991) Dispersal and connectivity in metapopulations. *Biol. J. Linn. Soc.*, **42**, 89–103.

Harrison, S. (1989) Long-distance dispersal and colonization in the Bay checkerspot butterfly. *Ecology*, **70**, 1236–43.

Harrison, S. (1991) Local extinction in a metapopulation context: an empirical evaluation. *Biol. J. Linn. Soc.* **42**, 73–78.

Harrison, S. and Quinn, J. F. (1989) Correlated environments and the persistence of metapopulations. *Oikos*, **56**, 293–8.

Harrison, S., Murphy, D. D. and Ehrlich, P. R. (1988) Distribution of the bay checkspot butterfly, *Euphydryas editha baynesis*: evidence for a metapopulation model. *Am. Nat.*, **132**, 360–82.

Harrison, S., Stahl, A. M. and Doak, D. (1993) Spotted owl update: US judge rejects Forest Service plan. *Conserv. Biol.* (in press).

Herben, T., Rydin, H. and Söderström, L. (1991) Spore establishment probability and the persistence of the fugitive invading moss, *Orthodontium lineare*: a spatial simulation model. *Oikos*, **60**, 215–21.

Lamberson, R. H., McKelvey, K., Noon, B. R. and Voss, C. (1992) A dynamic analysis of northern spotted owl viability in a fragmented landscape. *Conserv. Biol.*, **6**, 505–12.

Lande, R. (1987) Extinction thresholds in demographic models of territorial populations. *Am. Nat.*, **130**, 624–35.

Lande, R. (1988) Demographic models of the northern spotted owl (*Strix occidentalis caurina*). *Oecologia*, **75**, 601–7.

MacArthur, R. H. and Wilson, E. O. (1967) *The Theory of Island Biogeography*, Princeton University Press, Princeton, NJ.

McCauley, D. E. (1991) The effect of host plant patch size variation on the population structure of a specialist herbivore insect, *Tetraopes tetraopthalmus*. *Evolution*, **45**, 1675–84.

McKelvey, K., Noon, B. R. and Lamberson, R. H. (1992) Conservation planning for species occupying fragmented landscapes: the case of the northern spotted owl, in *Biotic Interactions and Global Change*, (eds P. M. Kareiva, J. G. Kingsolver and R. B. Huey), Sinauer, Sunderland, MA.

Murphy, D. D. and Ehrlich, P. R. (1980) Two new subspecies: one new, one on the verge of extinction. *J. Lepid. Soc.*, **34**, 316–20.

Opdam, P. (1990) Metapopulation theory and habitat fragmentation: a review of holarctic breeding bird studies. *Landsc. Ecol.*, **5**, 93–106.

Pulliam, H. R. (1988) Sources, sinks, and population regulation. *Am. Nat.*, **132**, 652–61.

Pulliam, H. R., Dunning, J. B. and Liu, J. (1992) Population dynamics in complex landscapes: a case study. *Ecol. Appl.*, **2**, 165–77.

Rolstad, J. (1991) Consequences of forest fragmentation for the dynamics of bird populations: conceptual issues and the evidence, in *Metapopulation Dynamics: Empirical and Theoretical Investigations* (eds M. E. Gilpin and I. Hanski), Academic Press, London, pp. 149–63.

Schoener, T. W. and Spiller, D. A. (1987) High population persistence in a system with high turnover. *Nature*, **330**, 474–7.

Sinsch, U. (1992) Structure and dynamics of a natterjack toad metapopulation (*Bufo calamita*). *Oecologia*, **90**, 489–99.

Sjögren, P. (1991) Extinction and isolation gradients in metapopulations: the case of the pool frog (*Rana lessonae*), in *Metapopulation Dynamics: Empirical and Theoretical Investigations* (eds. M. E. Gilpin and I. Hanski), Academic Press, London, pp. 135–47.

Stacey, P. B. and Taper, M. (1992) Environmental variation and the persistence of small populations. *Ecol. Applic.*, **2**, 18–29.

Taylor, P. D., Fahrig, L., Henein, K. and Merriam, G. (1993) Connectivity is a vital element of landscape structure. *Oikos*, **68**, 571–3.

Thomas, C. D. (1994) The ecology and conservation of butterfly metapopulations in fragmented landscapes, in *Ecology and Conservation of Butterflies* (ed. A. S. Pullin), Chapman and Hall, London.

Thomas, C. D. and Harrison, S. (1992) Spatial dynamics of a patchily-distributed butterfly species. *J. Anim. Ecol.*, **61**, 437–46.

Thomas, C. D. and Jones, T. M. (1993) Partial recovery of a skipper butterfly (*Hesperia comma*) from population refuges: lessons for conservation in a fragmented landscape. *J. Anim. Ecol.*, **62**, 472–81.

Thomas, C. D., Thomas, J. A. and Warren, M. S. (1992) Distributions of occupied and vacant butterfly habitats in fragmented landscapes. *Oecologia*, **92**, 563–7.

Verboom, J., Schotman, A., Opdam, P. and Metz, J. A. J. (1991) European nuthatch metapopulations in a fragmented agricultural landscape. *Oikos*, **61**.

Walters, J. (1991) Application of ecological principles to the management of endangered species: the case of the red-cockaded woodpecker. *Ann. Rev. Ecol. Syst.*, **22**, 505–23.

Warren, M. S., Thomas, C. D. and Thomas, J. A. (1984) The status of the heath fritillary butterfly, *Mellicta athalia* Rott., in Britain. *Biol. Conserv.*, **29**, 287–305.

Welsh, H. (1990) Relictual amphibians and old-growth forests. *Conserv. Biol.*, **3**, 309–19.

Whitlock, M. C. (1992) Nonequilibrium population structure in forked fungus beetles: extinction, colonization, and genetic variation among populations. *Am. Nat.*, **139**, 952–70.

Wootton, J. T. and Bell, D. A. (1992) A metapopulation of the peregrine falcon in California: viability and management strategies. *Ecol. Applic.*, **2**, 307–21.

Incorporating landscape pattern into conservation programs

13

Graham W. Arnold

13.1 INTRODUCTION

It is the purpose of this chapter to examine the extent to which conservation programs have been structured to provide the heterogeneity needed to sustain an organism, community or ecosystem. It is apparent that, at this point in time, very few attempts have yet been made. The reason for the gap between concepts and their application is because theoretical and empirical studies in this whole area have developed only very recently. It is often not clear just which concepts should be incorporated in a particular conservation program, especially when little is known about the autecology of the target species. The conservation programs are applied in either highly fragmented and/or human disturbed landscapes in which the natural heterogeneity has been replaced by a heterogeneity that differs temporally, spatially and in scale. The effects of these changes on ecosystem functions and the responses of organisms to these different landscape patterns are often equally unknown.

This said, land and resource managers have to make planning and management decisions. These are at the level of species, communities, reserves, landscapes, ecosystems and regions (Saunders, Hobbs and Arnold, 1993). In this chapter, a number of conservation programs are

Mosaic Landscapes and Ecological Processes.
Edited by Lennart Hansson, Lenore Fahrig and Gray Merriam.
Published in 1995 by Chapman & Hall, London. ISBN 0 412 45460 2

examined with emphasis given to the importance of heterogeneity in the design and implementation of the programs. The examples are diverse and are grouped into sections on conservation of individual species, conservation programs with reserves, conservation for species diversity in agricultural landscapes and conservation management of ecosystems at the regional level. This is followed by a discussion on management planning and the role of modeling. Finally, an assessment is made of whether any overall generalizations can yet be made about the ways in which heterogeneity should be incorporated into conservation programs.

13.2 CONSERVATION MANAGEMENT FOR INDIVIDUAL SPECIES

The majority of programs developed for individual species are set up either because the species is endangered or because it is a game species. They are all based on integrating different resource requirements of the species into a spatial pattern that the species can cope with and that will ensure long-term population survival. The critical components are the size of the patches of habitat (grain size), the homogeneity of the patches (grain evenness), the distribution of the patches (grain dispersion), the matrix surrounding the patches and the connectivity between the patches. The perception of the grain patterns in the landscape are species-specific and this is discussed in the examples.

13.2.1 THE CAPERCAILLIE (*Tetrao urogallus*) IN SWEDEN

The capercaillie is a species dependent on old growth forest but with a good ground layer of *Vaccinium myrtillus*. Angelstam (1992) states that there are three different kinds of critical area requirements for the amount of old forest that the capercaillie needs in order to maintain a stable population level:

1. 20–50 ha: the territory size of an individual cock at the lek;
2. 200–500 ha: the size of all the cock territories, i.e. a local lek population;
3. >10 000 ha: the size of an area required for a core area with a local viable population (three or four adjacent leks covering 2000 ha) that is unaffected by intruding generalist predators.

The grain size of the landscape mosaic is thus a major determinant of capercaillie population survival. However, if the clear-cuts in old growth forest are small, the capercaillie may perceive the forest landscape as one single forest instead of as a series of forest stands, i.e. it may tolerate the removal of a larger proportion of the old forest if clear-cuts are made sufficiently small (Rolstad and Wegge, 1987; Figure 13.1).

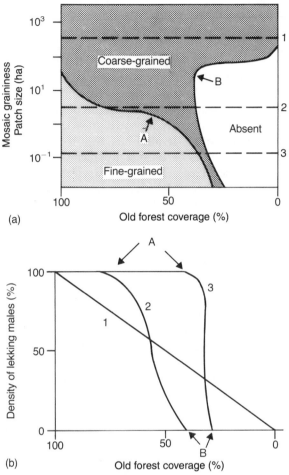

Figure 13.1 A predictive model of the grain response of capercaillie in relation to the percentage cover of old forest and the grain size of different landscapes. (a) The transition zone between a fine-grained male density independent of the proportion of old forest and a coarse-grained male density which tracks the amount of old forest-habitat use. (b) The critical values at which local populations are expected to go extinct due to an insufficient area of old forest remnants. 1, 2 and 3 depict male densities at three levels of landscape grain, i.e. coarse, intermediate and fine grained, respectively. (Source: Angelstam, 1992.)

13.2.2 THE RED GROUSE (*Lagopus lagopus scoticus*) IN SCOTLAND AND THE RUFFED GROUSE (*Bonasa umbellus*) IN MINNESOTA

The capercaillie is a single-patch user and is satisfied by one type of habitat to fulfil its daily, seasonal or lifetime demands. Other species are multi-patch users and require certain combinations of habitat type as well as certain densities of patches in the landscape (Angelstam, 1992). The red grouse is a good example. By managing heather moors in Scotland using fire, a patchwork of different successional stages of heather is maintained so that each territory has all the components needed (Hudson, 1986). Burning in a prescribed way can also increase the numbers of territories available and hence increase the population.

In Minnesota and Pennsylvania, logging in old growth forests is aimed at producing a chequerboard pattern (Figure 13.2) that maintains high densities of ruffed grouse. It also has resulted in an increased local diversity of breeding birds because of the niche created for edge species (Guillion, 1984; Yahner, 1984). One problem with this system is that it also results in higher densities of deer which can prevent forest regeneration (Alverson, Waller and Solheim, 1988).

13.2.3 AMERICAN MARTEN (*Martes americana atrata*) IN NEWFOUNDLAND, CANADA

The American marten prefers dense, mature coniferous forest or mixed forest with high overstorey density and it is becoming threatened through logging. This is a case where the creation of greater habitat heterogeneity in forests through logging is detrimental. In this case it is both grain size and grain evenness that are critical. Large clear-cuts are detrimental to marten populations, even regenerating clear-cuts are used infrequently or are avoided. Bissonette, Fredrickson and Tucker (1991) suggest that residual forest patches of >15 ha, and with patch shape values tending towards unity, are desirable landscape elements for martens. These need to be linked, preferably by riparian corridors. It is not possible to retain specific patches of old growth forest in perpetuity because of insect and fungal attacks. Thus landscape-level management is needed to maintain a shifting mosaic with predetermined proportions of each habitat element. To the extent that the habitat needs of core-sensitive species such as marten are known, the spatial arrangements of critical elements can be mimicked to provide for those needs under a higher disturbance regime.

13.2.4 RED-COCKADED WOODPECKER (*Picoides borealis*) IN TEXAS, USA

The endangered red-cockaded woodpecker provides another example where fragmentation from clear felling has a major detrimental impact on a species' survival. Conner, Snow and O'Halloran (1991) demonstrated that an alternative silvicultural approach might prevent the decline of the species. Seed-tree and shelterwood reproduction methods remove most trees but leave some residual pine trees to provide seeds and, in some cases, shelter for the next generation of pines.

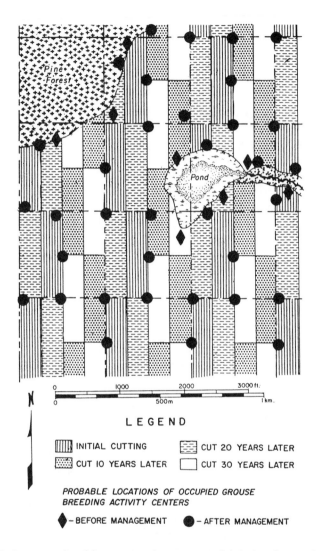

Figure 13.2 An example of how extensive commercial timber harvesting could be done in a 210 ha area of aspen or northern hardwood forest in a manner that should maintain good populations of ruffed grouse, woodcock and other forest wildlife. Each of these rectangular strips includes 4 ha, and 25 ha could be cut from a 2.6 km² area at one time. (Source: Guillion, 1984.)

The residual trees become potential cavity trees for the woodpeckers. The system also creates the open savanna habitat preferred by the woodpecker, and leaves some foraging habitat whilst reducing the hazard of southern pine beetles to the residual trees. Clumping of the residual trees creates larger openings that will enhance pine survival and growth following germination. It also mimics the clusters of mature pines that compose many of the current woodpecker colonies.

13.2.5 DEER AND GOLDEN-CHEEKED WARBLERS IN JUNIPER WOODLANDS, TEXAS, USA

Evergreen juniper woodlands have increased gradually throughout central Texas since the 1870s as a consequence of reduced fire frequency and heavy livestock grazing in the early years of settlement. White-tailed deer (*Odocoileus virginianus*) require managed habitats of juniper woodlands. Bryant (1991) recommends that stands of tall (> 3 m) juniper exceeding 20% canopy cover should be opened up in a design that includes 8-ha clearings of irregular shape. These clearings will provide foraging areas. Distance between clearings should be >200 m to give ample brush as security cover.

By contrast, the endangered golden-cheeked warbler (*Dendroica chrysoparia*) requires dense juniper with intermingled oaks. To maintain a breeding population requires preserving a minimum of 125 ha of dense stands. The trees need to be at least 50 years old for successful breeding. Thus the scale of juniper patches needed for deer and golden-cheeked warbler differ, and in a managed landscape different areas with different mosaic structures need to be maintained for different target species. Management priorities must depend on the conservation status of the different species both at the landscape and regional levels.

13.2.6 GRIZZLY BEAR (*Ursus ursus horribilis*) IN THE KOOTENAI NATIONAL FOREST, MONTANA, USA

Forestry management practices in this forest have profoundly affected the grizzly bear through changes in the habitat available (Garcia, 1985). The loss of habitat components requires the re-creation of patchiness within the framework of forest management. The program involves patch cutting of timber accompanied by burning and/or seeding of umbelliferous plants, prescribed burning, road-verge seeding of graminoids and legumes, and planting of berry-producing shrubs. The objective is to provide the diversity of food sources required by the bears at different times of the year. The patch size varies with the type of patch being created, from < 1 ha for reseeded patches to 30 ha for patches of burnt forest.

13.2.7 NORTHERN SPOTTED OWLS, OREGON, USA

The examples so far have been concerned with conservation of species at the landscape level. This example considers conservation of a species over the whole of its range. The Northern Spotted Owl is one of three subspecies of *Strix occidentalis* and ranges from south-western British Columbia south through the Coast Ranges and Cascade Range of Oregon and Washington and into the Klamath Mountains Physiographic Province, which includes south-western Oregon and the coastal mountains of north-western California.

Populations of the species have declined with the decline in the habitat that allows successful nesting, breeding and fledging of young, viz. old-growth forests and younger forests that include the structural characteristics of old-growth forest. The species has been the subject of intensive study with the objective of developing a habitat conservation plan. Murphy and Noon (1992) present the protocol of a model for the development of a reserve system covering a wide region. They point out that each habitat conservation plan is a unique combination of an understanding of the target species and of its habitats, and of the scientific framework that they present. Unfortunately, there are few species for which such detail is available.

The plan recognizes three critical life history phases: juvenile, sub-adult and adult. A key aspect is the forced juvenile dispersal, sometimes over long distances, in search of suitable territories and mates. The plan also had to allow selection, from potential reserve areas, of those reserves where pressures for other uses that could threaten the habitat could be overcome. Population viability analysis was used to determine reserve design characteristics (such as number of habitat patches, their sizes and their geographic distribution) that would reduce the chances of population extinction from natural stochastic phenomena.

A mapping procedure was developed to locate and then select an arrangement of habitat conservation areas as a system sufficient to allow the species a high likelihood of survival for 100 years. The design criteria relevant to this discussion are as follows:

1 Large habitat patches that support large populations of target species will tend to support that species for longer periods of time than will small patches that support fewer individuals.

2 Habitat patches that are continuous (less internally fragmented) in configuration tend to support species for longer periods than patches that are irregular.

3 Habitat patches that are sufficiently close together to allow dispersal tend to support target populations for longer periods than habitat patches that are far apart.

4 Habitat patches that are connected by habitat corridors, or that are set in a landscape matrix of a similar structure to the habitat patches, will allow a target species to disperse freely among patches and will support a species for longer periods than habitats not so situated.

13.2.8 CONCLUSIONS

The first six examples illustrate the relative importance of different components of landscape pattern to the conservation programs as perceived by their authors. In most cases there are not a lot of hard data on which to base the proposals.

Grain (patch) size is of greatest significance (capercaille, grizzly bear, white-tailed deer, golden-cheeked warbler, marten). Grain shape was considered important for white-tailed deer and marten. The red grouse, ruffed grouse, grizzly bear, white-tailed deer and red-cockaded woodpecker require a mosaic of different vegetation types whereas the capercaillie, golden-cheeked warbler and marten have less diverse, although specific, habitat requirements.

Although a program is designed for a target species, there will be other species that need similar landscape patterns and these species will also benefit. This may not always be desirable, as is the case with the ruffed grouse program.

The spotted owl program is one of very few that aim to conserve metapopulations of a species over the whole of its range. Another example is that of the Florida panther (*Felis concolot coryi*), but in this case it is based primarily on linking the reserves to allow dispersal and not on optimizing landscape patterns for the species.

13.3 MANAGEMENT FOR SPECIES DIVERSITY IN AGRICULTURAL LANDSCAPES

In the examples so far the conservation programs are all in multi-use forest or woodlands. In these managed landscapes the changes in landscape pattern are generally smaller than those in agricultural landscapes. Most of the native vegetation has been removed and the size and dispersion of the remaining areas creates a high probability that numbers of species will decline and genetic isolation will occur. Conservation programs in these landscapes should be aimed at maintaining or increasing species diversity and at linking populations into metapopulations by increasing the diversity of elements in the landscape and by connecting them.

13.3.1 INVERTEBRATES IN CROPLAND

Several recent studies have been concerned with conservation of invertebrates in these landscapes, with emphasis on the introduction of a system of linear or extended landscape elements like grass strips at the edges of fields, hedgerows, woodlots and wetlands, created and managed to overcome the isolation of animal communities. These habitat-linked projects have the effect of enhancing nature conservation and reducing pest problems in agricultural crops by natural predation (Mader, 1988; Thomas, Wratten and Sotherton, 1992). These introduced elements have been shown to attract populations of a wide range of invertebrate species and those species that are predators spread for some distance into the crops to reduce aphid numbers (Mader, 1988).

In a similar vein, Thomas, Wratten and Sotherton (1992) created grass habitats in cereal crops for the overwintering of arthropods which prey on cereal aphids. The refugees took the form of earth ridges sown with grasses bisecting cereal crops. Within three years the ridges supported high densities of polyphagous predators.

Dennis and Fry (1992) also examined the role of field margins influencing the distribution and abundance of polyphagous aphid predators in adjacent cereal crops in spring, and in nature conservation in supporting greater arthropod biodiversity in agroecosystems. Field margins increased general arthropod diversity on farmland: in summer, by providing a stable, complex habitat for species that would not survive in the farm landscape with the presence only of crop habitats; and in winter, by providing refuges for many arthropod species including those predators active in arable fields during the summer. They argue that the 'landscape' scale matrix of field margins may be vital for the effective dispersal of predators into crops and for achieving conservation goals. There is increasing evidence that the re-creation of heterogeneity in agricultural landscapes has the dual advantage of stabilizing insect pest populations and conserving arthropod species (e.g. Nentwig, 1989). Zanaboni and Lorenzoni (1989) found 13 species of Arenoida in the centre of a soybean field surrounded by hedges but only eight species in a similar field without any field margin. Likewise, Lagerlöf and Wallin (1993) found that the more diverse the flora in field margins, the higher the invertebrate diversity. Importantly, both for pest control in crops and for conservation, more alternative prey for polyphagous predator arthropods are found when the flora is diverse. Butterfly diversity also increases (Rands and Sotherton, 1986).

Non-pest insects, such as plant bugs (Heteroptera), Chrysomelidae (Coleoptera) and sourfly larvae (Symphyta) are also preferred food items for birds, e.g. partridge chicks. Thus maintaining and/or creating herb and grass species-rich field margins is important for the conservation of overall species diversity in agricultural landscapes.

The conservation of about 260 bee species found on the Wielkopolsko–Kujawska Lowland of western Poland depends on maintaining the mosaic structure of the landscape (Banaszak, 1992). This mosaic consists of meadows and residues of natural and seminatural communities as a faunal refuge system, and field crops as an additional nutritive system.

Banaszak estimates that no more than 75% of the landscape should be in crops if the bee species are to be conserved. Figure 13.3 shows that total bee density falls rapidly when less than 20% of the landscape is left as refuges.

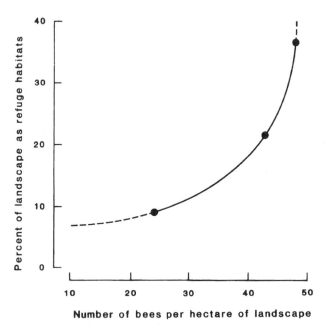

Figure 13.3 Dependence of bee density in agricultural areas of western Poland on the percentage of the landscape in refuge Kabitata. (Source: Banaszak, 1992.)

13.3.2 VERTEBRATES

The importance of hedgerows as wildlife habitat was demonstrated by Petrides (1942) in central New York State. Seventeen species of birds and 12 species of mammals were resident in the hedgerows. More recently, the diversity of the avifauna in hedgerows has been shown to depend not only on their structural diversity but also on the mosaic of woodlots and gardens in the immediate vicinity (Arnold, 1983). The presence of trees in the hedgerows significantly enhances the bird species diversity (O'Conner and Shrubb, 1986).

At the landscape level, Opdam (1990) argues that dispersal is the key to survival in fragmented agricultural landscapes. He says that 'landscape planners aiming at species conservation should consider and balance the advantages and the feasibility of habitat improvement, the extension of landscape patches and the construction of corridors simultaneously.'

Conservation strategies to retain a diverse avifauna in an agricultural landscape requires that both remnant size and heterogeneity need to be maximized. Greater habitat heterogeneity provides a more complete and more flexible range of the necessary resources in small areas for at least some species (Freemark and Merriam, 1986).

Various scales of patchiness are required to meet the different requirements for resources, and the distribution of these resources in space and time will influence the persistence of the remaining species of native animals (Lambeck and Saunders, 1993). These authors give guidelines for the restoration of patchiness in the central wheatbelt of Western Australia.

In heterogenous environments many species naturally occur in metapopulations (see Lambeck and Saunders, 1993). In reconstruction, provision of multiple representatives of given patch types will support multiple subpopulations. Creation of a variety of patch types provides resources that can be exploited at different times of the year and provides a range of resources for species which have different habitat requirements at different stages of their life cycle (section 13.2.6 and Chapter 6).

The re-creation of structural diversity in agricultural landscapes at the landscape level is currently being addressed in Australia, and particularly in the wheatbelt of south-western Australia (Hobbs, Saunders and Arnold, 1993; Hobbs et al., 1993). Here, the removal of over 90% of the original native vegetation, besides having an adverse effect upon the native flora and fauna (Saunders, Hobbs and Arnold, 1993), has resulted in an increase in salinization and soil erosion. Locally, groups of farmers are tackling land degradation by replanting perennial vegetation. In many areas this is being planned at catchment level with the dual objectives of restoring land productivity and nature conservation values.

13.3.3 CONCLUSIONS

The challenges for conservation programs in agricultural landscapes are great. The natural patterns in the landscape cannot be re-created. Existing elements have to be retained or enhanced in diversity. New elements that alter the dispersion of patches and connectivity between them have to be created. In doing this, it will rarely be possible to consider single target species, but working towards a specified objective is desirable (Hansen, Garman and Marks, 1993). In the central wheatbelt of Western Australia the objective could be to maximize the potential for survival of the suite of resident passerine bird species dependent on native vegetation. To achieve this, the revegetation strategy must be at the landscape level, within which patch size and diversity is determined by territory size, placement being such that the new patch's population is part of a metapopulation. This may need the creation of a connecting element.

13.4 CONSERVATION MANAGEMENT WITHIN RESERVES

Within managed landscapes larger remnant areas of native vegetation, be they publicly or privately owned, may require human management. In large reserves, the only management may be to allow natural processes to proceed. However, the long-term preservation of natural biota and ecosystem functions within small parks or reserves requires human intervention. In Texas state parks the management objective is to strive 'to make each park representative of the natural communities of which it was an integral part' (Hayes, Riskind and Park, 1987). Landscape heterogeneity is increased in order to enhance aspects of community structure and dynamics, which, due to size and configurational limitations, might not occur within the park. Such restorative and manipulative techniques to promote natural community heterogeneity in disjunct parklands may be characterized as patch-within-patch management. Hayes, Riskind and Pace (1987) go on to detail what is done in specific situations.

13.4.1 PRAIRIE RESTORATION

Different seed mixtures of local provenances are used. Seed mixtures are varied within each distinct microenvironment. Eight mixtures were used within one 80-ha restoration area. Once the vegetation communities are re-established, selective re-introduction of original faunal components of the prairie ecosystem are made, e.g. Prairie dogs (*Cynomys ludiovicianus*). This is an example of reconstruction, but based on recreating heterogeneity with a mosaic of plant communities.

13.4.2 HABITAT MANIPULATION TO PROMOTE NATURAL HETEROGENEITY

In addition to restoration of natural communities, management of small parks includes the manipulation of disturbed biotic communities to promote landscape heterogeneity. Fire is a major management tool. Management plans for specific parks include burn-grid maps delineating different burn cycles for sub-areas, based on fuel load, successional stage, natural and artificial fire breaks and topography. This is similar to the fire burning approach to management of the Uluru National Park, Australia (section 13.5.1) but on a different scale. In larger parks, wildfires are allowed which rarely exceed 150 ha due to environmental patchiness.

13.4.3 CREATION OF REFUGIA

Both the Commanche Springs pupfish (*Cyprinodon elegans*) and the Pecos gambusia (*Gambusia nobilis*) once thrived in the stenothermal spring runs surrounding marshes in west Texas. They are now endangered because of groundwater pumping. A refuge has been created in which the heterogeneity of the natural habitat of the species is mimicked. This includes a patchy distribution of both emergent and floating vegetation. The two species have successfully reproduced each year since the refuge was established.

13.4.4 ENHANCEMENT OF AVIAN HABITAT BY PROMOTING WOODLAND HETEROGENEITY

The golden-cheeked warbler and the black-capped vireo (*Vireo atricapillus*) have breeding ranges largely restricted to the Balcones Escarpment of the Edwards Plateau of central Texas. Both species are declining because of habitat loss and nest parasitism. Both are edge species inhabiting narrow ecotonal areas between oak–Asle juniper woodland and grasslands. Habitat manipulation is done mechanically to arrest succession within shrub patches. Under natural conditions these patches persist due to either edaphic factors or cyclic disturbances such as fire. Other manipulation, such as re-afforestation to replace trees killed by the oak wilt fungus, is done as required. The program applies to all 12 state parks in Texas within the breeding range of the two species. Management for the two species is community oriented in that numerous nontarget plant and animal species are also maintained by increased landscape heterogeneity.

13.4.5 MANAGEMENT OF DISTURBANCES

An essential feature of a natural disturbance regime is the variation in disturbance attributes (Table 13.1) such as size, timing, intensity and spatial location (Baker, 1992; Chapter 3). Baker points out that perpetuating disturbance variation has not been an explicit goal of either reserve design or management. Texas State Park management is thus innovative. Christensen (1990) lists five major ways to manage disturbances in natural areas, namely the use of surrogates, suppression, planned disturbance prescriptions, natural disturbance prescriptions and a natural disturbance regime.

The last option requires a regional approach or very large reserves. Baker (1992) stresses the importance, with managed disturbances of mim-

Table 13.1 Some attributes of individual disturbance patches, and the sets of disturbance patches that comprise a disturbance regime

Each disturbance patch

Type	Type of disturbance
Size	Land area disturbed
Shape	Measure by a shape index (cf. Austin, 1984)
Intensity	Physical force of a disturbance
Severity	Damage caused by a disturbance
Timing	Temporal setting of a disturbance
Spatial location	Spatial setting of a disturbance
Edge	Total length of the perimeter of a patch
Orientation	Compass direction of central axis of a patch

Spatial or temporal sets of disturbance patches

Type distribution
Size distribution
Shape distribution
Intensity distribution
Severity distribution
Timing distribution
Spatial distribution
Edge distribution
Orientation distribution

icking historical variation in disturbance sizes, as well as other attributes if possible. He cites the Knoza Prairie Research Natural Area as an example of a reserve in which natural disturbances are managed entirely by prescription (Marzolf, 1988) in a manner inconsistent with perpetuating landscape processes and form. The criticism is that the prescription is for fixed patch size and fixed times of the year for burning even though fire frequency is varied.

13.4.6 CONCLUSION

Maintenance and manipulation of landscape patterns with remnant areas is an essential part of any conservation program, but there is a need for clear objectives.

13.5 MANAGEMENT OF ECOSYSTEMS AT THE REGIONAL LEVEL

Conservation management is more difficult at the regional level because land tenure is rarely vested in a single authority at this scale. Whilst it is possible to obtain co-operation between various landowners in relatively small landscape areas (e.g. a water catchment), it is harder to obtain sufficient co-operation over a region with diverse tenure and land management goals to implement a general conservation program.

Murphy and Noon's (1992) program with the northern spotted owl illustrates this. The choice of reserves for the network is constrained by having to ascertain that the reserve will not be subject to some other land use in the future that degrades or eliminates it as owl habitat. The conservation goal in such situations is to provide the diversity of habitats required in a system of reserves spread through the region.

13.5.1 ULURU NATIONAL PARK, NORTHERN TERRITORY, AUSTRALIA

This large park contains 21 different land units and had been subjected to human-managed fire (patch burns) for 10 000 years, in addition to wildfire. The arrival of European settlers led to virtual elimination of patch burning, and as a consequence, large-scale wildfires occurred. The survival of wildlife species in the area depends on the irregular boundaries of small fires to provide a fine-grain mosaic of resources (Miller, 1982). A patch-burn strategy for fire management at Uluru was developed by the CSIRO (Saxon, 1984) which prescribes patch size and frequency of burning for each of the land units.

13.5.2 THE BOREAL FORESTS OF FENNOSCANDIA

These coniferous forests, or taiga, occupy 50×10^6 ha in Fennoscandia (Esseen et al., 1992). Forestry has a tremendous impact on the structure and function of the boreal ecosystem with consequent decreasing populations of many plant and animal species. Esseen et al. (1992) reviewed the changes in detail and then addressed the issue of conservation strategies for the future. They stress that the key guideline is to provide a natural mosaic of boreal habitats within a landscape framework. Diversity needs to be preserved at all scales and requires the preservation and maintenance of the processes causing diversity. The only way to do this is to preserve areas large enough to ensure operation of natural processes, such as the creation of old trees, snags, fallen logs and specific microclimates.

Natural disturbance creates much of the structural and functional diversity occurring in boreal forests. Thus, at the landscape level, disturbances such as fire that initiate secondary succession must be allowed to operate. At the local level, small-scale disturbances (tree falls, tree deaths, etc.) are needed to enhance microhabitat diversity. Achieving these goals requires the education of the forestry industry and politicians to produce an integrated forestry and nature conservation program.

An important component of the boreal landscape are the riparian communities. Apart from preserving some especially valuable areas, management of other areas with controlled use is essential (Nilsson,

1992). This requires the maintenance of fluctuating water levels which maintain high values of vegetation diversity and plant species diversity.

13.5.3 THE JARRAH AND KARRI FOREST, WESTERN AUSTRALIA

The dry sclerophyllous jarrah forest is dominated by jarrah (*Eucalyptus marginata*), but has a shrub understorey. It originally covered 5.3×10^6 ha in the south-west of Western Australia. Its prime uses are as a source of timber and as water catchments for a number of dams. It has a rich fauna comprising 100 bird and 31 mammal species. The karri forest is in the lower south-west with *Eucalyptus diversicolor* as the dominant tree species; it merges with the jarrah forest and has similar land uses. The majority of the forest areas are owned by the State of Western Australia.

The conservation management of these forests has evolved over the past 100 years. The current situation is that forest reserves have been set aside that cover as full a range as possible of the ecological types in the forests (Havel, 1989). These reserves are designated as management priority areas (MPAs) for flora and fauna. The management of individual reserves varies depending on the ecotype and on whether special management measures are needed for an endangered species or community.

These reserves are set in a matrix of forest that may either be virtually the same because it is water catchment but receives no priority management, or be in forest managed for timber. The timber harvesting is controlled so that a series of corridors of forest are left unlogged on a block by block basis forming a network of corridors of road, river and stream zones that connect with reserves (National Parks and Nature Reserves) (Figure 13.4). Research is continuing to determine the optimal distribution of these corridors and their long-term management (Wardell-Johnson and Christensen, 1992). The overall objective is the retention in perpetuity, as far as is possible in a multi-use land system, of the biological diversity of the forests.

This framework of conservation is being adopted within forests throughout Australia and elsewhere. As Hunter (1990) stresses,

> forests should be managed at a variety of scales. This will involve making silvicultural decisions on a landscape basis, not just stand by stand. Ideally, the management regimes should range from the very fine-scale management represented by selection cutting to the coarse-scale management effected by sizeable clearcuts. The maximum size of clearcuts should be determined by considering issues such as size of management unit, the home range requirement of large animals, aesthetics, nutrient loss and natural disturbance regime.

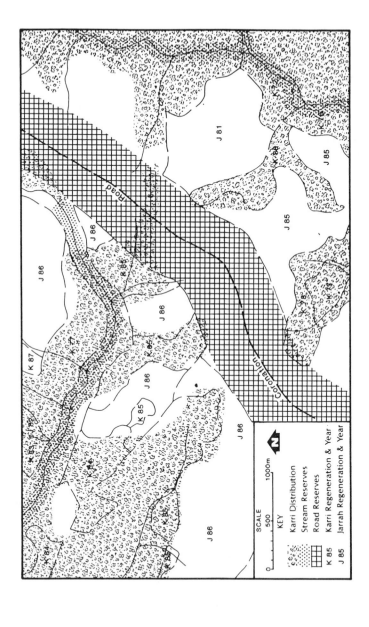

Figure 13.4 The spatial pattern of timber harvesting and regeneration in an area near Pemberton, Western Australia. (Source: Wardell-Johnson and Christensen, 1992.)

13.5.4 NORTHERN TERRITORY, AUSTRALIA

The native vegetation in the Northern Territory remains extensive and little fragmented so that there is still the opportunity to select a network of reserves using algorithms now capable of determining how to choose areas from an array of land units in a way which optimizes the representation of species or communities (e.g. Cocks and Baird, 1989; Margules, 1989).

The distribution and abundance of birds, and possibly other vertebrates, may be reasonably well associated at a local level with vegetation characteristics. At this level, a reserve network based on inclusion of all vegetation types might, theoretically, optimize conservation of most species. However, this approach does not take into account the needs of organisms requiring the resources of several communities to live or the presence of corridors to link the resource patches. There is another level to consider as well, and that is the temporal instability of many animal assemblages. Mobility is an important part of the ecology of many animal species in the Northern Territory and needs to be taken into consideration in developing strategies aimed at conservation of this fauna (Woinarski et al., 1992).

Woinarski et al. (1992) go on to illustrate the complexity of the problem with a number of examples. The survival of many rainforest species requires that their seed is dispersed amongst the 15 000 rainforest patches by frugivores, especially birds and fruit bats. The Gouldian finch (Erythruro gouldiae) appears to exploit landscape patchiness provided by episodic fires so that it is not possible to cater for this species in a system of static reserves. Distribution of populations of the magpie goose (Anseranas semipalmata) depend on broad-scale rainfall variability and local patchiness in rainfall events which means that there is no guarantee that population centers will be protected in a limited number of reserves.

Australian nectarivorous birds have very complex seasonal patterns of movement. Species which are obligate nectarivores must have flower production across substantial regions. Loss of critical habitat components may result in a crash of the entire population even from areas where local disturbance may be minimal (e.g. Franklin, Menkhorst and Robinson, 1989).

The problem is not unique to the Northern Territory. For example, some migratory passerines from Europe and North America are declining because of habitat changes in Africa and South America. The only solutions (partial) require the retention of as much as possible of the natural habitat of the species of concern in the areas it naturally spans.

13.5.5 CONCLUSIONS

Conservation programs for broad regions are feasible and some are being implemented. They have to recognize the need to preserve diversity at all scales and this requires the preservation and maintenance of the processes causing diversity (Chapter 2). The most difficult objective is undoubtedly retention of all the critical habitat components across a large region for bird species that are nomadic or migratory.

13.6 MANAGEMENT PLANNING AND MODELING

The conservation strategy within the forests of south-western Australia incorporates a variety of formal and informal planning procedures. The MPAs can be selected using models such as that of Margules (1989), whilst the design of the corridor linkages is informal in the sense that no criteria are yet available for determining the width and number of corridors needed for linking reserves. Noss and Harris (1986) formalized this approach in a conceptual scheme that evaluates not only habitat content within protected areas, but also the landscape context in which each reserve exists. Nodes of concentrated ecological value are linked into a functional network through the establishment of a system of interconnected multiple use modules (MUMs). The MUM network protects and buffers important ecological entities and phenomena, while encouraging movement of individuals, species, nutrients and energy across space and time. Noss and Harris (1986) present a network in Florida as an example (Figure 13.5).

On the design of inter-refuge corridors, Harrison (1992) suggests that they should contain enough suitable habitat for the target species to permanently occupy the corridor. The width is then that of a home range with a buffer to prevent deterioration of the habitat. This approach may be suitable for animal species that occupy non-exclusive home ranges and will not impede the movement of other dispersing individuals. However, the situation may be different for territorial species which will challenge intruders and, perhaps, block or slow movement. In this case the argument could be made that the corridor should be of a plant community through which the target species readily moves when dispersing or moving between resource patches, but which is unsuitable habitat for establishing territories.

13.6.1 THE SIX RIVERS NATIONAL FOREST, CALIFORNIA, USA

The Six Rivers National Forest covers 285 000 ha and is a major producer of commercial timber. The land managers required a strategy for habitat

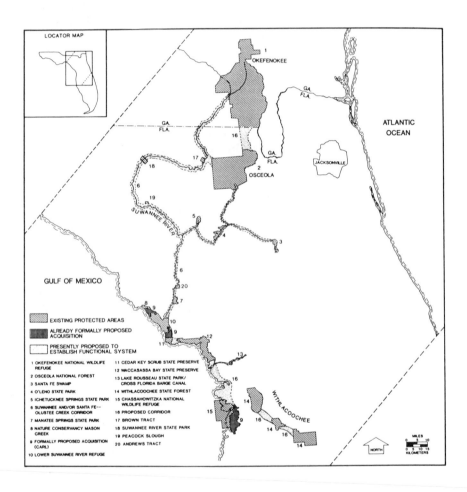

Figure 13.5 An example of a MUM network: a regional system of protected areas in north Florida and south Georgia, integrated by riparian and coastal corridors. (Source: Noss and Harris, 1986.)

management that would contribute to the long-term viability of wildlife populations. Toth, Solis and Marcot (1986) described a wildlife–habitat–

relationships model which provides the preliminary step in the development of a generalized approach for habitat diversity specific to this forest. The influence of habitat patch size and configuration on the 304 wildlife species was explored using a Wildlife Habitat Inventory Matrix Program (Figure 13.6).

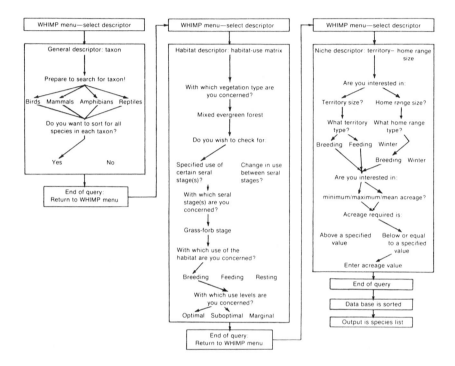

Figure 13.6 Logic flow of a query used with the wildlife habitat inventory matrix program (WHIMP) to sort species and habitat data in the wildlife habitat relationships (WHR) model. (Source: Toth, Solis and Marcot, 1986.)

Although more information is needed for some species, the results were used to develop management standards for the retention of special habitat components through time and guidelines for kinds, sizes, distributions and amounts of habitat patches through space.

13.6.2 BACHMAN'S SPARROW (*Aimophila aestivales*) IN THE SAVANNAH RIVER PINE PLANTATIONS, CAROLINA, USA

Pulliam, Dunning and Liu (1992) developed a spatially explicit simulation model of the Savannah River pine plantations and predicted the effects of landscape-level variation in habitat dispersion on the size and extinction probability of Bachman's sparrow in this region managed for timber production. In the model, habitat suitability and availability within the landscape changes annually as a function of timber harvest and management strategies. The life history characteristics of Bachman's sparrow used in the model include dispersal, survivorship and reproductive success.

The model showed that the probability of extinction was very sensitive to adult and juvenile survivorship and that the presence of habitat types that serve as permanent sources of dispersers increases the total population size in the landscape, and lowers the probability of extinction.

The model should allow the development of an improved pattern of timber harvest in the 21-year rotation that allows the sparrow to colonize newly created patches of prime habitat more readily. This can be done by designing a network of contiguous, suitable habitat patches scattered throughout the region devoted to pine production. Such a network ought to increase greatly the total population size in the managed area. Pulliam, Dunning and Liu (1992) point out that models such as this one must be tailored to meet specific local conditions of landscape characteristics, and must use locally accurate inormation of habitat-specific demography and dispersal behavior.

13.6.3 COOK–QUENTIN WATERSHED, WILLAMETTE NATIONAL FOREST, OREGON, USA

The Cook–Quentin watershed covers 3318 ha and is multiple use forest land in the west Cascades. Hansen, Garman and Marks (1993) used simulation modeling to assess the long-term (140 years) effects of four management strategies on 51 bird species and on timber production. Knowledge about the bird species was limited to habitat suitability and life-history attributes were used as surrogates for detailed demographic data. The watershed was mapped to show the areas suitable for each species. The species were ranked (1, 2, 3) for likely responsiveness to landscape change for each of eight life-history traits (reproductive effort, nest type, nest height, minimum territory size, seral stage association, microhabitat association, response to edge and response to patch size). The ranks were added.

The simulation model used calculates several landscape metrics at each time step and classifies habitat suitability for each bird species. Survival of populations of a species were assumed to be related to the area of habitat suitable and the sensitivity of the species to changes in the landscape. The authors suggest that land managers, after weighing the assumptions and limitations of the methodology, can at least use the information as a guide to the trade-off between timber production and conservation of avifauna (in this case). Hansen, Garman and Marks

(1993) suggest that managers, as an experiment, implement one or more prescriptions and monitor the responses of habitat and the population responses of select species.

13.6.4 CONCLUSIONS

The examples given cover the range in level of detail that can be used in developing conservation programs using logical frameworks (Chapter 12). No judgment can be made of the merits of the different approaches since the information input needed differs in each case. What would be interesting would be to use the models of WHIMP and Hansen and co-workers for the Bachman's sparrow and compare the predictions with those from the model of Pulliam and co-workers.

The formal modeling approach has a role to play where the autecology or habitat requirements of a species is known (Chapter 5). However, where there is little information then expert systems may have a role to play. Marcot (1986) cautions on the cost, need and utility of such systems and, as yet, there are no examples of this use in the development of management plans for nature conservation. Rather, expert systems have been used to identify suitable habitat for a given species in areas where this is not known.

13.7 GENERAL DISCUSSION

In natural landscapes there is both spatial and temporal heterogeneity (Southwood, 1977) at several scales (Kotliar and Wiens, 1990; Chapter 1). Grain size (patchiness) in the landscape mosaic is determined not only by the geological, hydrological, edaphic and disturbance patterns in the landscape and their effects on plant community development but also by the perception of the mosaic by individual animal species (Chapter 9). What is perceived as a grain by a large herbivore may be a mosaic of many different grain types to a spider. Thus trying to incorporate the concepts of landscape heterogeneity into conservation programs requires setting specific conservation objectives and then determining the landscape pattern to aim for.

Where the objective is to manage landscapes or regions to retain biotic diversity, the retention of as many areas as possible of each of the ecological types present in the region is essential. These areas need to be managed to maintain the patterns and processes causing diversity (Esseen *et al.*, 1992). This includes disturbances such as fire, and ensuring, as far as is possible, that the areas burnt are within the natural distribution range in size with natural frequency and seasonality, e.g. the Uluru National Park (Chapter 3).

In landscapes where these natural processes cannot occur because of fragmentation or because of the risk to human activities or of causing extinction of a species, there is still a need to simulate successional stages through management intervention as is done in Texas parks. This is necessary to conserve plant and animal communities. The question of specific conservation programs to retain either plant or animal communities has received little attention but needs addressing. Some programs for conservation of a particular species also conserve those members of the community that have similar habitat requirements. For example, fire management to conserve the endangered noisy scrub birds (*Atrichornis clamosus*) in the Two People's Bay Nature Reserve in Western Australia benefits the rare western bristlebird (*Dasyornis longirostris*) and Western whipbird (*Psophodes nigrogularis*) which have similar habitat requirements (Smith, 1985). The examples given of management of the patchiness and patch types for individual animal species illustrate a number of points. One is the need for species-specific information on the habitat requirements and the size and spatial distribution of the habitat patches needed to sustain a population. Since many animal species that live in heterogeneous environments occur in metapopulations, multiple representatives of given patch types are needed to support multiple subpopulations. Then there is need for some species for temporal variation in habitat types depending on their life history stage (e.g. spotted owl), or on the seasonal changes in food and/or habitat requirements due to reproduction (grizzly bear). There can be no hard and fast rule about the value of heterogeneity when considering conservation of individual species.

The restoration of invertebrate diversity in agricultural landscapes by increasing the patchiness or heterogeneity of the landscape is feasible. Restoration of vertebrate diversity, especially of less vagile species, requires the creation of patches in spatial arrangements that allow recolonization from sources. Preferably, these sources would be natural ones of quality habitat of the target organism. However, re-introductions to create new sources may well be necessary. Such sources would need to be at the nodes of networks of habitat patches.

Sufficient examples have been given to show that the development and maintenance of landscape patterns is central to many conservation programs. Conflicts arise in multiple-use managed landscapes between primary production and nature conservation. The various modeling approaches described allow the effects of given management strategies to be predicted. Validation of predictions will be essential for developing land managers confidence in investing in them.

What has to be recognized in all conservation programs (and few specifically do so) is the temporal dynamics of the natural landscape patterns (Noss and Harris, 1986; Chapter 6). Within managed forest landscapes, temporal dynamics can be incorporated in conservation programs. It appears impossible to do so in programs in agricultural landscapes where conservation depends primarily on relics of the native vegetation, but long-term viability of conservation programs at the landscape level requires that ways of simulating the space–time mosaics must be found.

13.8 SUMMARY

The role of spatial and temporal heterogeneity in planning conservation programs at the levels of species, landscapes, ecosystems and regions is considered. At the species level, examples are given of management plans to meet specific grain size, type and spatial juxtapositions for individual species in managed forest landscapes. At the landscape, ecosystem and regional level the importance of retaining both the range of ecotypes and natural disturbance processes, or simulating them is stressed. It is suggested that simulation models or expert systems coupled with geographic information systems will be increasingly used to produce land management plans for sustainable conservation of wildlife coupled with sustainable forestry and agriculture.

REFERENCES

Alverson, W. S., Waller, D. M. and Solheim, S. L. (1988) Forests too deer: Edge effects in Northern Wisconsin. *Conserv. Biol.* **2** (4), 348–58.

Angelstam, P. (1992) Conservation of communities – the importance of edges, surroundings and landscape mosaic structure, in *Ecological Principles of Nature Conservation. Applications in Temperate and Boreal Environments* (ed. L. Hansson), Elsevier Applied Science, London, pp. 9–70.

Arnold, G. W. (1983) The influence of ditch and hedgerow structure, and area of woodland and garden on bird numbers on farmland. *J. Appl. Ecol.*, **20**, 731–50.

Austin, R. F. (1984) Measuring and comparing two-dimensional shapes, in *Spatial Statistics and Models* (eds G. L. Gaile and C. J. Wilmott), D. Reider Publ. Co., Boston, MA, pp. 293–312.

Baker, W. L. (1992) The landscape ecology of large disturbances in the design and management of nature reserves. *Landsc. Ecol.* **7**, 181–94.

Banaszak, J. (1992) Strategy for conservation of wild bees in an agricultural landscape. *Agric., Ecosys. Environ.*, **40**, 179–92.

Bissonette, J. A., Fredrickson, R. J. and Tucker, B. J. (1991) American marten: A case for landscape-level management, in *Wildlife and Habitats in Managed Landscapes* (eds. J. E. Rodiek and E. G. Bolan), Island Press, Washington, DC, pp. 117–34.

Bryant, F. C. (1991) Managed habitats for deer in juniper woodlands of

West Texas, in *Wildlife and Habitats in Managed Landscapes* (eds J. E. Rodiek and E. G. Bolan), Island Press, Washington, DC, pp. 59–75.

Cocks, K. D., and Baird, I. A. (1989) Using mathematical programming to address the multiple reserve selection problem: an example from the Eyre Peninsular, South Australia. *Biol. Conserv.* **49**, 113–30.

Conner, R. N., Snow, A. E. and O'Halloran, K. A. (1991) Red-cockaded woodpecker use of seed-tree/shelterwood cuts in eastern Texas. *Wildl. Soc. Bull.,* **19**, 67–73.

Christensen, N. L. (1990) Variable fire regimes on complex landscapes: ecological consequences, policy implications, and management strategies. Paper presented at the International symposium on fire and the environment: ecological and culture perspectives, March 20–24, 1990 Knoxville, Tennessee.

Dennis, P. and Fry, G. L. A. (1992) Field margins: can they enhance natural enemy population densities and general arthropod diversity on farmland? *Agric. Ecosys. Environ.,* **40**, 95–115.

Esseen, P. A., Ehnström, B., Ericson, L. and Sjöberg, K. (1992) Boreal forests – the focal habitats of Fennorscandia, in *Ecological Principles of Nature Conservation. Applications in Temperate and Boreal Environments* (ed. L. Hansson), Elsevier Applied Science, London, pp. 252–325.

Franklin, D. C., Menkhorst, P. W. and Robinson, J. L. (1989) Ecology of the Regent Honeyeater *Xanthanzya phrygia. Emu,* **89**, 140–54.

Freemark, K. E. and Merriam, H. G. (1986) Importance of area and habitat heterogeneity to bird assemblages in temperate forest fragments. *Biol. Conserv.,* **6**, 115–40.

Garcia, E. R. (1985) Grizzly bear direct habitat improvement on the Kootenai National Forest, in *Proceedings of the Grizzly Bear Habitat Symposium* (eds G. P. Contreras and K. E. Evans), Intermountain Research Station, Ogden, UT, pp. 185–189.

Guillion, R. F. (1984) *Managing Northern Forests for Wildlife,* The Ruffed Grouse Society, Misc. Journal Series, Minnesota Agric. Stn, St Paul.

Hansen, A. J., Garman, S. L. and Marks, B. (1993) An approach for managing vertebrate diversity across multiple-use landscapes. *Ecol. Applic.,* **3** (3), 481–96.

Harrison, R. L. (1992) Toward a theory of inter-refuge corridor design. *Conserv. Biol.,* **6**, 293–5.

Havel, J. J. (1989) Conservation in northern jarrah forest, in *The Jarrah Forest – A Complex Mediterranean Ecosystem* (eds J. Dell, J. J. Havel and N. Malajczuk), Kluwer Academic Publishers, pp. 379–99.

Hayes, T. D., Riskind, D. H. and Pace, W. L. (1987) Patch within-patch restoration of man-modified landscapes within Texas state parks, in *Landscape Heterogeneity and Disturbance* (ed. M. G. Turner), Springer-Verlag, New York, pp. 173–98.

Hobbs, R. J., Saunders, D. A., Lobry de Brun, L. and Main, A. R. (1993)

Changes in biota, in *Reintegrating Fragmented Landscapes: Towards Sustainable Production and Nature Conservation* (eds R. J. Hobbs and D. A. Saunders), Springer-Verlag, New York, pp. 65–106.

Hobbs, R. J. Saunders, D. A. and Arnold, G. W. (1993) Integrated landscape ecology: a Western Australian perspective. *Biol. Conserv.,* **64**.

Hudson, P. (1986) *Red Grouse – The Biology and Management of a Wild Game Bird*, The Game Conservatory Trust, Bourne Press, Bournemouth.

Hunter, M. L. (1990) *Wildlife, Forests and Forestry. Principles of Managing Forests and Biological Diversity*, Prentice Hall, Englewood Cliffs, NJ.

Kotliar, N. B. and Wiens, J. A. (1990) Multiple scales of patchiness and patch structure: a hierarchical framework for the study of heterogeneity. *Oikos,* **59**, 253–60.

Lagerlöf, J. and Wallin, H. (1993) The abundance of arthropods along two field margins with different types of vegetation composition: an experimental study. *Agric. Ecosys. Environ.,* **43**, 141–54.

Lambeck, R. J. and Saunders, D. A. (1993) The role of patchiness in reconstructed wheatbelt landscapes, in *Nature Conservation 3: Reconstruction of Fragmented Ecosystems* (eds D. A. Saunders, R. J. Hobbs and P. Ehrlich), Surrey Beatty & Sons, Chipping Norton, NSW.

Mader, H. J. (1988) Effects of increased spatial heterogeneity on the biocenosis in rural landscapes *Ecol. Bull.,* **39**, 169–79.

Marcot, B. G. (1986) Use of expert systems in wildlife-habitat modelling, in *Wildlife 2000. Modelling Habitat Relationships of Terrestrial Vertebrates,* (eds J. Verner, M. L. Morrison and C. J. Ralph), pp 145–50.

Margules, C. R. (1989) Introduction to some Australian developments in conservation evaluation. *Biol. Conserv.,* **50**, 1–11.

Marzolf, R. (1988) Konza prairie research natural area of Kansas State University. *Trans. Kansas Acad. Sci.,* **91**, 24–9.

Miller, G. T. (1982) *Living in the Environment*, Wadsworth, Belmont, CA.

Murphy, D. D. and Noon, B. R. (1992) Integrating scientific methods with habitat conservation planning: reserve design for northern spotted owls. *Ecol. Applic.,* **2**, 3–17.

Nentwig, W. (1989) Augmentation of beneficial arthropods by strip-management II. Successional strips in a winter wheat field. *J. Plant. Dis. and Protect.,* **96**, 89–99

Nilsson, C. (1992) Conservation management of riparian communities, in *Ecological Principles of Nature Conservation: Applications in Temperate and Boreal Environments* (ed. L. Hansson), pp. 352–72.

Noss, R. F. and Harris, L. D. (1986) Nodes, networks and MUMs: preserving diversity at all scales. *Environ. Manage.,* **10**, 299–309.

O'Connor, R. J. and Shrubb, M. (1986) *Farming and Birds*, Cambridge University Press, Cambridge.

Opdam, P. (1990) Dispersal in fragmented landscapes: the key to survival, in *Species Dispersal in Agricultural Habitats* (eds R. G. H. Bunce and D. C. Howard), Belhaven Press, London, pp. 3–17.

Petrides, G. A. (1942) Relation of hedgerows in winter to wildlife in central New York. *J. Wildl. Manage.,* **6**, 261–79.

Pulliam, H. R., Dunning, J. B. and Liu, J. (1992) Population dynamics in complex landscapes. *Ecol. Applic.,* **2**, 165–77.

Rands, M. R. W. and Sotherton, N. W. (1986) Pesticide use on cereal crops and changes in the abundance of butterflies on arable farmland in England. *Biol. Conserv.,* **36**, 71–82.

Rolstad, J. and Wegge, P. (1987) Distribution and size of capercaillie leks in relation to old forest fragmentation. *Oecologia (Berl.),* **72**, 389–94.

Saunders, D. H., Hobbs R. J. and Arnold, G. W. (1993) The Kellerberrin project on fragmented landscapes, a review of current information. *Biol. Conserv.,* **64**, 185–92.

Saxon, E. C. (ed.) (1984) *Anticipating the Inevitable: A Patch-Burn Strategy for Fire Management at Uluru (Ayers Rock – Mt Olga) National Park,* CSIRO, Australia.

Smith, G. T. (1985) Fire effects on populations of the noisy scrub-bird (*Atrichornis clamosus*), Western bristle bird (*Dasyarnis longirostris*) and Western whipbird (*Psophodes nigrogularis*), in *Fire Ecology and Management in Western Australian Ecosystems* (ed. T. R. Ford), WAIT, Curtin, Western Australia, pp. 95–101.

Southwood, T. R. E. (1977) Habitat, the templet for ecological strategies. *J. Anim. Ecol.,* **46**, 337–65.

Thomas, M. B., Watten, S. C. and Sotherton, N. W. (1992) Creation of 'island' habitats in farmland to manipulate populations of beneficial anthropods: predator densities and species composition. *J. Appl. Ecol.,* **29**, 524–31.

Toth, E. G. Solis, D. M. and Marcot, B. G. (1986) A management strategy for habitat diversity: using models of wildlife habitat relationships, in *Wildlife 2000. Modelling Habitat Relationships of Terrestrial Vertebrates* (eds J. Verner, M. L. Morrison and C. J. Ralph), University of Wisconsin Press, pp. 139–44.

Wardell-Johnson, G. and Christensen, P. (1992) A review of the effects of disturbance on wildlife of the karri forest, in *Research on the Impact of Forest Management in South-west Western Australia,* Dept Conserv. Land Manage. Occasional Paper 2/92, pp. 33–58.

Woinarski, J. C. Z., Whitehead, P. J., Bowman, D. M. J. S. and Russell-Smith, J. (1992) Conservation of mobile species in a variable environment: the problem of reserve design in the Northern Australia. *Global Ecol. and Biogeogr. Letters,* **2**, 1–10.

Yahner, R. H. (1984) Effects of habitat patchiness created by a ruffed grouse management plan on breeding bird communities. *Am. Midl. Nat.,* **111**, 409–13.

Zanaboni, A. and Lorenzoni, G. G. (1989) The importance of hedges and relict vegetation in agroecosystems and environment reconstitution. *Agric., Ecosys. Environ.*, **27**, 155–61.

Summary
Ecology of Mosaic Landscapes: Consolidation, Extension and Application

Landscape ecology has become an accepted subdiscipline within ecology. Certain authors (e.g. Wiens *et al.*, 1993) even suggest that traditional ecology should be replaced by landscape ecology. However, there still are large areas of traditional ecology that have not yet been considered within landscape ecology and there are significant problems for testing theoretical ideas in landscape ecology.

Theories in landscape ecology make many simplifying assumptions but landscape ecological theory is still clearly more complex than ecological theory for homogeneous systems. Landscape theories have to be tested under the spatial conditions assumed by the theory. The most accepted method for testing theories in ecology is through experimentation. Landscape experiments at the actual scale of the substrate and the process involved often are impossible and the experimental units must be scaled down (Forney and Gilpin, 1989; Ims, Rolstad and Wegge, 1993). However, the results of such small-scale experiments are of uncertain validity when extrapolated to natural systems because, for example, there may be threshold effects when moving from one scale to another (Chapter 1). In addition, there may be significant biological differences between systems that can be studied experimentally and natural systems. For example, enclosure experiments such as those of Ims, Rolstad and Wegge (1993) do not consider predation effects, and predation may change the dynamic pattern of the focal species (e.g. Hanski *et al.*, 1993).

Since experimentation with adequate replication for statistical power at the landscape scale is in general not possible, we must find alternative

Mosaic Landscapes and Ecological Processes.
Edited by Lennart Hansson, Lenore Fahrig and Gray Merriam.
Published in 1995 by Chapman & Hall, London. ISBN 0 412 45460 0.

means of testing landscape theories. One approach is to use the theory or model to make a critical prediction (e.g. a predicted contrast in population dynamics in two different kinds of landscapes), and to test that prediction in an experiment designed to look at this single critical prediction (e.g. Fahrig and Paloheimo, 1988). The important aspect of this type of test is that it must be a 'strong' prediction in the sense that detection of a single case that does not match the prediction is sufficient to invalidate critical aspects of the model, because it is unlikely that replication will be feasible. If the critical test does not invalidate the model, we can then feel more confident in using the model to make other, more extensive predictions concerning landscape effects. However, it is important to note that the constraint of large spatial scale implies that in experimental field tests we will not be able to obtain sample sizes normally considered appropriate for inference.

Some form of field test at the appropriate scale is nevertheless necessary. Highly disaggregated, spatially explicit simulation models of species in landscapes are becoming increasingly common, and there is a tendency to use these models as though they were a complete description of the real system without testing their predictions (e.g. Pulliam, Dunning and Liu, 1992). Embedded within these models are assumptions about which processes are most important in the real system and many processes are of necessity left out. There is no more basis for believing (without field tests) predictions from such models than predictions from simpler analytical models.

An alternative approach to starting with models (analytical or simulation) is to use another research paradigm in landscape ecology, such as inductive generalization, as suggested cautiously by Wiens (Chapter 1). If so, we may need to gather data from a larger variety of landscape structures and try to extract generalizations from these empirical data. Such large-scale observational studies in which the observations are spread thinly across a large number of landscapes, may in fact also be a more efficient means than experimentation for testing theories at the landscape scale (Eberhardt and Thomas, 1991).

The problems of complexity, scale and idiosyncracies in landscape ecology must be assessed and carefully evaluated. We should not invoke landscape explanations unless they are really needed. Therefore a major area of future research should be to evaluate which ecological processes can be well-explained at large scales by conventional ecological theory assuming spatio-temporal homogeneity. If landscape considerations really are necessary, ways of making inferences and predictions have to be evaluated thoroughly. An important integral part of this re-evaluation may be a fundamental consideration of the nature of evidence and the acceptability of evidence relevant to particular environmental problems and their consequences.

A major facet of present work in landscape ecology centers on individual movements, especially dispersal, and the dynamics of single-species populations which often are spatially divided. However, there are few single-species populations which exist independently of other populations in natural environments. Chapters in this volume by Ilkka Hanski, Henrik Andrén and Judith Bronstein demonstrate the potential importance of multispecies interactions at the landscape scale. It is therefore important to to determine the circumstances under which the spatial patterns of interactions with competitors, predators, disease and mutualists need to be understood in order to understand the dynamics of focal populations.

We also need to know more about landscape ecology of multispecies communities. However, current landscape ecology does not seem to include even the rudiments of any theoretical treatment of multispecies communities in the landscape. An important aspect of such a landscape community approach would be to assess the functional completeness and species proportions of ecological communities in various landscape compositions and configurations. Certain species disappear from simplified landscapes and the dynamics of surviving species may be severely affected (e.g. Bengtsson, 1989; as discussed in Chapter 9).

Pest, game and endangered species live in heterogeneous environments; even outbreaks of pest species in monocultures usually depend on refugial habitats in a mosaic (Frank, 1956; Southwood and Jepson, 1961). Therefore, applied ecology relies heavily on a presently insufficient theory of landscape ecology. This problem is especially apparent for work in conservation and biodiversity management. Recent research in conservation sometimes, but certainly not always, adopts a landscape perspective but there often seems to be an incomplete consideration of the scaling problem; local resource levels and distribution, and habitat heterogeneity are often examined in detail, when the problem is actually on the landscape or regional scale. Thus there appears to be a need in conservation research for theory and methods for selecting the pertinent scale for a problem. A useful contribution in this regard would be a comparison of population dynamics of a variety of species across a variety of scales (e.g. Holling, 1993). Studies should be framed in terms of alternative hypotheses relating processes at different scales to population dynamics. If it were found that the most accurate predictions could be made for most species at the same scale (or small range of scales), the problem of management for multispecies communities would be greatly simplified. Conservation and biodiversity research are flourishing; incorporating landscape perspective into the burgeoning research in this area could also allow testing and development of landscape ideas and encourage the gathering of databases needed for further generalizations.

REFERENCES

Bengtsson, J. (1989) Interspecific competition increases local extinction rate in a metapopulation system. *Nature,* **340**, 713–15.

Eberhardt, L. L. and Thomas J. M. (1991) Designing environmental field studies. *Ecol. Monogr.,* **61**, 53–73.

Fahrig, L. and Paloheimo, J. E. (1988) Effect of spatial arrangement of habitat patches on local population size. *Ecology,* **69**, 468–75.

Forney, K. A. and Gilpin, M. E. (1989) Spatial structure and population extinction: a study with *Drosophila* flies. *Conserv. Biol.,* **3**, 45–51.

Frank, F. (1956) Grundlagen, Möglichkeiten und Methoden der Sanierung von Feldmausplagegebieten. *Nachrichtenblatt des deutsches Pflanzen-schutzdienstes,* **8**, 147–58.

Hanski, I., Turchin, P., Korpimäki, E. and Henttonen, H. (1993) Population oscillations of boreal rodents: regulation by mustelid predators leads to chaos. *Nature,* **364**, 232–5.

Holling, C. S. (1993) Cross-scale morphology, geometry, and dynamics of ecosystems. *Ecol. Monogr.,* **62**, 447–502.

Ims, R. A., Rolstad, J. and Wegge, P. (1993) Predicting space use responses to habitat fragmentation: Can voles *Microtus oeconomus* serve as an experimental model system (EMS) for capercaillie grouse *Tetrao urogal-lus* in boreal forest? *Biol. Conserv.,* **63**, 261–8.

Pulliam, H. R., Dunning, J. B. and Liu, J. (1992) Population dynamics in complex landscapes: a case study. *Ecol. Applic.,* **2**, 165–77.

Southwood, T. R. E. and Jepson, W. F. (1961) The fruit fly – a denizen of grassland and a pest of oats. *Ann. Appl. Biol.,* **49**, 556–66.

Wiens, J. A., Stenseth, N. C., Van Horne, B. and Ims, R. A. (1993) Ecological mechanisms and landscape ecology. *Oikos,* **66**, 369–80.

Index

Page numbers in **bold** refer to figures and page numbers in *italic* refer to tables.